AF825975

NUCLEAR FUSION

Half a Century of Magnetic Confinement Fusion Research

NUCLEAR FUSION

Half a Century of Magnetic Confinement Fusion Research

C M Braams

and

P E Stott

Taylor & Francis
Taylor & Francis Group
New York London

Taylor & Francis is an imprint of the
Taylor & Francis Group, an informa business

Published in 2002 by
Taylor & Francis Group
270 Madison Avenue
New York, NY 10016

Published in Great Britain by
Taylor & Francis Group
2 Park Square
Milton Park, Abingdon
Oxon OX14 4RN

© 2002 by Taylor & Francis Group, LLC

No claim to original U.S. Government works
Printed in the United States of America on acid-free paper

International Standard Book Number-10: 0-7503-0705-6 (Hardcover)
International Standard Book Number-13: 978-0-7503-0705-5 (Hardcover)
10 9 8 7 6 5 4 3 2

This book contains information obtained from authentic and highly regarded sources. Reprinted material is quoted with permission, and sources are indicated. A wide variety of references are listed. Reasonable efforts have been made to publish reliable data and information, but the author and the publisher cannot assume responsibility for the validity of all materials or for the consequences of their use.

No part of this book may be reprinted, reproduced, transmitted, or utilized in any form by any electronic, mechanical, or other means, now known or hereafter invented, including photocopying, microfilming, and recording, or in any information storage or retrieval system, without written permission from the publishers.

Trademark Notice: Product or corporate names may be trademarks or registered trademarks, and are used only for identification and explanation without intent to infringe.

Library of Congress Cataloging-in-Publication Data

Catalog record is available from the Library of Congress

Visit the Taylor & Francis Web site at
http://www.taylorandfrancis.com

To Lieke and Olga

Thanks for your patience and encouragement

Contents

Figures

The authors gratefully acknowledge the following organizations and individuals for granting permission to reproduce figures and illustrations. Copyright in these materials remains with the original publishers: EFDA-JET (figures 5.4, 6.1, 6.5, 7.2, 8.1, 8.5, 8.6, 9.1 and 9.3), Elsevier Scientific (figures 2.1, 2.2, 2.3, 2.4, 2.7, 3.1, 4.8, 4.11, 4.12, 4.13, 5.3, 6.6a and the figure in box 2.4), Euratom-UKAEA Fusion Association Culham (figures 1.5, 1.8, 2.10 and 8.8), General Atomics (Dr T N Carlstrom for figure 5.7), International Atomic Energy Agency (figures 3.3, 3.5, 3.9, 3.12, 3.13, 4.1, 4.2, 4.3, 4.5, 4.14, 5.2, 5.5, 5.8, 5.10, 6.6b, 6.9 and 7.1), International Thermonuclear Experimental Reactor (ITER) (figure 9.4), Japan Atomic Energy Research Institute (figure 8.3), Lawrence Livermore National Laboratory (figures 3.4, 3.8, 4.7 and 4.10), Max-Planck-Institut für Plasmaphysik, Garching (Dr H Wobig for figures 2.8 and 5.9 and Professor F Wagner for figures 5.12, 7.3 and 8.7), National Institute for Fusion Science (figure 5.11), Nature Publishing Group and Dr N J Peacock (figure 5.6), Oxford University Press and Dr J Wesson (figures 6.2 and 6.7), Princeton Plasma Physics Laboratory (figures 5.1, 6.3, 8.2 and 8.4) and Trilateral Euregio Cluster (TEC) (figure 6.4). Every effort has been made to contact all copyright holders. The authors will be pleased to make good in any future editions any errors or omissions that are brought to their attention.

Some figures have been redrawn or modified by the authors, in particular figures 1.1 and 1.2 after Raeder J *et al.* 1986 *Controlled Nuclear Fusion* (Chichester: Wiley); figure 1.4 after Alfvén H 1950 *Cosmical Electrodynamics* (Oxford: Clarendon Press); figure 1.9 after Glasstone S and Lovberg R H 1960 *Controlled Thermonuclear Reactions* (Princeton: D van Nostrand); figures 2.5 and 2.6 after Rose D J and Clark M 1961 *Plasmas and Controlled Fusion* (Cambridge: MIT Press); figure 3.6 after Kawabe T 1983 *Varenna Course on Mirror-Based and Field-Reversed Approaches to Magnetic Fusion* **1** 43; figure 3.7 after Kumazawa R 1983 9th Conf. Plasma Physics and Controlled Nuclear Fusion Research, Baltimore, 1982, *Nucl. Fusion Suppl.* (Vienna: IAEA) **1** 566; figure 3.10 after Mather J W and Williams A H 1966 *Physics of Fluids* **9** 2080; figure 4.4 after Baker D A and Quinn W E 1981 *The Reversed-Field Pinch,* chapter 7 in *Fusion* **1A** ed. E Teller (New York: Academic Press); figure 4.9 after Momota H 1992 *Fusion Techn.* **21** 2309; figure 6.8 after various sources; figure 9.2 after Moir R W 1972 4th Conf. Plasma Physics and Controlled Nuclear Fusion Research, Madison, 1971, *Proceedings Series* (Vienna: IAEA) **3** 315 and 1975 5th Conf. Plasma Physics and Controlled Nuclear Fusion Research, Tokyo, 1974, *Proceedings Series* (Vienna: IAEA) **3** 589.

Figures 1.3, 1.6, 1.7, 2.9, 3.2, 3.14, 4.6, 4.15 and those in boxes 2.1, 2.3, 5.2 and 7.2 were produced by the FOM Institute Rijnhuizen, which gladly waives its copyrights.

Preface

Physicists who first cast an eye on the problem of the peaceful use of thermonuclear energy, around the middle of the twentieth century, soon realized that to try something and see what might happen would get them nowhere until they understood more of plasma physics. But although plasmas occur widely in nature, indeed most of the known matter in the universe is in the plasma state, this area of physics had been largely neglected. When relativity and quantum mechanics had presented themselves in the early years of the twentieth century as the areas where the greatest advances in physics were to be made, the unexplored land between Newtonian mechanics and Maxwellian electromagnetism was simply overlooked. Astronomers and gas discharge physicists were the first to become aware of this omission and to take steps to correct it and, when the thermonuclear problem came up, they were among the first to bring their expertise to bear on the subject. In the course of the 1950s this developed into a recognized new branch of physics and as this unfolded, controlled thermonuclear research, or fusion research as it came to be known, evolved from basic to applied plasma physics. As it turned out, of all the magnetic field configurations investigated for their capability to confine a hot plasma, the tokamak became the leading contender, to the point where we can now proceed to the construction of a test reactor and to contemplate the next step: a prototype commercial reactor.

In this book, we attempt to sketch the different paths followed by fusion research from initial ignorance to present understanding, be it the understanding of why a particular scheme would not work, or why it was more profitable to concentrate, at least for the time being, on the mainstream—tokamak development. The authors have been actively involved in these events from 1957 to 1987 (CMB) and from 1962 to the present time (PES). We do not regard ourselves as historians, but we hope that our account will help future historians as well as scientists in other fields to find their way through this difficult terrain. We also hope that our present and former colleagues will, when reading the book, recall the excitement with which all of us witnessed what went on in this challenging field of discovery. And most of all we wish to help those new in the field or preparing to enter it, to see it in a broad perspective and to realize that the course of fusion will not have been run even when the first demonstration reactor has come into operation.

In the course of writing this book we have drawn on the vast volume of published material relating to fusion in scientific journals and elsewhere, as well as on unpublished material and on the reminiscences of our colleagues. We have tried to give an accurate and balanced account of the development of fusion research that reflects the relative importance of the various lines that have been pursued and gives credit to the contributions from the many institutions in the countries that have engaged themselves in fusion research. However, compressing half a century of research into a book of manageable length has not been an easy task, so inevitably there will be issues, topics and contributions that some readers might feel deserved more detailed treatment.

We would like to thank all of our colleagues who have helped and advised us in many ways. In particular we are indebted to Folker Engelmann, Masami Fujiwara, Viktor Golant, the late Igor Golovin, Evgeniy Gusakov, Ian Hutchinson, Atsuo Iiyoshi, Sigeru Mori, Vyacheslav Strelkov, Masaji Yoshikawa and Kenneth Young who took a great deal of time and trouble to read an early draft of the book and who gave constructive criticism and valuable suggestions for its improvement. We are grateful also to Rosario Bartiromo, Michael Bell, Bas Braams, Hardo Bruhns, Niek Lopes Cardozo, Chris Carpenter, James Drake, Umberto Finzi, Alan Gibson, Richard Gill, Günther Grieger, Hans-Jürgen Hartfuss, Richard Hawryluk, Bick Hooper, John How, Jan Hugill, Otto Kardaun, Winfried Kernbichler, Bo Lehnert, Anthony Leonard, Jonathan Lister, James Luxon, Akko Maas, the late Charles Maisonnier, Dale Meade, Tobin Munsat, Katsunori Muraoka, Per Nielsen, Noud Oomens, Ronald Parker, Carol Phillips, Kseniya Razumova, Jan Rem, Michael Roberts, Chris Schüller, Richard Siemon, Shigeru Sudo, Tatsuo Sugie, Vladimir Tereshin, Paul Thomas, Vladimir Voitsenya, Friedrich Wagner, Hans Wilhelmsson, Horst Wobig, Alan Wootton and many others who have helped us to check specific points of detail, generously provided figures and assisted in other ways. A special thanks to John Navas, Simon Laurenson and the staff of Institute of Physics Publishing for their patience and encouragement. We would like to thank Piet van Kuyk and Wim Tukker who produced or edited most of the figures.

CMB acknowledges the hospitality he enjoyed at the FOM Institute 'Rijnhuizen', The Netherlands after his retirement and the support he received from its scientific staff and its technical and administrative services.

The content of this book and the views expressed therein are the sole responsibility of the authors and do not necessarily represent the views of the European Commission or its services.

Prologue

Is a fusion reactor an interesting proposition, from a scientific, technical, social and economic point of view? When physicists started to look seriously into this question around the middle of the twentieth century, surely no one expected that by the turn of the millennium the answer would still not be an unequivocal yes or no. '*Fusion will be there when society needs it*', said Lev Artsimovich, and thirty years after his untimely death no better answer can yet be given.

Thermonuclear fuel burns at a temperature of some hundreds of million Kelvin and needs to be confined by either a magnetic field or, in micro-explosions, by inertia; we restrict ourselves here to the former method. The physics of magnetized plasma has turned out to be far more difficult than anyone who early on elected to work in this field appreciated. The key problem is the energy lost by the plasma to its surroundings, which must be compensated by the fusion energy yield for reaction conditions to be maintained.

Radiative energy loss scales as the square of the density, as does the reaction yield; their ratio is independent of the size of the reactor. Only a very restricted class of reactions between light nuclei is able to sustain the ignited or nearly ignited state required for net energy gain, the most reactive fuel being a mixture of deuterium and tritium. But instabilities of the state, far from thermodynamic equilibrium, in which the thermonuclear fuel must be brought to yield its treasure, at worst may cause the plasma to escape as a whole from its confinement and at best result in anomalous particle and energy transport across the field, which comes in addition to the unavoidable normal transport caused by particle collisions. As a consequence, the energy balance scales favourably with size. Thus, like any system that has to maintain a certain temperature to sustain its metabolism, such as a star or a warm-blooded animal, the fusion reactor must have a minimum size; fusion research is about determining where precisely this minimum lies.

There is no doubt that with the knowledge gained so far one could construct an energy-producing device, but to make it competitive requires further research and development. Theory must provide more elaborate and precise models to describe the plasma's seemingly erratic behaviour. Likewise, fusion technology must tread on uncharted ground to reach the level of safety and reliability below which the whole effort is not even worth trying. Above all, experiments inevitably become increasingly complex and costly as thermonuclear conditions are approached. Yet, the decisive evidence must come from large projects, and approval for these hinges as much on political judgements concerning the needs of societies as on the scientific and technical progress.

How large is the smallest feasible and economically attractive reactor and how small is the largest one that can be accommodated in a future power distribution system? If by current knowledge there is a gap between the two, can it be bridged by further research and development? A variety of confinement systems have been investigated and, so far, only the tokamak and the stellarator show sufficient promise to warrant a large-scale reactor development programme. The tokamak is universally regarded as our present best choice, but it has so much in common with the stellarator that the development of one in no way pre-empts future decisions in favour of the other. There is an ongoing debate about whether one should now make the step to an energy-producing tokamak test reactor for both scientific and technological studies, or dwell longer upon the physics and the materials science in order to test new ideas that could eventually make for a smaller and better reactor. Within the research community, there is broad support for a bold step forward and an intense international collaboration has resulted in the design of a large machine that can be confidently expected to serve its purpose. The governments concerned have engaged in serious negotiations, so when the next step will be made, how ambitious it will be and which countries will make that step, either alone or in an international collaboration, is in their hands.

It is difficult to predict how societies will ultimately judge the safety of the reactor and its fuel cycle. There are good prospects that tritium burning reactors can be made intrinsically safe, so that no foreseeable accident could pose an immediate threat to the lives of those living outside the industrial premises, nor force them to evacuate their homes. Detailed risk analyses of fully elaborated reactor designs must confirm this expectation. But whether even a highly affirmative technical judgement will eventually overcome lingering fears of any kind of nuclear energy will depend to a large extent on how populations in different countries perceive their energy needs and the security of supplies, as well as environmental values, in years to come. To some fusion scientists, such considerations make it desirable or even imperative to consider fuels other than the deuterium–tritium (D–T) mixture favoured by most, even if less inflammable 'advanced' fuels require higher temperatures, plasma pressures and energy containment times than D–T. There exists no confinement scheme that can be confidently advanced for this purpose at the present stage; mirror systems and field-reversed configurations need to be further investigated in this context, but few fusion scientists appear ready to discard D–T burning and to opt for the uncertain prospects of other fuels.

Fusion research stepped into the open at the Second United Nations International Conference on the Peaceful Uses of Atomic Energy, held at Geneva in 1958. Immediately, the United Nations International Atomic Energy Agency (IAEA) in Vienna undertook to play a coordinating role in the rapidly expanding international effort, mainly by organizing a series of conferences on plasma physics and controlled fusion and by publishing the journal *Nuclear Fusion*, both of which continue to play the leading role in their field. Initially, the emphasis in both was on plasma physics whilst speculation about fusion reactors was considered premature, but by the time of the third IAEA Conference, held in August 1968 in Novosibirsk, the time seemed ripe for more directed efforts towards reactor development. This led to the INTOR and ITER design studies for an international tokamak reactor, carried out under the IAEA.

The magnetic fields in which a thermonuclear plasma may be confined have open or closed magnetic surfaces and can be pulsed or in a steady state. Within each of the four classes so defined, there exist a variety of specific field configurations that have been studied, if not because they were thought to be suitable for a reactor, then because of their scientific interest. As potential reactors the steady-state closed systems and in particular the tokamak have moved ahead of the other schemes. In the first two chapters we discuss the scientific background and the early work that was revealed at the Geneva Conference in 1958. Next we deal with the open systems, steady state and pulsed, and with the alternative toroidal systems. This will set the scene for the mainstream of fusion research; in the second half of the book we follow the tokamak and its close relative, the stellarator, in their development from table-top experiments to the proposed International Thermonuclear Experimental Reactor (ITER). Finally, we give a brief overview of reactor studies and in the Epilogue look back at what half a century of dedicated scientific effort has taught us and look forward to the next steps that are now being contemplated.

Chapter 1

The road to Geneva

1.1 The scientific roots

1.1.1 Fusion energy in stars

With thermodynamics, nineteenth-century physics and chemistry seemed to have achieved their ultimate unifying theory and there was little doubt that its validity would extend over the neighbouring fields of astronomy and geology as well. Its basic principle, conservation of energy, therefore required that the Sun's radiation be accounted for in terms of an equivalent source. Chemical fuels were inadequate and Helmholtz found that gravitational contraction could account for only some 20 million years of the Sun's radiation output. Kelvin estimated the time it would have taken the Earth to cool down to its current state at one hundred million years, and although geologists could accept this figure—their own estimates also tended to converge towards it—even half a billion years was not enough for Charles Darwin to comply with his evolution theory. Kelvin had convinced the geologists that the Earth could not have existed for an indefinite length of time, as had been their accepted doctrine. But he kept revising his figures downwards, and when he eventually arrived at an estimate of 20–40 million years they could no longer stay with him. The physicists for whom it had been heresy to question the axiom of the immutability of the atom or to contemplate the existence of as yet unknown forms of energy, had overplayed their hand. Surely, thermodynamics would stand the test of time, but there would turn out to be 'more in Heaven and Earth'. Around the turn of the century, Becquerel demonstrated that the physics of Helmholtz and Kelvin was incomplete. Radioactivity at once turned out to be the key, both to determining the ages of rocks and to explaining the Earth's hidden source of heat. Now the ball was in the astronomers' court. With physics, geology and biology agreeing that the age of the Earth was in fact several billion years [1], the question was from where the Sun, which clearly could not be younger than the Earth, obtained its heat.

Einstein's equivalence of mass and energy shed a new light on the problem and the thought arose that 'atomic transmutations' might be the source of solar and stellar energy. Immediately after Aston had measured the mass defect of helium relative to four hydrogen atoms with his mass spectroscope, Arthur Eddington, one of the leading astronomers of his time, put forward the conversion of hydrogen into helium [2]:

> *A star is drawing on some vast reservoir of energy by means unknown to us. This reservoir can scarcely be other than the sub-atomic energy which, it is known, exists abundantly in all matter; we sometimes dream that man will one day learn how to release it and use it for his service.... I think that the suspicion has been generally entertained that the stars are the crucibles in which the lighter atoms, which abound in the nebulae, are compounded into more complex elements.*

This was in 1920. Russell had by then established a connection between the liberation of energy and the temperature inside a star and when Eddington refined the existing model for the stellar interior by including radiation pressure and radiative energy transport, he found that the observational data—the Herzsprung–Russell diagram in which each star is represented by a point on a luminosity versus surface temperature diagram—suggested a central energy source, rather than a more uniform source like gravitation [3]. Eddington concluded that the energy release must set in at a critical temperature, for which his model yielded 40 million Kelvin (MK) (it was later adjusted to 15–20 MK). He considered fusion of hydrogen into helium the most plausible reaction, although there was a problem in that the Sun was supposed to contain only a few per cent hydrogen. A few years later, Atkinson and Houtermans invoked Gamov's theory of barrier penetration, suggesting the fusion of hydrogen into helium through successive proton and electron captures in as yet unidentified heavier nuclei, followed by α-decay [4].

At that time, the nucleus was thought of as an assembly of protons and electrons, or of alpha particles built up from these constituents. So, precisely which reactions might be involved remained unclear. In an astounding development, nuclear physics in hardly a decade matured to the point where it was possible to identify the appropriate thermonuclear reactions and to calculate their reaction rates. Important stepping stones were the introduction of particle accelerators in 1930, the discovery of deuterium and of the neutron, both in 1932, and the theory of beta decay in 1934 [5]. In 1936, Atkinson suggested the beta decay of colliding protons as the first step towards helium production via the proton–proton chain:*

* Here, the three hydrogen isotopes are represented by the nuclear physicist's symbols p, d and t. Later on, when our emphasis shifts from the nuclear processes to the fuels burnt and ashes produced, we shall use the chemical symbols H, D and T, in accordance with most recent fusion literature.

$$p + p \rightarrow d + e^+$$

$$p + d \rightarrow {}^3He$$

$$^3He + {}^3He \rightarrow {}^4He + 2p$$

The reaction rate was calculated by Bethe and Critchfield [6]. Von Weizsäcker and Bethe independently recognized the CNO cycle [7]:

$$^{12}C + p \rightarrow {}^{13}N \rightarrow {}^{13}C + e^+$$

$$^{13}C + p \rightarrow {}^{14}N$$

$$^{14}N + p \rightarrow {}^{15}O \rightarrow {}^{15}N + e^+$$

$$^{15}N + p \rightarrow {}^{12}C + {}^4He$$

as the one for which Atkinson and Houtermans had been looking and, again, Bethe calculated the reaction rate. In 1939 he wrote a classic paper [8], in which he concluded that the p–p chain and the CNO cycle are equally probable at a solar temperature of 16 MK, and that the latter would be dominant at the assumed 19 MK. A review of the subject in which several side branches of these reaction sequences are taken into consideration shows that, on present knowledge, conditions in the Sun favour the p–p chain [9].

The subject of energy release in stars is, of course, intimately related to that of nucleosynthesis. Aside from the early reviews by von Weizsäcker [10], who clearly posed the outstanding problems, and Bethe, who quantitatively evaluated the relevant reaction rates, articles by Burbidge *et al.* and Cameron stand as landmarks in the development of this field [11].

Although burning ordinary hydrogen can comfortably sustain a main-sequence star's radiation flux for billions of years, the process is utterly inadequate for earthly purposes: the reaction rates and, hence, the power production per unit volume are too low for economic purposes. Even at a pressure as high as that in the Sun, the average power density is 0.27 W/m^3 against some MW/m^3 required for a commercial reactor.

Stars differ from prospective fusion reactors in two other aspects: confinement and radiative equilibrium. In stars, the nuclear fuel is held together by gravity. As to radiation, bremsstrahlung in a fusion reactor counts as an immediate plasma heat loss because it has a mean free path for reabsorption far in excess of the dimensions of any conceivable earthly magnetic-confinement reactor.* In stars, on the other hand, this reabsorption length is small compared with the star's size, so that radiation from the centre diffuses slowly towards the photosphere and the reaction temperature can be sustained with the power density referred to above.

* The situation is different in inertial confinement of pellets compressed to extremely high densities, a subject not dealt with in this review.

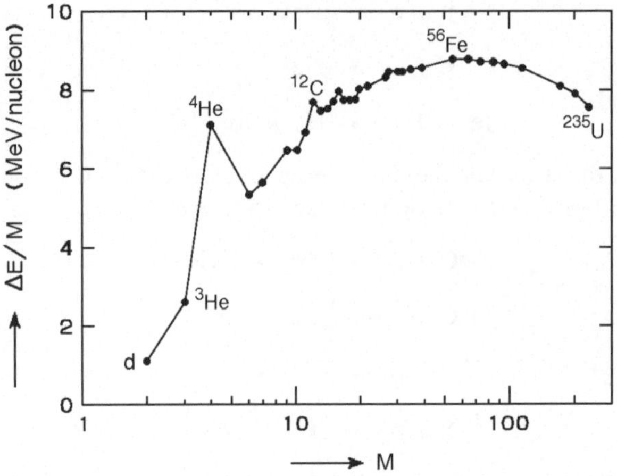

Figure 1.1. Binding energy or mass defect per nucleon, $\Delta E/M$, of stable nuclei as a function of nuclear mass number M. Nuclear energy is released when reactions lead to an increase in the binding energy, which is negative potential energy.

1.1.2 Fusion reactions on Earth

When astrophysics and nuclear physics pointed towards the conversion of nuclear binding energy into heat,* the thought immediately arose [12] that this '*latent power*' could be controlled '*for the well-being of the human race—or for its suicide*'. Hydrogen presented itself as the first candidate for nuclear fuel, because the repulsive Coulomb-barrier varies as the product of the nuclear charge numbers of the reactants. And since, among the lightest nuclei, ^{4}He stands out with as much as 7.1 MeV binding energy per nucleon (figure 1.1), thoughts turned towards the conversion of hydrogen isotopes into helium.

When systematic investigations of nuclear reactions became possible by the introduction of particle accelerators, it appeared that in particular the loosely bound deuteron gave prolific reactions on various target nuclei [13]. The reaction may involve the formation of a compound nucleus or proceed through a mechanism called 'deuteron stripping'. In either case, one constituent of the deuteron is trapped in the target nucleus and the other flies off, carrying most of the reaction energy (the difference between the reaction products' and the reactants' binding energies) as kinetic energy. But when later considered as a source of nuclear energy, these reactions were named 'fusion', because the alliteration with 'fission' accentuates the difference and because the more descriptive term 'thermonuclear reaction' in people's minds was associated with the hydrogen bomb.

* Commonly referred to as 'energy production'.

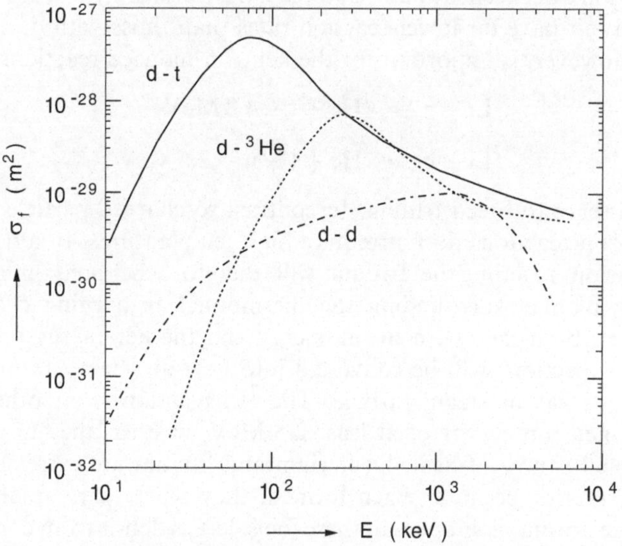

Figure 1.2. Cross-section, σ_f, for the d–d, d–t and d–^{3}He fusion reactions as a function of kinetic energy, E, of the relative motion of the colliding nuclei. The curve marked d–d indicates the total cross-section for the two reactions d + d mentioned in the text.

Most of what is now called fusion research aims at exploiting deuteron break-up by letting deuterium react with itself, with tritium, or with ^{3}He. The latter are produced in d–d reactions, but may also be supplied as fuel along with the deuterium. The reactions in question are:

$$d + d \rightarrow t + p + 4.0\,\text{MeV}$$

$$d + d \rightarrow {}^3\text{He} + n + 3.3\,\text{MeV}$$

$$d + t \rightarrow {}^4\text{He} + n + 17.6\,\text{MeV}$$

$$d + {}^3\text{He} \rightarrow {}^4\text{He} + p + 18.3\,\text{MeV}$$

of which the first and second have approximately equal probability (figure 1.2). The high cross-section and the high energy yield of the d–t reaction make it the favourite candidate for terrestrial fusion.

To confuse the terminology still further, even a fission, or spallation reaction like

$${}^{11}\text{B} + p \rightarrow 3 \times {}^4\text{He} + 8.7\,\text{MeV}$$

has been labelled 'fusion', which must now be understood to include all energy-producing reactions between light nuclei.

Reactions between lithium isotopes and normal hydrogen have been considered but have far lower reaction rates than those with d, t, and ^{3}He. Lithium, however, is important for the neutron-induced reactions

$$^6\text{Li} + \text{n} \longrightarrow {}^4\text{He} + \text{t} + 4.8\,\text{MeV}$$

$$^7\text{Li} + \text{n} \longrightarrow {}^4\text{He} + \text{t} + \text{n} - 2.5\,\text{MeV}$$

which can serve to breed tritium. In contrast to charged-particle reactions, these neutron reactions do not require high temperatures. In a deuterium–tritium fusion reactor, the tritium will therefore be bred in a lithium-containing blanket surrounding the thermonuclear burning chamber. In this blanket, both the d–t neutron energy and the net energy produced in the lithium reactions will be converted into heat at a temperature suitable for driving a gas or steam turbine. The ^{7}Li reaction is endothermic and requires a neutron energy of at least 2.5 MeV, whereas the ^{6}Li reaction is an additional source of heat. Deuterium and lithium are extremely rare in the Sun's interior because, when formed, they react very rapidly. But on Earth, there are ample supplies of these fuels, left as debris from cosmological processes and supernova explosions of early massive stars. Some 150 ppm* of natural hydrogen is deuterium, and even with this small amount one litre of ocean water contains as much energy as three hundred litres of gasoline. Such comparisons triggered wild fantasies of virtually inexhaustible energy supplies. Also lithium, which has an abundance of 0.17 ppm in ocean water and is available in widely scattered salt deposits and brines, can support any foreseeable demand from fusion energy for centuries.

The initial supply of tritium can be extracted from fission reactors, whereupon a fusion reactor fitted with a lithium-containing blanket can breed its own fuel, each triton burned producing one neutron and this, in turn, producing one triton from ^{6}Li. Inevitable neutron or tritium loss can be compensated by first letting fast neutrons from the d–t reaction interact with ^{7}Li, which gives a triton plus a slow neutron, and then capturing the latter in ^{6}Li. Alternatively, one may use (n, 2n) reactions with beryllium or a heavy element, such as lead, as a blanket material to generate more neutrons. Finally there is the possibility to use fissionable materials as neutron multipliers. But although the tritium–lithium cycle was already considered in the 1950s [14], thoughts on fusion energy were initially focused on deuterium, with tritium and helium-3 only as intermediate products [15], even though the temperature† required for d–d reactions is of the order of 100 keV, against 20 keV for d–t.

* Parts per million.

† Generally, we use MKS units, except that, henceforth, temperatures are mostly expressed in eV or keV. The unit eV corresponds with 11.6×10^3 K. With this convention, Boltzmann's constant becomes $k = 1.6 \times 10^{-19}$ J/eV $= 1.6 \times 10^{-16}$ J/keV.

The world became acutely aware of the devastating power of nuclear energy when it learned of the two fission weapons exploded over Japanese cities towards the end of the Second World War and of the first thermonuclear explosions in 1952. But there was also the fission reactor, which was seen as a great hope for cheap and abundant energy. Physicists who had worked on the bomb were among the first, along with the astrophysicists, to tackle the problem of completing the quartet by developing a fusion reactor [16].

Although it is relatively easy to accelerate nuclei to the energies required for fusion and to demonstrate the reactions by shooting particle beams onto solid targets, it was quickly realized that this would not be a feasible way to generate energy. Most of the accelerated nuclei give up their energy to electrons in the target and relatively few approach the target nuclei sufficiently close to undergo fusion. The energy gain from fusion does not match that invested in accelerating the much larger number of nuclei that are stopped in the target. Shooting a beam into a plasma is, under special conditions (box 8.1), a viable option, but thoughts on fusion turned primarily towards reactions in a thermal plasma.

1.1.3 The origins of plasma physics*

The word 'plasma' entered the vocabulary of physics through Irving Langmuir's description of the positive column of a gas discharge [17]. One interpretation is that he saw an analogy with blood plasma in that both were carrying particles, but it seems more plausible that he thought of the Greek word for 'to mould', because the luminous substance tended to take the shape of its glass container [18]. The plasma, an example of which is the light-emitting gas in a fluorescent lamp, is electrically quasi-neutral and highly conductive, hence nearly electric-field-free (in contrast to the region in front of the cathode—the cathode fall—in which there is a steep potential gradient). The study of atomic, molecular, and surface phenomena determining the state of ionization in these discharges had started with the discovery of the electron and would develop into a mature science in the years before the Second World War [19]. But the collective motions of ions and electrons, which are what plasma physics in the narrow sense is about (a wider sense includes the ionization phenomena) were only beginning to be studied towards the end of the pre-war period.

Langmuir introduced two important concepts in plasma physics—the Debye screening distance and the plasma frequency. He applied the Debye–Hückel theory of electrolytes to describe the space-charge sheath in front of an electrode and called the thickness of this sheath the Debye

* The reader who does not wish to go too deeply into plasma physics may choose to skip the boxes.

Box 1.1 Debye length and plasma frequency

The basic parameters characterizing collective phenomena in a non-magnetized plasma are the Debye length

$$\lambda_d = (\varepsilon_0 k T_e / n_e e^2)^{1/2}$$

which is normally very small, and the plasma, or Langmuir frequency

$$\omega_p = (n_e e^2 / \varepsilon_0 m_e)^{1/2}.$$

Here, n_e is the density, T_e is the temperature, e is the charge and m_e is the mass, all of the electrons, while k is Boltzmann's constant and ε_0 is the permittivity of free space. The product of λ_d and ω_p equals the electron thermal velocity.

If, in the positive column of a gas discharge, n_i is the density of an ion species with charge number Z_i, the deviation from neutrality, $(\sum Z_i n_i - n_e)/n_e$, is of the order of $(\lambda_d/a)^2$, where a is the radius of the column. The Debye length is also called the Debye screening distance because it is a measure for the thickness of the layer outside which a positive charge immersed in a plasma is screened by a locally increased electron density—or a negative charge by a decreased electron density. In this layer, the electrostatic field, determined by Poisson's equation, makes equilibrium with the pressure gradient in the electron gas. The plasma frequency appears when the electrostatic field and the electron mass together govern electron-density oscillations. If there is a magnetic field, the gyroradius and the gyrofrequency (see box 1.2) also come into play.

Since $m_i \gg m_e$, the ions may often be thought of as only a fixed neutralizing background for high-frequency fluctuations in the electron gas, particularly in classical gas discharges where $T_i \ll T_e$. In this case, one may speak of an electron plasma. In the same vein, electrons in solid conductors or semi-conductors can behave as a plasma. For a complete description of low-frequency phenomena in ionized gases one needs to introduce also the ion Debye length and the ion plasma frequency, λ_{di} and ω_{pi}, in which case the corresponding electron parameters defined above are written as λ_{de} and ω_{pe}.

length. This appears also as a characteristic length in the description of the positive column (box 1.1). Langmuir also explained [20] the high-frequency noise observed by Penning [21] as emitted by oscillations resulting from collective interactions of ions and electrons through self-generated electric fields.

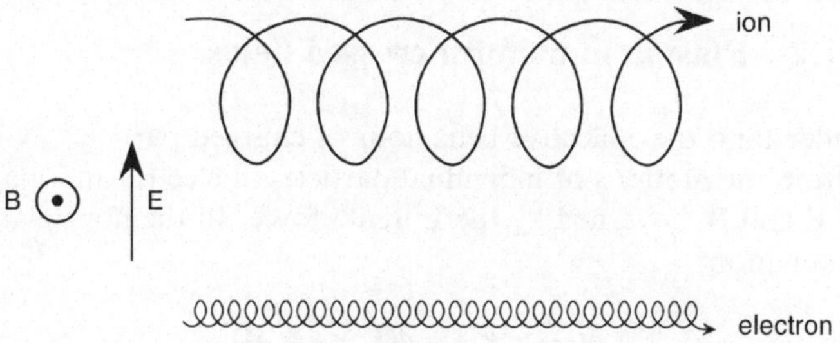

Figure 1.3. The $\mathbf{E} \times \mathbf{B}$ drift of ions and electrons (not to scale). An ion gains speed and the radius of curvature of its orbit increases when it moves in the direction of $\mathbf{E}$; the inverse holds for an electron. The magnetic field points out of the page.

Electrostatic waves* were found to occur in electronic tubes and were exploited in devices like klystrons and magnetrons. This subject was elaborated particularly in the wartime development of radar systems [22]. In the same context, gas discharges found applications in switching devices like spark gaps and thyratrons. And before that time, Heaviside and others had already studied the transmission of radio waves through and their reflection from ionized layers in the upper atmosphere [23].

Collective interactions involving magnetic fields were first recognized in 1934 by Bennett [24]. In its original version, the Bennett-pinch was a pair of counter-streaming relativistic electron and ion beams whose relatively small random transverse velocities were contained by the self-magnetic field of the beams. Tonks [25], who apparently was unaware of Bennett's work, considered the opposite case of a magnetically constricted plasma current carried by ions or electrons with an axial drift velocity small compared with the thermal speed, and came to essentially the same result (box 1.4). Before we continue with the action of a magnetic field on a plasma, we must say something about the motion of individual particles.

An electron or an ion in a uniform and constant magnetic field will describe a spiral orbit, the radius and the frequency of which are designated with the prefix gyro, cyclotron or Larmor—although Larmor did not study the orbits of free electrons [26]. Electrons spiral clockwise and ions anticlockwise when viewed in the conventional direction of the field. If in addition there is an electric or a gravitational field perpendicular to the magnetic field, the particles respond not by falling in the direction of this force but by moving sideways, that is in the third perpendicular direction (figure 1.3). In crossed electric and magnetic fields, **E** and **B**, positive and negative charges drift in the same direction with a velocity equal to E/B (box 1.2). So if these fields extended to infinity, a plasma would move

* Conventionally, an electric field is called electrostatic if it is associated with an electric charge (not necessarily constant), and electrodynamic if it is induced by a time-varying magnetic field.

Box 1.2 Plasma in uniform crossed fields

To understand the collective behaviour of charged particles, one may start from the motions of individual particles in electric and magnetic fields, **E** and **B**, governed by the Lorentz force. In the non-relativistic approximation

$$m\frac{\mathrm{d}\mathbf{v}}{\mathrm{d}t} = \mathbf{F}_\mathrm{L} = q(\mathbf{E} + \mathbf{v} \times \mathbf{B})$$

where m, q and $\mathbf{v}$ are the mass, charge and velocity of the particle. In uniform and constant crossed fields ($\mathbf{E} \perp \mathbf{B}$), this equation is satisfied by $\mathbf{v} = \mathbf{v}_\mathrm{d}$, where $\mathbf{v}_\mathrm{d}$ is the drift velocity

$$\mathbf{v}_\mathrm{d} = \mathbf{E} \times \mathbf{B}/B^2$$

which is independent of mass and charge. J J Thomson's velocity selector was based on this relation, which reappears as the drift of a plasma in crossed electric and magnetic fields. Superimposed on this drift may be a random velocity in the form of a spiralling motion with arbitrary radius and pitch, that is with arbitrary parallel and perpendicular velocities, and with angular frequency

$$\omega_\mathrm{c} = qB/m.$$

The reader who is familiar with linear differential equations will recognize a special solution of the inhomogeneous, and the general solution of the homogeneous equation.

sideways without a current being induced. But in a gravitational field the ions and electrons move in opposite directions, producing precisely the current density, **j**, which by the $\mathbf{j} \times \mathbf{B}$ force supports the plasma against the force of gravity.

A charged particle in a circular orbit has a magnetic moment—the ring current times the surface area. This moment is directed against the field, so it is diamagnetic. And as a paramagnetic material is attracted by a magnet, so is the diamagnetic plasma repelled by a magnetic field. More precisely, a charged particle experiences a force directed against the gradient of the field strength. The component of this force perpendicular to the magnetic field causes a sideways drift, but the parallel component repels the particle from the stronger field. So wherever field lines converge or diverge, the particle will be pushed back from the stronger field.

The complicated motions that arise in non-uniform and time-dependent fields were studied first in astrophysical contexts. Not surprisingly, Scandinavian physicists, who had personal experiences of the beauty and marvel of aurora borealis, played a leading role in studying the phenomenon. Birkeland fired electron beams at magnetized spheres to simulate the behaviour of solar-wind particles in the Earth's field. Inspired by these 'terrella' experiments, Störmer made extensive calculations of such orbits [27]. He distinguished allowed and forbidden regions and found impact zones explaining the geographical location of auroras. Alfvén developed the concept of a guiding-centre orbit [28], about which the particle gyrates with approximately constant magnetic moment (box 1.3, figure 1.4).

Another interesting aspect of the motion of a single particle is the ponderomotive force exerted on electrons by radiofrequency waves (box 3.3). This is proportional to the square of the amplitude, so that it appears as a non-linear effect in plasma–wave interactions, notably in laser-produced plasmas; it has also played a role in fusion research as a possible means for helping to confine a plasma. It was first addressed, in different parts of the world, in the context of work on high frequency generating tubes but was soon considered for thermonuclear applications as well [29].

To study particle orbits in given fields is only one step towards understanding plasma behaviour. We have already seen some collective effects in which particles alter, if not shape, the fields through space charges and currents. A more general treatment had to be based on statistical mechanics. The first point that needed clarification was how the concept of a collision as an instantaneous exchange of momentum translates to what happens in a plasma, in which a particle at any time interacts with many others through long-range forces. Building upon the Liouville and Boltzmann equations which describe the evolution of the electron velocity-distribution function, Landau [30] derived a formula for the collision term, while Vlasov [31] proposed to include the smoothed self-fields of the particles. Landau [32] predicted collisionless damping of waves in plasma by resonant particles with thermal velocity near the phase velocity. The kinetic theory of gases and its application to plasmas was reviewed in Chapman and Cowling's book *The Mathematical Theory of Non-Uniform Gases* [33].

Viewed from a different angle, the plasma may be seen as a conducting fluid. In 1950, the Swedish astrophysicist Hannes Alfvén, who received the Nobel prize for physics in 1970, published his book *Cosmical Electrodynamics*, in which he argued that, on astronomical scales, electrical forces are usually small, but that magnetic forces can have a dominant effect on fluid motion [34]. Hence, the name hydromagnetic or magneto-hydrodynamic (MHD) theory. In fact, Alfvén showed that cosmical plasmas can often be considered as ideal conductors—with negligible resistivity. This led him to the concept of frozen-in magnetic fields [35] and to his prediction of the existence of MHD waves (box 7.2). These are now known as Alfvén waves and their phase velocity is called the Alfvén velocity. MHD shock waves were studied by de Hoffmann and Teller [36]. The idea of a magnetic field moving around with a conducting fluid—from which magnetic field lines derive an identity—was

Box 1.3 Guiding-centre motion

The magnetic moment of a particle with perpendicular velocity $v_\perp$ relative to the direction of the field

$$\mu = mv_\perp^2/2B$$

is constant to a high approximation while the motion is affected by the space and time derivatives of the field strength. This allows the motion to be treated as the superposition of a gyration and the motion of the guiding centre—the centre of curvature of the gyration.

In a constant field, the kinetic energy $mv^2/2$ is also constant, so that the particle exhausts its parallel energy when travelling towards a stronger field—a magnetic mirror. Reflection occurs where the guiding-centre velocity parallel to the magnetic field, $v_{||}$, equals zero, that is where

$$B_{\mathrm{refl}} = mv^2/2\mu.$$

In analogy with the cross-field drift caused by a perpendicular electric field (box 1.2), a perpendicular gradient of the magnetic field strength causes a guiding centre drift in a direction perpendicular to both **B** and the gradient, again with approximately constant magnetic moment. And a similar drift is caused by curvature of field lines

$$v_{\mathrm{d}} = \frac{mv_\perp^2}{2qB^3}\mathbf{B} \times \nabla B + \frac{mv_{||}^2}{qB^2R^2}\mathbf{R} \times \mathbf{B}$$

where **R** is the radius of curvature of the field line at the guiding centre. The first term may be understood as a response to the time-averaged force on a particle gyrating in a non-uniform field, whereas the second arises from the centrifugal force resulting from its motion along curved field lines. In a toroidal vacuum field, the gradient is connected with the curvature and the drift becomes

$$\mathbf{v}_{\mathrm{d}} = \frac{w_\perp + 2w_{||}}{qB^2R^2}\mathbf{R} \times \mathbf{B}$$

where $w_\perp$ and $w_{||}$ stand for the kinetic energy in the perpendicular and the parallel velocity.

In a time-dependent field, the constancy of μ implies that the perpendicular energy varies with B while the parallel energy stays constant. This can be exploited as a heating mechanism—magnetic pumping.

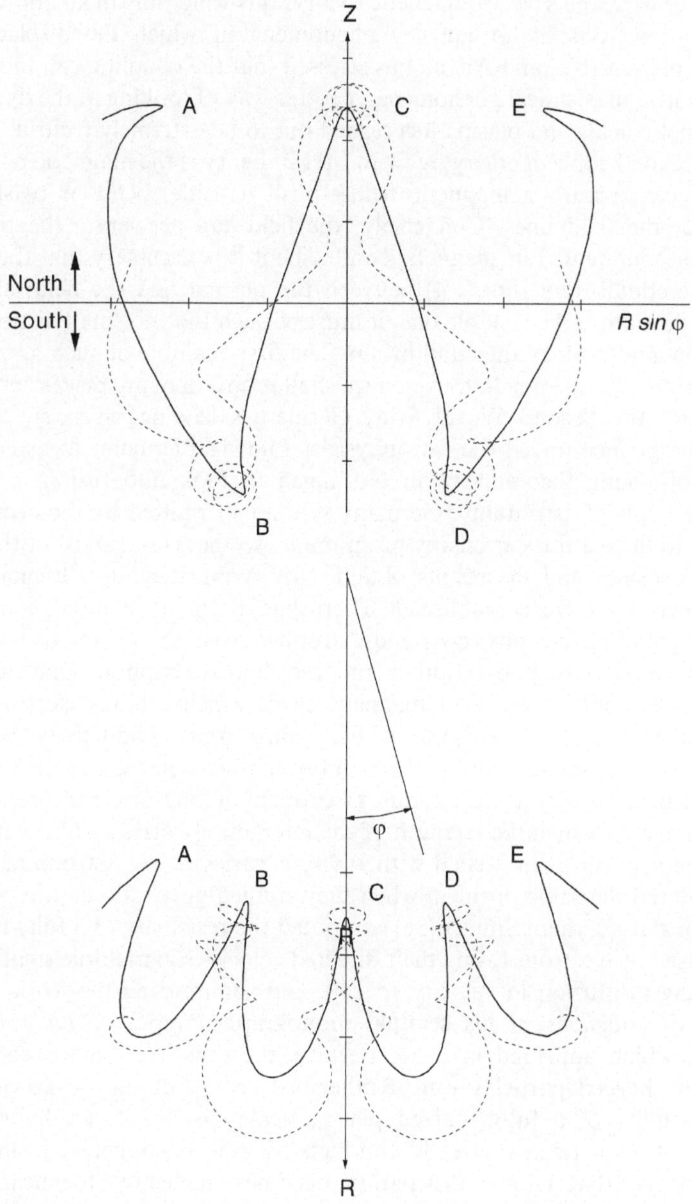

Figure 1.4. Motion of a charged particle in the Earth's dipole field, calculated by numerical integration (broken lines) and by guiding centre theory (solid lines). R, φ and Z are cylindrical coordinates. Above: projection on the $\varphi = \pm\pi/2$ plane through the polar axis, below: projection upon the equatorial plane ($Z = 0$). The orbit displays both repeated reflections from the magnetic mirrors in the northern and southern hemispheres (points A–E), and azimuthal drift due to the gradient of the field strength and the curvature of the field lines.

foreign to Maxwell's electromagnetic theory. It is valid only in an approximation that neglects high-frequency phenomena in which the displacement current plays a role, and Alfvén has stressed that the coupling can indeed be broken, also in cosmical phenomena. Yet this way of looking at the dynamics of magnetic fields and plasma has turned out to be extremely fruitful.

A related concept emerging from MHD theory is the magnetic pressure. Plasma can disturb a magnetic field—push it aside, bend or twist it or constrict the field lines. Conversely, the field can accelerate the plasma, shape or confine it. For magnetic confinement it is necessary that there is a pressure equilibrium (box 1.4) between the plasma and the field; and for the equilibrium to be a stable one, it must be such that no small disturbance can grow and destroy the equilibrium. The first example of such a pressure equilibrium was the pinch, to which we shall return in many places hereafter.

After the Second World War, Germany's leading physicist, Werner Heisenberg, encouraged the astrophysicist Ludwig Biermann to assemble a group of young theoreticians in Göttingen to look into fusion from the vantage point of astronomy. Germany was not permitted by the occupying powers to have a nuclear energy programme, so the work stayed at the level of basic science and the results obtained by Arnulf Schlüter, Reimar Lüst and others were freely published. There had been much discussion, both among gas discharge physicists and astrophysicists, about how to reconcile the individual particle description and the fluid description. One question was, for example, whether a magnetic field, which inhibits electron flow perpendicular to it, gives rise to a highly anisotropic conductivity. Schlüter developed a two-fluid theory [37] which brought light in such controversies.

We have already noted Landau's treatment of long-distance interactions in a plasma, in which the strength of the interaction varies as $1/r^2$ while the number of particles in a shell with radius r varies as r^2. Astronomers had encountered the same problem when they studied gravitational interactions between stars. Lyman Spitzer [38] computed the resistivity of a fully ionized gas caused by electrons losing their directed velocities in multiple small-angle scattering ('diffusion in velocity space'), and summarized the work on the effects of collisions in his seminal monograph *Physics of Fully Ionized Gasses,* which appeared in 1956. Because the cross-section for collisions between charged particles—the Rutherford cross-section—scales as T_e^{-2}, the resistivity of a fully ionized plasma scales as $T_e^{-3/2}$. (A hydrogenic plasma of $T_e = 1.5\,\mathrm{keV}$ already conducts as well as copper at room temperature, so that Ohmic dissipation becomes ineffective for heating as thermonuclear temperatures are approached.)

Also in 1956, Richard Post [39] published a survey of the fusion-relevant unclassified physics known at that time. Like Spitzer, he had to stop short of revealing secrets of the thermonuclear programme in which, as it turned out later, both were deeply involved. The constancy of the magnetic moment and the action of a magnetic mirror were well known at that time and Fermi [40]

Box 1.4 Magnetic pressure

By summing up the contributions of individual particles to the local current density in a steady-state plasma, one finds the diamagnetic current density associated with a temperature or density gradient,

$$\mathbf{j} = (\mathbf{B} \times \nabla p)/B^2.$$

In a macroscopic picture, this is the current through which, in static equilibrium, the magnetic field exerts a balancing pressure perpendicular to **B**:

$$\mathbf{j} \times \mathbf{B} = -\nabla p.$$

In a rectilinear magnetic field the sum of the plasma pressure, $p = n_e kT_e + \sum n_i kT_i$, and the pressure or energy density, $B^2/2\mu_0$, of the magnetic field is uniform; in a curvilinear field there is a magnetic tension along **B** of the same magnitude which provides the restoring force in Alfvén waves. The parameter β is defined as the ratio

$$\beta = 2\mu_0 p/B^2$$

and when this refers to a magnetically confined plasma, p is meant to be the maximum pressure while B is usually measured outside the plasma. In a curvilinear system, there is some arbitrariness in where precisely B is measured, so that one may even find instances where β is said to exceed one.

The pressure equilibrium between the self-field of a current in a plasma and the plasma pressure is known as the Bennett relation; in Tonks' version it reads

$$I^2 = 2 \times 10^7 \left(N_e kT_e + \sum N_i kT_i \right)$$

where I is the discharge current in amps, N_e is the number of electrons per metre length of the plasma column and T_e is the electron temperature (in eV), while N_i and T_i refer to the various ion species that may constitute the plasma. In a sharp-boundary model, in which the plasma is separated from the field by a thin current sheath, one finds this result by posing an arbitrary plasma radius, calculating the magnetic field and the plasma pressure, and noting that the radius drops out of the equation for the pressure equilibrium. Its validity can also be demonstrated, however, for arbitrary pressure profiles and the corresponding profiles of current density.

had even proposed repeated reflections between opposite mirrors moving towards each other as a mechanism for accelerating particles in cosmic magnetic fields. Yet secrecy prevented mirrors as means for confining plasma from being mentioned in Post's review article and Spitzer's book. Despite such limitations, these works defined the scientific background against which the groups that began to look into the subject around that time started their work. For those who entered the new field with a background of knowledge about individual particles and their motions, and naïvely thought that it would be just a small step to come to grips with their collective behaviour, the plasma had surprises in store. Alfvén, Spitzer and Post were the guides who led the emerging fusion community to the new way of thinking. (Post made up for his earlier omission by publishing in 1987 a monumental review of mirror confinement, and in 1995 he looked back at twentieth century plasma physics [41]).

1.2 In and out of secrecy

1.2.1 Programmes taking shape

Although the physics community was at first divided about the possibility of exploiting nuclear energy on earth—Rutherford [42] had called it '*moonshine*'—speculation on the subject abounded from the days when it was suspected that nuclear processes might be important for the stars. Remarkably, cold fusion was among the first schemes that were considered. Already in 1926, German chemists had reported the production of helium in hydrogen-loaded palladium, but before long they had to retract their claim [43]. Tandberg, a Swedish engineer, nevertheless continued along this line until the 1930s [44]. This topic returned briefly to prominence in 1989 with the much-publicized claims that cold fusion had been observed in electrolytic cells, but the results did not stand up to close scrutiny [45].

Magnetic confinement of hot plasma was actively pursued in 1938 by Kantrowitz and Jacobs [46] from the US National Advisory Commission for Aeronautics, who attempted heating a plasma by radio-frequency power in a toroidal magnetic field, but had to give up for lack of support. And in 1937, the German physicist Fritz Houtermans had made plans for work in the Ukrainian Physico-Technical Institute in Kharkov before he became one of Stalin's political prisoners.

In Britain, thoughts on thermonuclear reactors began to take shape shortly after the war. In 1946, George P Thomson and Moses Blackman at Imperial College in London had registered a patent for a thermonuclear reactor. This was based on magnetic forces—the pinch effect—for the confinement of electrons and on electrostatic forces, resulting from the space charge of these electrons, for the confinement of ions. In a torus of

Figure 1.5. Two of Thonemann's early tori at the Clarendon Laboratory, Oxford (*circa* 1949–50). The torus on the left is of Pyrex with a water-cooled coil wound around it to provide a toroidal magnetic field. The plasma current (about 10 A) was induced via RF coils and a transformer core (most of which has been removed for clarity in the picture). The torus on the right is of copper and was used to study the pinch effect at currents up to 2 kA.

major radius $R_0 = 1.3$ m and minor radius $a = 0.3$ m, a ring current of 0.5 MA was to have been established by betatron acceleration and to be maintained by a travelling wave fed by 3 GHz magnetrons. Ions would be heated by collisions with electrons and, with a postulated space-charge voltage of 0.5 MV, their temperature could run up to several hundred keV. The fuel would be deuterium and the fusion energy yield 9 MW(th), to be exploited as heat or as neutrons for breeding fissile materials. This proposal provoked much discussion within British atomic energy circles, but little action so that it would not be until 1948 [47] that Alan Ware could start experimental fusion research at Imperial College. The patent was classified, so the details were not made public at that time, and three years later this work was moved for security reasons to Associated Electrical Industries (AEI) in Aldermaston.

A parallel initiative had been started in 1946 in the Clarendon Laboratory at Oxford University by Peter Thonemann, who built a series of glass tori in which alternating current discharges could be induced electromagnetically with initially a 5 MHz and later a 100 kHz power source (figure 1.5). But only when the radio frequency source was used to drive a direct current in the discharge (an arrangement similar to the current drive that will be discussed in section 7.3) was the pinch effect clearly seen. From 1953 on, Thonemann used an iron-core pulse transformer, in which a unidirectional primary current pulse was driven by a switched capacitor bank and the plasma loop acted as the secondary winding (figure 1.6), the technique adopted earlier by Ware as well as by Russians and Americans. After

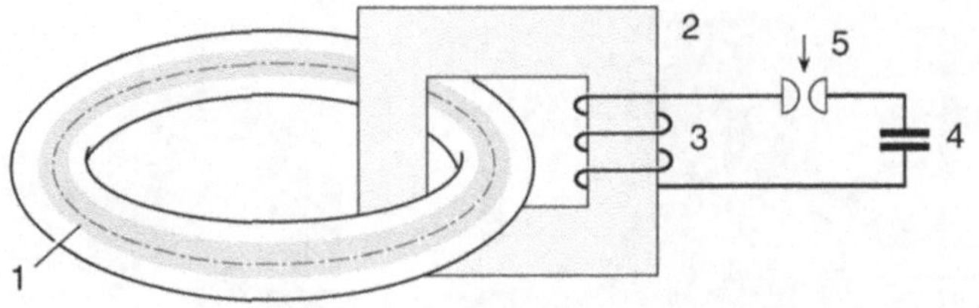

Figure 1.6. Transformer-driven toroidal discharge. The plasma loop, 1, acts as the secondary winding of a transformer, 2, whose primary winding, 3, is fed by a capacitor bank, 4, discharged through a spark gap, 5.

Thonemann's work had been transferred to Harwell, it rapidly expanded to culminate in the famous ZETA experiment. A third British programme developed in the mid-1950s in the Atomic Weapons Research Establishment, also at Aldermaston but quite separate from the AEI programme referred to earlier. The AWRE fusion work was originally concerned with imploding shock waves and linear pinches, but later expanded in other directions as well. These separate elements of the UK fusion research were brought together at the new Culham Laboratory established in the early 1960s.

Meanwhile, both Ware and Thonemann [48] had published papers on the subject which until 1956 would remain the most relevant places in the literature. Since all this work became classified in 1951, little was heard about it until 1956, except that in 1951 an Austrian physicist working in Argentina moved his benefactor, President Perón, to announce a breakthrough in fusion research. The scientific merits of this work were subsequently disproved, but Perón did manage to draw the attention of both scientists and government officials in the USA to the subject and so became the catalyst that activated their fusion research [49].

Fusion energy had been discussed in Los Alamos already before the end of the war, and some of the basic physics was recognized at that time, but it had remained dormant until an ambitious classified programme was launched by the Atomic Energy Commission (AEC) in 1952 and 1953. This started with the stellarator, conceived by Lyman Spitzer at Princeton as a steady-state magnetic field with closed magnetic surfaces on which the field lines had 'rotational transform' (section 2.4). Spitzer, himself a theoretician, attracted further mathematical support from Martin Kruskal and Martin Schwartzschild. Next, J L (Jim) Tuck, an Englishman who had already worked in Los Alamos during the war and had returned there from Oxford, initiated work on pinches. He started with an inductively driven discharge, named Perhapsatron, but before long Los Alamos had linear pinches as well. Richard Post, who first worked at Berkeley but then moved to Livermore, proposed mirrors. Initially, the Berkeley Laboratory's director, Herbert York, had considered a cylindrical system with an axial magnetic field plugged at the ends with radio-frequency fields, but this soon gave way to the mirrors. Later, Livermore broadened its programme with pinch work by Sterling Colgate and a relativistic-electron ring, Astron, invented by Nicholas Christofilos, while Oak Ridge entered the field with a study by Albert Simon of magnetized arc discharges, later to be followed by work on injection of ions into mirror machines [50]. Edward Teller, then already a senior physicist, and Admiral Lewis Strauss, chairman of the US AEC, were influential supporters of the American programme.

In the Soviet Union, Oleg Lavrentiev—a soldier in the far-east army without even a high-school diploma—appears to have been the first to draw the attention of his government to controlled thermonuclear reactions [51]. The letters he wrote in 1949 and 1950 were passed on to Igor Tamm, who in turn asked Andrej Sakharov (both of them worked in the nuclear weapons programme) to look into the problem.* They saw that Lavrentiev's suggestion to confine ions in electrostatic fields (section 3.5) was rather naïve, and developed their own ideas about toroidal magnetic thermonuclear reactors (MTR) [52]. In May 1951 the Soviet Government formally launched the controlled fusion programme and established a Council on MTR headed by Kurchatov, director of the Institute of Atomic Energy (IAE) in Moscow—which was later named after him. Lev Artsimovich became responsible for the experimental programme in the IAE, while Mikhail Leontovich directed theoretical studies. Leontovich was a gifted teacher, whose school brought forth eminent theoreticians who rose to prominent positions in their country. Kurchatov also drew institutes in Moscow (besides the IAE), Kharkov, Leningrad, Sukhumi, and later Novosibirsk into the work on plasma physics and controlled fusion which, like that in the US and Britain, was initially kept secret.

The early experimental work in the IAE was organized in three groups under Andrianov, Golovin, and Yavlinsky. Artsimovich first stimulated work on fast linear pinches, done by Filippov in Andrianov's group. Inductive discharges in strong toroidal fields as proposed by Sakharov and Tamm—known as tokamaks from 1959 on—initially were a common programme of Golovin's and Yavlinsky's groups. Osovets studied confinement by the time-averaged magnetic pressure exerted by a travelling wave. Later, when Vedenov and his theoretical colleagues had understood ponderomotive forces (box 3.3), RF confinement took the form of repelling the axial flow of a plasma in a straight field. In 1956, after Budker [53] had pointed out that drift orbits would lead to enhanced diffusion in toroidal systems and

* In 1950, the Harwell theoretician Klaus Fuchs was arrested for espionage on behalf of the Soviet Union and another theoretician, Bruno Pontecorvo, defected to Moscow. They knew about the work by Thonemann and Ware, and likely passed on information about this before it was published. There is, however, no evidence that this had any influence on Tamm and Sakharov, whose first ideas show little resemblance to the British work.

had proposed confining high-energy ions between magnetic mirrors, M S Ioffe formed a group to study such 'adiabatic traps', and started the PR series of experiments that would eventually run to PR-7. Artsimovich had become concerned about current-driven instability and tended to favour Ioffe's mirror experiments, but when he grew convinced of the potential of toroidal discharges, he encouraged Yavlinsky to attract a strong group of young physicists to this work.

The hierarchy in the IAE was complicated in that Igor Golovin was until 1958 a deputy to Kurchatov, which gave him an opportunity to promote a large mirror facility, OGRA, against the will of Artsimovich who wanted to conduct initial experiments on a smaller scale first. Kurchatov then formed an independent group under Golovin to build OGRA, but after Kurchatov's death in 1960, Artsimovich became the undisputed leader of all fusion research in the Soviet Union. He held important positions in the IAE, the Academy of Sciences and Moscow University which, however, did not keep him from taking direct charge of the tokamak group when in 1962 Yavlinsky died in a plane crash. By his deep insight in both the theoretical and the experimental aspects of the problem, his sober-minded and critical assessment of results and his sharp judgement of what were the most promising developments, Lev Artsimovich grew to become the world's leading authority on fusion research. He was a gifted lecturer and author of one of the most comprehensive books [54] on fusion research that were to appear in the early 1960s.

1.2.2 Looking behind the curtain

The first UN Conference on the Peaceful Uses of Atomic Energy, *Atoms for Peace*, convened in Geneva in 1955 upon the initiative of US President Eisenhower. Here, fusion was mentioned only in the opening address of the president of the conference, Homi Bhabha, who ventured his much-quoted prediction:

> *It is well known that atomic energy can be obtained by fusion processes as in the H-bomb and there is no basic scientific knowledge in our possession today to show that it is impossible for us to obtain this energy from the fusion process in a controlled manner. The technical problems are formidable, but one should remember that it is not yet fifteen years since atomic energy was released in an atomic pile for the first time by Fermi. I venture to predict that a method will be found for liberating fusion energy in a controlled manner within the next two decades. When that happens the energy problem of the world will truly have been solved forever for the fuel will be as plentiful as the heavy hydrogen in the oceans.*

These words stimulated laboratories around the world to take up the subject. Several contributions to the 1957 Venice, 1958 Geneva and 1959 Uppsala

Conferences report work started around 1956. In France, for example, where the fusion research that Meunier had started in 1952 had been abandoned when he left Fontenay-aux-Roses, a new programme was initiated in 1956 by Hubert, with support from Vendryès, Yvon and Francis Perrin. In Japan, the Universities of Tokyo, Nagoya, Osaka and Kyoto started pioneering research [55]. Special committees of the Japanese Atomic Energy Commission and the Science Council of Japan charted out a course in which first the scientific ground would be explored before the Japan Atomic Energy Research Institute (JAERI) would become involved. This led to the foundation of the Institute of Plasma Physics, affiliated to Nagoya University, in 1961. For its first twelve years this institute was directed by Kodi Husimi, a man of dignity and refinement who after his retirement continued to serve as the president of the Science Council.

Well before the advent of fusion research, gas discharge physics had been an active field of study, both in universities and in industrial laboratories. The highest energy densities were found in arcs, struck between metal or carbon electrodes. But in a classical arc, the degree of ionization is low and the ion temperature cannot rise much above the temperature of the neutral gas, which in turn is cooled by the surroundings. To overcome this barrier one had to resort to entirely different techniques. As a first step, the pinch effect would isolate a fully ionized plasma from the wall; the next step would be to bend the discharge tube into a torus so as to avoid heat loss to the electrodes. Naturally, linear and toroidal pinches were among the first schemes to be considered for the production of thermonuclear plasmas.

The simplest way to produce a high-current arc is to discharge a high-voltage capacitor bank through a low-inductance circuit into a pre-ionized gas. Under these conditions the formation of the plasma involves shock-wave heating and adiabatic compression, both of which are more efficient than Ohmic (or Joule) heating at elevated temperatures. Moreover, shock waves tend to heat the ions preferentially. So, although most groups saw this scheme as only a stepping stone towards toroidal discharges, one could envisage also very fast, very high-density linear pinches being established and conditions for fusion energy gain being reached on a time scale shorter than those of instabilities and axial heat loss.

When the first experiments on magnetic confinement were initiated, it had already been recognized that the pressure equilibrium could be unstable [56]. As a first step towards a more comprehensive MHD-stability theory, Kruskal and Schwartzschild [57] described the flute instability of a plasma supported against gravity by magnetic pressure, as well as the kink instability*

* Unstable deformations with amplitude varying as $\exp[i(m\theta + kz)]$ in cylindrical coordinates or $\exp[i(m\theta + n\phi)]$ in toroidal coordinates are called sausage instabilities if $m = 0$ and kink instabilities if $m = 1$.

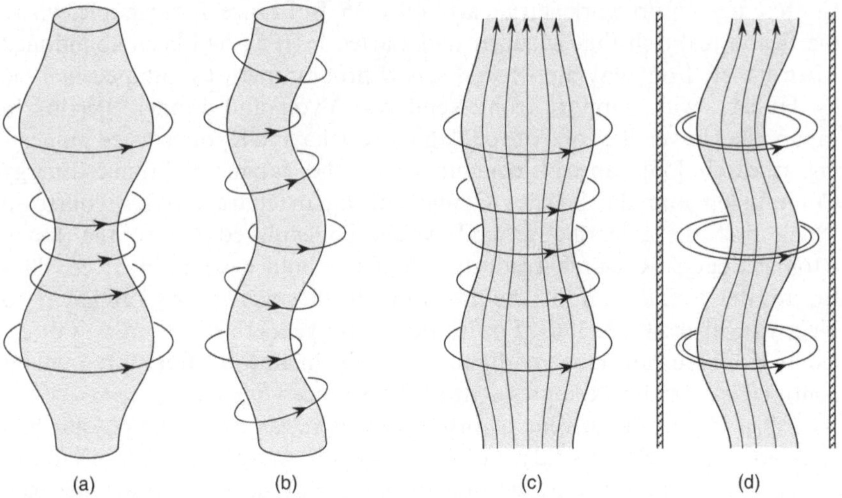

Figure 1.7. Sausage and kink instabilities. (a) and (b) Unstabilized, the field is stronger at the neck of the sausage than at the bulge and it is stronger at the inside of the kink than at the outside. (c) The sausage is stabilized by a trapped axial field; both the azimuthal field of the plasma current and the trapped field are strongest in the neck, but the former goes as r^{-1}, while the latter goes as r^{-2}. The kink is stabilized by the stress of the trapped field, combined with the compression of the pinch field between the plasma and the metal wall.

(figure 1.7) of a pinched discharge (box 1.5). Their paper had escaped censorship, but further studies of pinch stability in the US, Britain and the Soviet Union, which included the effects of trapped magnetic fields, metal shields, finite boundary layers and arbitrary distortions, were classified.

The first glance behind the curtains of secrecy was given on the occasion of a Soviet state visit to Britain in April 1956. As a member of the delegation, Kurchatov [58] was allowed to present a lecture at Harwell, where he surprised the world by bringing the Russian experiments on discharges in straight tubes, filled with deuterium or some other gas, into the open. Kurchatov reported neutrons [59], but made it clear that they were produced by deuterons colliding with other deuterons—possibly adsorbed at the electrodes or at the wall—after having been accelerated over several tens of kilovolts, rather than by true thermonuclear reactions. Neutron pulses were always coincident with bursts of hard (up to 300–400 keV) X-rays. Kurchatov expressed doubt about the possibility of keeping the pinched column from coming in contact with the walls and attached particular interest '*to methods in which stationary processes may be employed*', without elaborating on this subject. In the discussion [60] that followed the lecture it became clear that they were also working on toroidal pinches, but he may as well have been referring to the mirror devices on which his laboratory had just started experimental work.

Box 1.5 Stability of the pinched plasma column

The simplest form of a pinch, a straight plasma cylinder carrying an axial current whose azimuthal field balances the plasma pressure, is first of all unstable against sausage deformations. The field is strongest in the constrictions, squeezing the plasma farther into the bulges. A trapped axial field resists both the contraction and the expansion because its strength varies as the square of the radius, whereas the confining field varies linearly. In a kink deformation the driving force comes from the crowding of the azimuthal field lines on the inside and the thinning out on the outside. The stabilizing force is the tension of the trapped axial field, or the increase of its energy associated with the stretching.

The earliest theoretical work on these 'stabilized pinches' was based on a normal-mode analysis and dealt with a sharp-boundary model in which the plasma current was restricted to an infinitely thin layer. It showed that a trapped field would be effective for stabilizing short wavelengths while a conducting wall outside the plasma would suppress long-wavelength deformations by image currents or, what amounts to the same, by the field of the plasma current being compressed between the plasma and the wall.

So far, no toroidal effects had been considered, other than the requirement that the pitch of a disturbance must be an integer fraction of the length of the cylinder for the endplanes to match when it is bent into a torus or toroid. But then the equilibrium is lost, so one first had to see how this could be restored before one could turn to the stability of toroidal systems. Shafranov described the equilibrium of the toroidal pinch, after which he and Grad independently derived the general equation for the equilibrium in a closed system from which subsequent stability analysis had to start.

A series of papers published in *Atomnaya Energya* [61] described the Russian pinch work in considerable detail. They were summarized by Artsimovich and Golovin in an extraordinary session of the Stockholm Symposium on Electromagnetic Phenomena in Cosmical Physics of the International Astronomical Union in August of 1956 [62]. Artsimovich reviewed the evidence for the occurrence of both sausage and kink instabilities. He estimated the ion temperature at 100 eV, thus excluding a thermonuclear origin of the observed neutron flux. In an impressive performance, he then went on to present Shafranov's paper, which extended earlier theoretical work on the kink instability by including an axial (B_z) magnetic field.

Golovin described more details of the experimental work. Artsimovich's paper opened with an introduction in which he argued that the energy balance of a reacting plasma is determined by the lifetime of an ion in it. From this he derived a criterion for power generation in a D–D reactor: $10^{-15} n\tau \approx 1$ at $T \approx 10^8$ K, where n is the density of the plasma in cm^{-3}, τ is the lifetime of an ion in the system in seconds and T is the temperature. This is, in essence, the criterion derived a few years earlier by Lawson from Harwell, but not published because of secrecy until 1957. There is no trace of this result in the original Russian papers, so one must assume that it reflected how Artsimovich himself had thought about the thermonuclear problem. Yet he never claimed credit for having derived the Lawson criterion independently.

At this Stockholm conference, the only *acte de présence* of fusion research in the Western countries was by R S (Bas) Pease from Harwell, who derived the pinch current (1–2 MA) at which bremsstrahlung makes equilibrium with Ohmic dissipation [63]. Meanwhile, pressure towards declassification built up, both in the scientific community and in public opinion, and some fusion work began to trickle into scientific journals also west of the Iron Curtain.

Harwell responded to Kurchatov's challenge by producing a series of papers for the January 1957 issues of the *Philosophical Magazine* and the *Proceedings of the Physical Society*. Here, they established the Lawson criterion and the Pease current, described theoretical work on stability and on MHD shocks, and published the first photographs (figure 1.8) of an unstable toroidal pinch. Elsewhere, W B Thompson published a brief theoretical introduction to the thermonuclear problem [64].

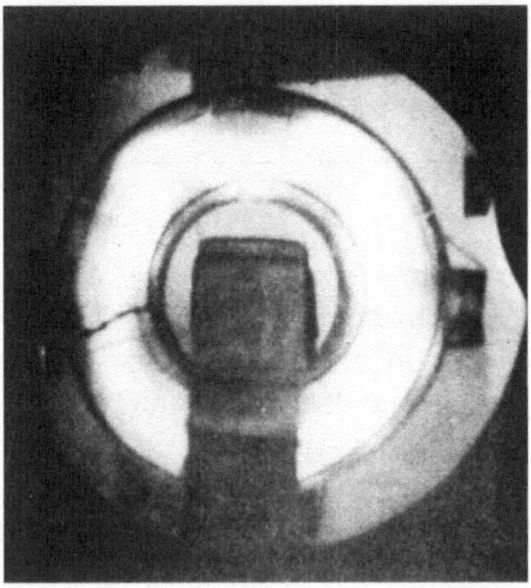

Figure 1.8. Photograph of an unstable toroidal pinch discharge with current 1.3 kA in a xenon plasma. The glass tube has a major radius of about 0.3 m.

When a plasma is confined in a magnetic field, MHD instability may be driven by two sources of free energy, the pressure gradient and the magnetic energy associated with the deviation from the vacuum field. The second cause had been studied—in an astronomical context—by Lundquist [65], who had derived an energy principle for a non-vacuum field in a pressureless conducting fluid (one in which the current is everywhere parallel to the magnetic field). Several groups had undertaken to extend this treatment, to take account of the free energy residing in the plasma pressure. While most of the earliest work was concentrated on the kink instability of the pinch [66], Rosenbluth and Longmire [67] had studied flute instabilities in geometries with curved field lines, recognizing the unfavourable curvature in mirror devices and throwing a new light on the sausage instability of the pinch. Further work along these lines eventually led to the general energy principle laid down in the definitive 1957 BFKK (Bernstein, Frieman, Kruskal and Kulsrud) [68] paper, but independently derived in essence by Kadomtsev, by Hain, Lüst and Schlüter and by Berkowitz *et al.*

Not yet having their own channels of communication, fusion physicists from East and West set out to invade the Third International Conference on Ionization Phenomena in Gases, held in Venice in June 1957 [69]. Groups from Harwell and from the Centre d'Etudes Nucléaire in Saclay, France, reported on highly unstable toroidal pinches, but Harwell saw the first signs of stabilization by an 'axial' or toroidal magnetic field. Aldermaston, Berkeley and Saclay also had work to report on linear pinches, not unlike that disclosed earlier by the Russians, except that Aldermaston ventured to claim a possible thermonuclear origin for their neutrons. Both Shafranov and the Göttingen group discussed toroidal equilibria; the latter had derived a general stability theory and proposed magnetic pumping as a heating mechanism. Papers from Los Alamos, Harwell and CERN (Geneva) discussed runaway electrons, a phenomenon caused by fast electrons having a longer mean-free path for collisions with ions than slow ones, so that above a critical field strength—the Dreicer limit—there develops a beam of fast electrons. Los Alamos, Harwell and Aachen expanded the theory of the B_z-stabilized pinch; Kruskal from Princeton presented his theory of adiabatic invariance of the magnetic moment.

The next step in the declassification process was preceded by a series of papers on equilibrium and stability from the Göttingen group, who were not restricted by rules of secrecy, in *Zeitschrift für Naturforschung* [70]. Then on 25 January 1958 the British journal *Nature* [71] published a series of papers that represented the outcome of British–American exchanges of information and negotiations about declassification [72]. Optimism ran high in Britain and the United Kingdom Atomic Energy Authority (UKAEA) had been anxious to open up their thermonuclear file. Harwell had recently installed the large toroidal pinch apparatus ZETA (Zero-Energy Thermonuclear Assembly, a name alluding to fission terminology in which a low-power critical assembly was called a zero-energy pile), and had neutron fluxes suggestive of a temperature of 0.4 to 0.5 keV. In the US, Los Alamos had comparable results from their Perhapsatron S-3 toroidal pinch, which gave neutron yields consistent with 0.5 keV ion temperatures. There were also reports about toroidal pinch work in the AEI laboratory at Aldermaston, as well as stabilized linear pinch work from Los Alamos. In the same *Nature* issue, Spitzer sounded a warning by pointing out that the ion temperatures claimed for toroidal pinches could not be explained on the basis of Ohmic heating, and elsewhere [73] Livermore disclosed neutron-spectroscopic measurements on a linear pinch which showed that deuterons had been accelerated axially to energies over 50 keV, presumably by sausage instabilities.

Despite these critical notes, the ZETA results gave rise to great expectations. A journalist enticed Harwell's director Sir John Cockroft to say that he was ninety per cent certain that the neutrons had a thermonuclear origin. But a few months later, in May 1958, nuclear physicists from Harwell [74] reported anisotropic neutron spectra showing the presence of deuteron streams and invalidating the thermonuclear claim. About the same time, the Americans disclosed much of their non-pinch work in the Spring Conference of the American Physical Society (APS) [75]. Spitzer presented the stellarator and discussed the main methods—figure-eight tubes and helical windings—to produce rotational transform. Albert Simon introduced the Oak Ridge approach towards trapping injected ions in a mirror machine and Post described the Livermore pulsed-mirror or pyrotron line of experiments. Details were withheld for the forthcoming second Atoms for Peace Conference, however. Thus, the ball was again in the Soviet court, but their side deferred further declassification until the conference.

1.2.3 The road to travel

Let us recall at this point what could be said around that time, based upon sources in the open literature such as the books by Alfvén and Spitzer and the 1956 review article by Post, about the problem of building a thermonuclear reactor.

By considering the yield-to-loss ratio,* R, in a pulsed reactor system—which did not rely on ignition—Lawson had derived general criteria for the temperature, density and confinement time required for a reactor to become energetically self-sustaining (box 1.6). The reaction-rate parameters as known at this time are shown in figure 1.9 [76].

In a steady state, τ assumes the character of an energy-confinement time, τ_E. (In early work particle loss went much faster than heat conduction, so that the energy loss was primarily the energy carried off by the particles and τ_E was essentially the particle confinement time, τ_p.) The energy confinement time is

* Elsewhere, we use the symbol R to denote the major radius of a torus.

Box 1.6 Lawson's criteria

The fusion power density generated in a D–D plasma is

$$P_f = \tfrac{1}{2} n_i^2 \langle \sigma v_i \rangle E_{DD}$$

where $\langle \sigma v_i \rangle$ is the reaction cross-section averaged over the relative velocity of colliding ions and E_{DD} is the mean reaction energy of the two primary D–D reactions ($E_{DD} = 3.65\,\text{MeV}$). In a D–T plasma with equal amounts of D and T,

$$P_f = \tfrac{1}{4} n_i^2 \langle \sigma v_i \rangle E_{DT}$$

where $E_{DT} = 17.6\,\text{MeV}$.

The bremsstrahlung power loss scales as

$$P_B \propto n_e^2 T_e^{1/2}$$

so that the ratio of the reaction yield to the bremsstrahlung loss is a function of the temperature only. For net energy production, this ratio has to be at least in the order of one, the precise requirement depending on the fuel cycle and the energy conversion efficiencies. This defines a lower limit for the operating temperature.

In a pulsed system, the energy balance must also account for the thermal energy invested in the plasma,

$$W = \tfrac{3}{2}(n_i k T_i + n_e k T_e) \approx 3nkT.$$

If τ is the duration of the pulse, the yield-to-investment ratio, R, is a function of $n\tau$:

$$R = \frac{\tau P_f}{W} = \frac{n\tau f(T)}{3kT}.$$

Lawson concentrated on pulsed systems and drew curves of constant $n\tau$ in an R versus T diagram, but plots of $n\tau$ versus T or of $nT\tau$ versus T for constant R are also called Lawson diagrams. For a pulsed reactor with an efficiency of 33% for conversion of heat to electricity, Lawson gave criteria both for the minimum temperature and for the minimum $n\tau$ value at the optimum temperature:

$$T \geq 20\,\text{keV and } n\tau \geq 10^{22}\,\text{m}^{-3}\,\text{s at } T \approx 100\,\text{keV for DD}$$

or

$$T \geq 3\,\text{keV and } n\tau \geq 10^{20}\,\text{m}^{-3}\,\text{s at } T \approx 30\,\text{keV for DT.}$$

For an energy-producing reactor it is of course not sufficient that the system can sustain its own energy demand; to arrive at an economically acceptable level of circulating power, it must exceed Lawson's $n\tau$ limit by nearly one order of magnitude.

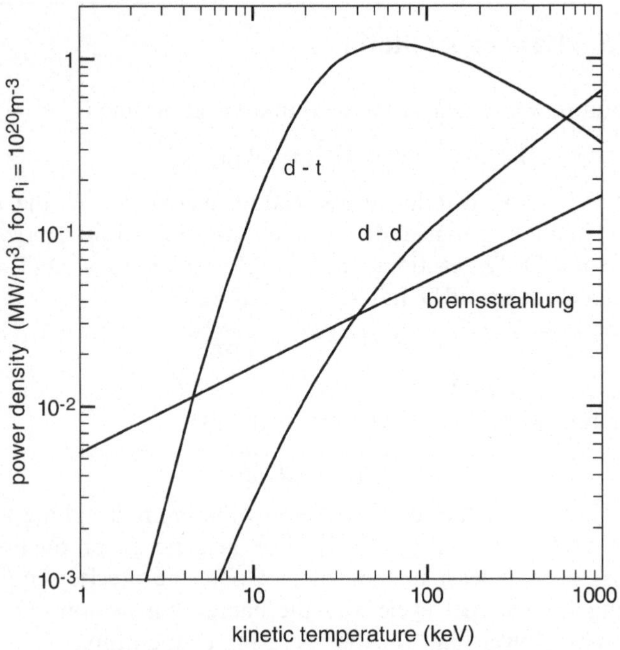

Figure 1.9. Power density of the fusion reaction in pure deuterium and in a 50% D plus 50% T mixture as a function of plasma temperature compared with bremsstrahlung.

related to the heating power P_H needed to maintain the plasma at the temperature T through $P_H = 3VnkT/\tau_E$, where V is the volume of the plasma and the factor 3 expresses that the ions and the electrons both have $\frac{3}{2}kT$ as mean energy. Then, the power loss scales as nT/τ_E and the power yield as $n^2\langle\sigma v\rangle E_r$, so that R is proportional to $n\tau_E f(T)$. Neglecting all losses other than bremsstrahlung and assuming equal ion and electron temperatures, Post estimated the ignition point to lie at 35 keV in deuterium and 4 keV in a 50 : 50 deuterium–tritium mixture. To find the optimum values for $n\tau$ (or $n\tau_E$) and T, one must of course specify the mode of operation, pulsed or steady state, and the efficiencies of various energy conversion steps in more detail. Moreover, the way in which τ_E depends on n and T affects the optimization. But whether τ is taken to be a pulse duration, a particle-confinement time or an energy-confinement time,* the outcome is not greatly different, so that Lawson's figure $n\tau = 10^{14}\,\text{cm}^{-3}\,\text{s}$ ($10^{20}\,\text{m}^{-3}\,\text{s}$ as we quote it), stuck in people's minds and became the milestone against which progress in fusion research was measured, regardless of the precise meaning of τ.

* For the present discussion, energy confinement is of course what counts. But the charm of the $n\tau$ criterion is its global nature, which obviates the need to define precisely whether τ represents energy transport, particle loss or pulse duration.

In Alfvén's terminology, magnetic confinement is an example of a field 'frozen' in a plasma, although in this case one should perhaps say frozen out. Such a state will decay towards thermodynamic equilibrium, in which the temperature and the density of the plasma are uniform and the field has the vacuum configuration consistent with the currents in external coils; the decay time is given in box 1.7. The time constant characterizing this decay would then figure as the energy-confinement time in a Lawson criterion. Assuming this classical loss rate, one can calculate that a D–D reactor operating at $T \approx 50\,\text{keV}$ would need $Ba \approx 1\,\text{T m}$ (Tesla-meter), with the additional requirement that B has to satisfy the pressure balance (the optimum temperatures under these conditions are about a factor 2 lower than in Lawson's pulsed model). For a D–T reactor at 15 keV this figure would be only $10^{-1}\,\text{T m}$, but to confine 3.5 MeV α-particles with $B\rho = 0.27\,\text{T m}$, one would still need Ba in the order of 1 T m.

All this seemed well within the range of technical feasibility. It was clear, however, that classical confinement could not be taken for granted.

Box 1.7 Classical loss rate

To estimate the decay time of a plasma column in a magnetic field, one first calculates the E/B drift velocity (box 1.2) associated with the diamagnetic current (box 1.4)

$$\mathbf{v}_\text{d} = -\frac{\eta \nabla p}{B^2}$$

where η is the resistivity of the plasma, and obtains as an estimate of the plasma or particle confinement time

$$\tau_\text{p} \approx \frac{B^2 \Lambda^2}{\eta p}$$

Λ being the scale length of the pressure profile. As a first estimate, one may put $\Lambda \approx a/2.4$, where a is the plasma radius. (The factor 2.4 is appropriate for a Bessel function profile in a plasma cylinder; it would be $\pi/2$ for a cosine profile in a plane slab.)

The Spitzer resistivity of a hydrogenic plasma is given by

$$\eta \approx 3 \times 10^{-8} T_\text{e}^{-3/2}\ \text{ohm-m} \quad [T_\text{e}\ \text{in keV}]$$

so that, if this 'classical' resistive particle diffusion were the dominant loss mechanism, a thermonuclear plasma with $T_\text{e} \approx T_\text{i} \approx T$ and $p = 2n_\text{i}kT$ would be confined with

$$n\tau_\text{p} \approx 1.8 \times 10^{22} (Ba)^2 T^{1/2}.$$

Switching from an MHD to a particle point of view, one can show that, in binary collisions, the guiding centres of the two particles make equal or opposite jumps, depending on whether the charges are opposite or equal. Thus, ion–electron collisions (responsible for Spitzer resistivity in the MHD picture) give rise to equal diffusion rates for ions and electrons (ambipolar diffusion). Classical diffusion may be shown to correspond with

$$D_{\rm ci} = \frac{\eta(p_{\rm e} + p_{\rm i})}{B^2},$$

which has an $nB^{-2}T^{-1/2}$ dependence.

This classical picture of cross-field diffusion, however, fails to account for the actually observed losses of magnetically confined plasmas. Anomalous loss caused by collectively generated field disturbances was first encountered in the context of the wartime development of ion sources for isotope separators [77]. Bohm proposed a threatening semi-empirical formula for the diffusion coefficient

$$D_{\perp\rm B} = \frac{kT}{16eB}.$$

The numerical factor is 1/10 in Bohm's paper, but is usually quoted as 1/16, in agreement with the companion papers co-authored by Burhop and Massey. This 'drain diffusion' was suspected of being connected with microscopic plasma turbulence (box 3.3) and was long feared to be an intrinsic property of magnetized plasma, even though Simon [78] had shown that, at least in arc discharges like those in the isotope separators, which were what Bohm diffusion was really about, the loss rate could be explained by assuming that the process was not really an instance of cross-field diffusion of the plasma, but came about by ions diffusing radially across the magnetic field and electrons being driven out axially by the resulting electric field. But when in later years stellarators once more displayed anomalous loss, such an explanation was not immediately at hand and in this area Bohm's (or by then rather Spitzer's) conjecture was already living its own life. In fact, we shall see in later chapters that something akin to it, albeit at a far lower level, continues to play a role in discussions about anomalous loss in tokamaks and stellarators.

Bohm diffusion as first hypothesized would lead to a staggering requirement of $B^{3/2}a > 100\,\mathrm{T}^{3/2}\,\mathrm{m}$ for a D–T reactor* and to even an order of magnitude more for D–D. So, where Lawson had set the goal and Spitzer's resistivity had defined the shortest possible path along which to reach it, Bohm had placed a seemingly insurmountable roadblock. Until the issue was finally settled at the 1968 IAEA Conference in Novosibirsk, Bohm diffusion haunted fusion research like an evil spirit.

* Strictly, the requirement would be: $B^{3/2}a \approx 3.4 \times 10^{-10}(n\tau)^{1/2}T/\beta^{1/2}$. With reasonable estimates for a D–T reactor, like $n\tau = 10^{21}$, $\beta = 0.1$ and $T = 15\,\mathrm{keV}$, this would translate into $B^{3/2}a \approx 500\,\mathrm{T}^{3/2}\,\mathrm{m}$. (Note: T represents *temperature*, T the unit of Tesla.)

Chapter 2

Geneva 1958

In the 1955 'Atoms for Peace' conference in Geneva, the nuclear powers had placed their efforts towards civilian applications of fission energy before the world. When preparing the second conference of this kind, which would be held in September 1958, again in Geneva, they decided to yield to the pressure to declassify their fusion research and to display some of their most significant experiments in working condition in the exhibition on the UN premises.

2.1 Fast linear pinches or Z-pinches

As we saw, the most important results obtained with fast linear pinches had already been disclosed prior to the Conference and in this respect Geneva brought no great surprises. The most elaborate experiments were reported by the Institute of Atomic Energy* in Moscow [1], where Filippov had pushed these discharges to the highest currents and energies, and had introduced discharge tubes with metal walls in order to reduce the influx of impurities. This group had observed neutrons already in the summer of 1952 and the neutron yield had gone up to 3×10^9 per pulse. The emission appeared to originate from a region near the anode where the current sheath first collapsed upon the axis of the tube. Axial plasma jets emerged from this spot [2], as witnessed by craters formed on the anode and by erosion of a diaphragm placed halfway between the anode and the cathode. The neutron spectrum was strongly asymmetric, indicating an acceleration mechanism. Filippov's metal-wall device became known in later years as the 'plasma focus'.

* The Institute of Atomic Energy in Moscow was renamed 'I V Kurchatov Institute of Atomic Energy' in 1960 in honour of its first director. It is generally known as the Kurchatov Institute and we will hereafter adopt this name throughout.

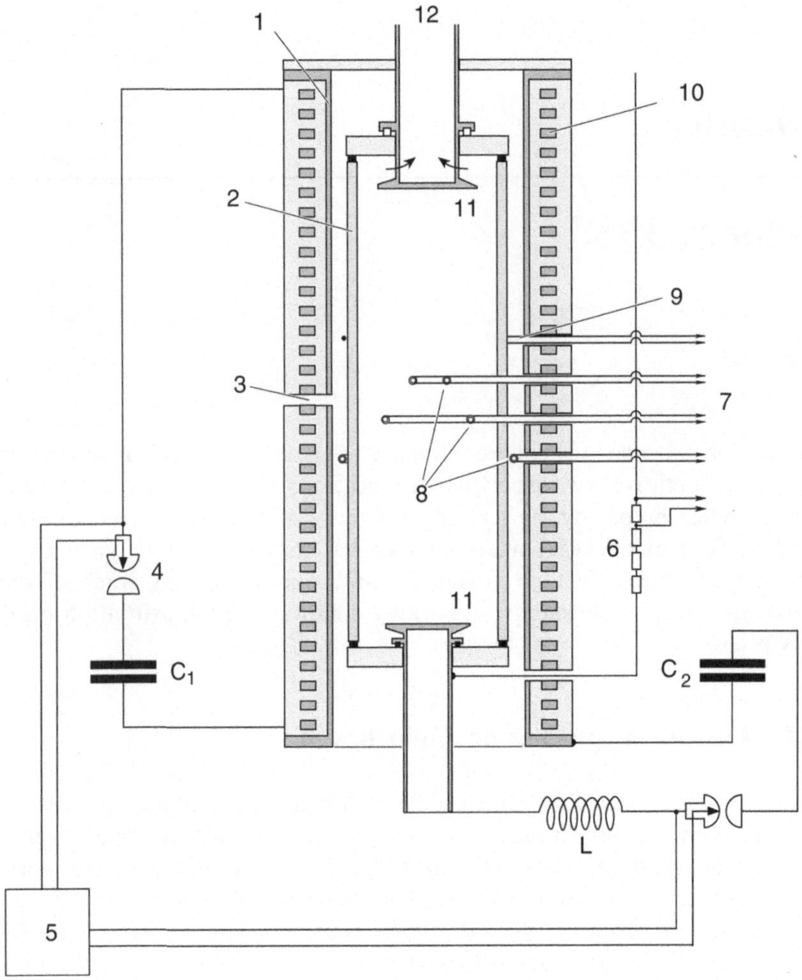

Figure 2.1. Pinch discharge tube: 1, coaxial return conductor; 2, porcelain or glass tube; 3, slot for photography; 4, spherical spark gap; 5, trigger control; 6, voltage divider; 7, connection to oscillograph; 8, Rogovski belts for measuring current distribution; 9, turns for measuring plasma diamagnetism; 10, longitudinal field coil; 11, copper or stainless steel electrodes; 12, vacuum outlet.

It was suggested, however, that the neutrons produced in one of their Pyrex-wall chambers (figure 2.1) might be of thermonuclear origin. In his review of the work in the USSR, Artsimovich [3] reported that a temperature in excess of 0.3 to 0.4 keV was reached in a 500 kA discharge in deuterium. (The more detailed paper [4] quotes a yield of 5×10^7 neutrons at 1.5 MA.) Since Doppler broadening of spectral lines from impurities failed to yield

Table 2.1 Linear pinches presented at the Geneva 1958 Conference

Location	Device (author)	Diameter (cm)	Length (cm)	V_{bank} (kV)	$\tau_{1/4}$ (μs)	I_{max} (MA)	B_z (T)	Remarks (neutrons/discharge)
IAE (Kurchatov	(Andrianov)	40	90	40	8	0.5		
Institute), Moscow	(Andrianov)	40	50	100	7	1.3		(5×10^7)
	(Andrianov)	17–18	80–100	40		0.2		Various diagnostics
	(Andrianov)	42	70	30	7	0.5		Plasma focus (3×10^9)
	(Komelkov)	60	17.5	30	7	0.9		Current distribution
	(Komelkov)	18.5	4.7	40	2	1.4–1.8		Fast current rise $(<10^6)$
	(Komelkov)	19	15–25	40	3	0.9		Fast photography
	(Golovin)	18–23*	80		≤900	0.3	2.7	B_z-stabilization
Livermore	Screw-dynamic	10	20	70	3	0.25	3	Helical return current (10^5)
	Triax**	10/20	50	30	3	1.3		(2×10^4)
	Triax**	2.5/20	50	20	4	1.3	0.02	(2×10^5)
Los Alamos	Columbus II	10	30	100	3	0.5	0.05	Asymmetric neutron spectrum (6×10^8)
	Columbus S-4	13	61	20	6	0.2	0.2	Current distribution
	(Baker)	15*	600	3	40	0.07	0.05	$m = 1$ instability
	(Baker)	90*	360	3	200	0.6	0.2	Highly unstable
GA, San Diego	(Kerst)	4–7	30	8	5	0.1–0.4	2	Current profiles
AWRE, Aldermaston	(Curran)	15	28	25	4	0.54	0.2	Low inductance (10^7)
Fontenay	(Aymar)	28	90	35	5	0.2	0.15	Cinematography (10^7)
KTH, Stockholm	(Siegbahn)	30	60	50	6	0.4	0.07	(10^6)

* Metallic discharge tube; others are glass or ceramic.

** Sheet pinches; the diameters given are those of the inner and outer metal cylinders.

unambiguous deuteron temperatures, he compared observable variables such as currents, voltages and magnetic fields with their computed values to lend credence to the ion temperature predictions produced by the same calculations. He concluded that the neutron emission was in large measure of thermonuclear origin, but went on to say:

> *At present, however, proof or disproof of the thermonuclear origin of a small burst of neutron emission is hardly of such importance as to warrant special attention in discussions on this subject. This is why I do not consider it necessary to insist that in the above-mentioned experiments thermonuclear reactions were actually observed. The question of whether a given neutron belongs to the noble race of descendants of thermonuclear reactions or whether it is the dubious offspring of a shady acceleration process is something that may worry the pressmen but at the present stage in the development of our problem it should not ruffle the composure of the specialists. When the number of neutrons arising in a single discharge pulse reaches a value in excess of 10^{12} then all doubts as to the origin of this effect will vanish.*

This set the tone for further discussions on 'true' and 'false' neutrons, but did not keep others who saw neutrons, nor Artsimovich himself some years later, from setting out their thermonuclear stakes.

Some parameters of the linear pinch devices reported at Geneva are listed in table 2.1. Neutron yields were mostly in the range 10^5–10^7 per discharge. Where neutron spectra were obtained, there was always at least some asymmetry in the axial direction, indicative of directed rather than thermal ion velocities. Moreover, the neutron pulse was often accompanied by hard X-rays, indicating that at least some of the electrons had

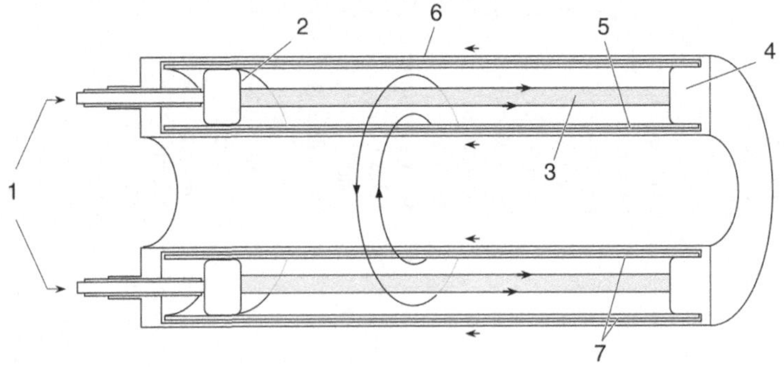

Figure 2.2. Sheet pinch or hard-core pinch. The plasma current flows in a hollow cylinder, the return current divides itself between the inner and outer metal cylinders. 1, current feed; 2, electrode; 3, plasma; 4, electrode; 5, inner metal cylinder; 6, outer metal cylinder; 7, insulators.

superthermal energies. But where electron temperatures were measured or derived from the electrical conductivity of the plasma, they were found to be as low as 5–10 eV.

The Berkeley laboratory also had a modified pinch device, the sheet pinch (figure 2.2), in which the plasma current ran between two concentric metal cylinders. This was thought of as an approximation to a current-carrying flat sheet of infinite extent, that would be stable against the characteristic pinch instabilities. The discharges indeed turned out to be more reproducible than normal stabilized pinches, but they were not free from small-scale instabilities and spurious neutron emissions [5].

2.2 Steady-state mirror confinement

The idea of confining particles between magnetic mirrors (box 2.1) had occurred in at least two places: Berkeley and Moscow. In 1951/1952, Post [6] started experimental work in the Radiation Laboratory of the University of California at Berkeley, but this work was later transferred to Livermore. Independently, Budker [7] started a theoretical study of mirror confinement in the Kurchatov Institute in Moscow, which gave rise to experimental work under Golovin and Ioffe.

In steady-state mirror machines, the 'magnetic bottle' or 'magnetic trap' (in US or Russian terminology) may be filled with hot plasma by a variety of means like injection of fast particles, an ion-magnetron, RF heating, or a plasma gun. Trapping of individual ions, as practised in Oak Ridge [8] and Moscow, requires a special provision, because when the magnetic moment is preserved a charged particle entering the confinement region from outside will leave it again at once. The first beam-trapping experiments were done with molecular D_2^+ or H_2^+ ion beams which disintegrated into D^+ or H^+ ions and D or H atoms by collisions with residual gas, with particles already trapped between the mirrors, or with particles in an arc especially provided to intersect the injected beam. From the outset, the Oak Ridge laboratory mostly made use of deuterium beams, whereas the initial experiments in Moscow were performed with hydrogen. Other early proposals to trap injected beams suggested either a rising mirror field to make the ions withdraw from the injector (Post), or RF fields at the ion-cyclotron resonance to increase the magnetic moment of ions injected through the mirrors—the Sinelnikov scheme [9].

The Oak Ridge and Moscow groups reported on large steady-state mirror devices: the Direct Current Experiment, DCX [10] (figure 2.3), and OGRA (figure 2.4). Contrary to the interpretation '**Ar**tsimovich–**Go**lovin', which the editors of the Geneva Proceedings [11] had been led to accept, OGRA stood for '*odin* ***gram*** *neitronov v sutki*' (one gram of neutrons per day), which is what the Soviet team had secretly hoped to produce in the

Box 2.1 Mirror confinement

If the magnetic field strength has a gradient along the direction of the field, the Lorentz force, averaged over one gyration period, has a component directed from the stronger towards the weaker field. If the parallel component of the velocity is not too large compared with the perpendicular component (the gyration), a particle travelling along a field line may be trapped between local field maxima—which are therefore called magnetic mirrors.

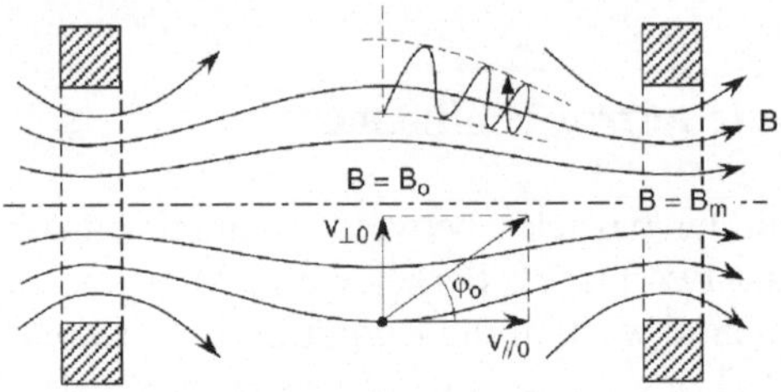

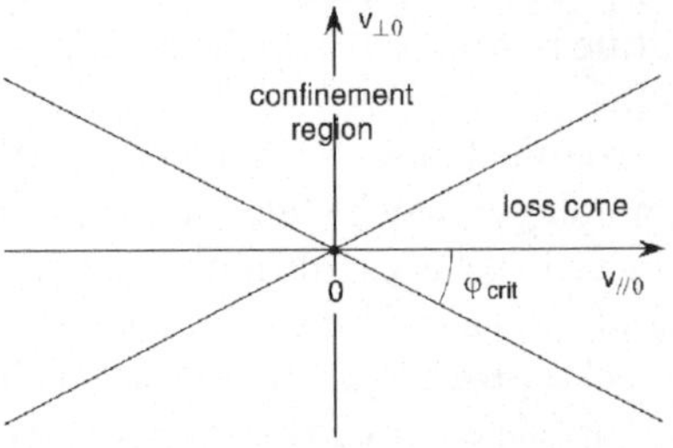

The upper part of the figure shows a magnetic mirror field and the coils that produce it; the lower part shows the confinement region, $\varphi_{\text{crit}} < \varphi_0 < \pi/2$, and the loss cone, $\varphi_0 < \varphi_{\text{crit}}$, in velocity space. The components of the velocity parallel and perpendicular to $\mathbf{B}$ at $B = B_0$ are named $v_{\perp 0}$ and $v_{\parallel 0}$. The magnetic moment of a particle with velocity v and pitch angle φ—the angle between its velocity and the magnetic field lines—equals $mv^2 \sin^2 \varphi/2B$; its constancy implies that $\sin^2 \varphi/B = \sin^2 \varphi_0/B_0$, where φ_0 and B_0 are the values of φ and B at the midplane. Thus, the value of B at the point of reflection, B_{refl}, is $B_{\text{refl}} = B_0/\sin^2 \varphi$, and if the particle is to be reflected before it reaches the mirror, where $B = B_{\text{m}}$, it is necessary that $\sin^2 \varphi_0 > B_0/B_{\text{m}}$. This defines the critical pitch angle. The quantity B_{m}/B_0 is called the mirror ratio, M, so we see that a particle is confined between the mirrors if:

$$\sin^2 \varphi_0 > \sin^2 \varphi_{\text{crit}} = M^{-1}$$

whence $\varphi_0 < \varphi_{\text{crit}}$ is called the loss cone in velocity space.

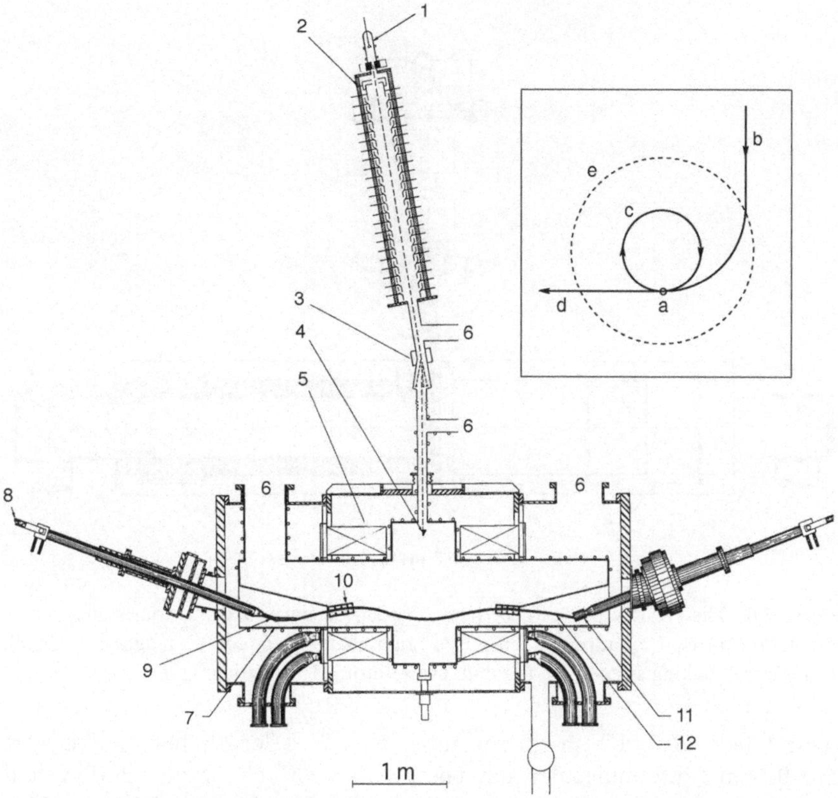

Figure 2.3. The DCX mirror experiment. 1, molecular ion source; 2, accelerator tube; 3, deflection magnet; 4, D_2^+ beam; 5, coil; 6, vacuum outlet; 7, high-vacuum liner; 8, current feed; 9, cathode; 10, baffle; 11, anode; 12, arc. Insert: end-on view showing particle trajectories: a, arc; b, D_2^+ beam; c, trapped D^+ ions; d, D^0 atoms; e, field boundary.

machine.* DCX had already been in operation; OGRA had only recently been assembled by a group under Golovin [12]. Both employed injection of molecular ions. In DCX a carbon arc was drawn along an eccentric field line and the beam was directed so that, upon dissociation of D_2^+ ions in the arc, the D^+ ions would describe circular orbits centred upon the axis of symmetry. In OGRA, dissociation of H_2^+ was to proceed on the residual gas and, eventually, on the plasma formed in the trap. In either apparatus, the main experimental goal was to achieve 'burnout', a state in which the plasma would become sufficiently dense to be opaque to neutral particles from the wall, so that the neutral gas and the ensuing charge-exchange losses would disappear from the interior of the plasma [13]. DCX had a

* This would have been a momentous achievement indeed, requiring a deuterium beam of several A_{eq} and yielding a continuous power of nearly 10 MW.

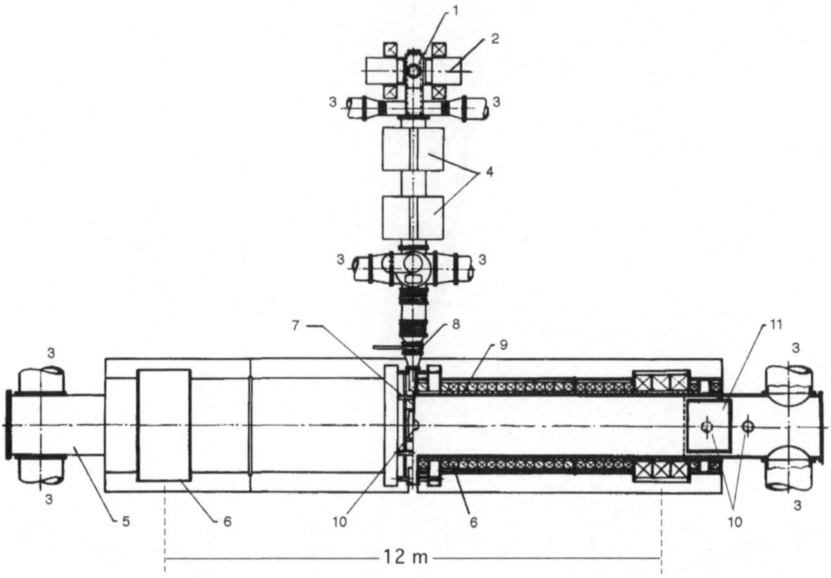

Figure 2.4. The OGRA mirror experiment. 1, 200 kV ion source; 2, source magnet; 3, evacuation ports; 4, quadrupole lens; 5, vacuum chamber; 6, coil; 7, magnetic channel; 8, shutter; 9, baking system; 10, titanium evaporator; 11, electrode.

central field $B_0 = 1\,\mathrm{T}$, a mirror ratio $M = 2$, a length between mirrors $L = 0.76\,\mathrm{m}$, and molecular ion energy $E = 600\,\mathrm{keV}$, while OGRA had $B_0 = 0.5\,\mathrm{T}$, $M = 1.6$, $L = 12\,\mathrm{m}$ and $E = 200\,\mathrm{keV}$. Although the Oak Ridge group worked mainly with deuterium, they reported a 10 ms decay time, determined by charge exchange, for H^+ ions injected as H_2^+, and could not achieve burnout of their neutral gas with the 200 μA beam available and their base pressure of 5×10^{-7} mTorr ($\sim 7 \times 10^{-8}$ Pa).

In Ioffe's PR-2 ion magnetron experiment, a radial electric field extracted ions from a central arc column in a magnetic mirror field. This had produced a density of $10^{15}\,\mathrm{m}^{-3}$ of 250 eV ions in one experiment [14] and a fast-ion confinement time of a few milliseconds in another [3]. A similar geometry, but an entirely different plasma, had been studied in the rotating plasma devices Homopolar at Los Alamos and Ixion at Berkeley [15]. The idea was that the centrifugal pressure exerted by the rotating plasma on the magnetic field would make the field bulge out so as to increase the mirror ratio and, moreover, would provide an additional barrier (the potential of the centrifugal force) against escape of ions through the axial mirror. With either a metallic conductor (Homopolar) or a plasma column (Ixion III) as central electrode, these groups observed rotational velocities of several 10^4 m/s. The discharge behaved as a capacitive circuit element, indicating that rotational energy was stored in the plasma; a fraction of

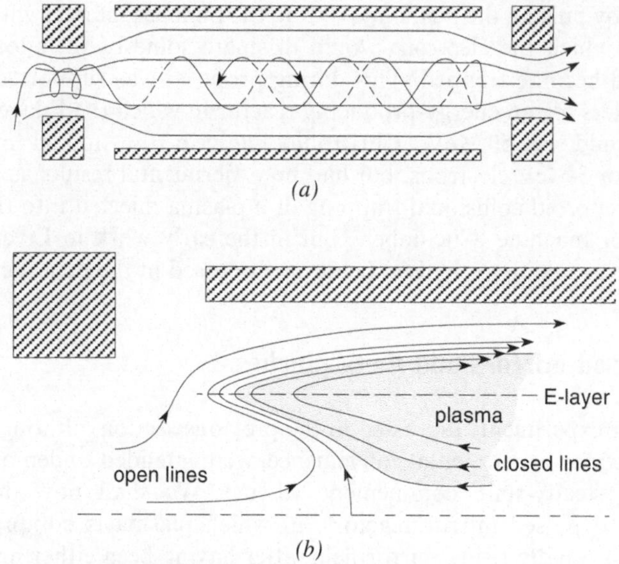

Figure 2.5. The ASTRON experiment. (a) Mirror coils and field lines without E-layer. After entering through the left-hand mirror, electrons must lose axial velocity to be reflected against the right-hand mirror and become trapped. (b) Detail of the configuration with E-layer. The field inside this layer is reversed, producing a plasma in closed magnetic surfaces, embedded in a mirror field which acts as a divertor.

the stored energy could be recovered upon short-circuiting. No measurements demonstrating containment of thermal plasma energy were reported, however. Neither could neutron bursts from Ixion be explained as being of thermonuclear origin.

Another proposal for a steady-state mirror device was Astron (figure 2.5), conceived in 1953 and launched in 1957 by Christofilos at Livermore [8]. Nick Christofilos was a remarkable inventor without a formal training in physics, who had already established a reputation by proposing the strong-focusing principle that has come to dominate nuclear accelerator and beam-line design. In the Aston concept, a cylindrical layer of high-energy electrons encircling the axis of symmetry and oscillating axially between the mirrors—the E layer—would modify the original mirror field sufficiently to reverse the magnetic field on axis. Thus, field lines in the core would be closed and the confinement geometry would turn into a set of toroidal magnetic surfaces with highly elongated minor cross-section embedded in a set of mirror-like open magnetic surfaces (this was later called a Field Reversed Configuration). The electron beam would ionize the gas present initially and heat the resulting plasma to thermonuclear temperature, all the while maintaining its cylindrical shape. The difficulty of trapping a beam after it had passed through one mirror would be overcome by pulsing the beam and letting the bunches interact with resistive circuits in which the electrons would dissipate some of their longitudinal energy and become trapped. This daring proposal was supported by a list of parameters for an energy-producing reactor in which the E layer electron energy would be 50 MeV. Christofilos described a model experiment designed for 3 MeV electrons, but had no experimental results as yet.

Post reported collisional trapping of a plasma injected into the steady-state mirror machine 'Cucumber', but in the early work in Livermore the emphasis was mainly on pulsed fields, as discussed in the next section.

2.3 Pulsed mirrors and theta pinches

The mirror experiments discussed in the previous section, although in some cases pulsed for experimental convenience, were intended to demonstrate an essentially steady-state confinement scheme. We shall now discuss the approach to pulsed mirror reactors, in which plasma is compressed and heated by a rapidly rising mirror field, after having been either produced *in situ* by some form of pre-discharge or injected by a plasma gun, reaction products being exhausted and fresh fuel provided between pulses.

The experiments by Post's group in Livermore in time developed into the steady-state low-beta devices that we discussed in the previous section, but the early versions were true compression experiments. This group reported a series of experiments in which plasma was injected into a weak mirror field, which then, on a time scale of several hundred microseconds, was increased to its final value in order to compress and heat the plasma [16]. With large compression ratios, initial plasma densities of 10^{17}–$10^{18}\,m^{-3}$ were increased, in Post's estimate, to about $10^{20}\,m^{-3}$. Energies of escaping electrons, measured by absorber techniques, yielded electron temperatures of 10–20 keV; ion energies were not measured directly, but were believed to be around 1 keV and in some cases 2–3 keV. The plasma escaped through the mirrors with a decay time of a few ms, which could be explained by Coulomb scattering, but energetic electrons were observed to escape radially at an anomalously fast rate. In a three-stage compression experiment, a plasma was compressed in one containment region and then twice, by means of asymmetrically energized mirror coils, translated axially through one mirror to a smaller volume with a stronger mirror field. Each translation caused the plasma to separate from neutral gas emitted by the plasma source, so that in the third compression stage a relatively pure plasma with small charge-exchange losses was obtained. The compressed plasma was too small in volume, however, to yield a measurable neutron output.

Mirror compression was carried an important step forward by simplification of the scheme and improvement of the technique of energy storage

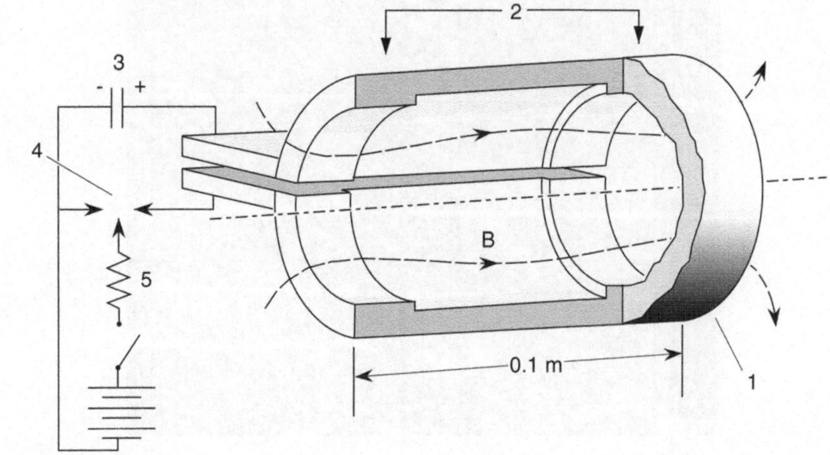

Figure 2.6. Cutaway view of a theta pinch. 1, single-turn coil; 2, mirror sections; 3, capacitor bank; 4, triggered spark gap; 5, high resistance.

in high-voltage, low-inductance capacitor banks. Experiments on magnetic-shock heating and fast magnetic compression were performed at the Naval Research Laboratory (NRL) [17] in Washington DC, at Livermore [18] and at Los Alamos [19]. The emphasis in this work was not so much on mirror confinement as on the processes of plasma formation and compression in rapidly rising magnetic fields. The essential characteristic of a theta pinch (figure 2.6) is a rapidly rising longitudinal magnetic field, which if the plasma is sufficiently pre-ionized induces an azimuthal current in it. This more or less excludes the magnetic field from the interior and thus provides a high-beta equilibrium between the plasma and the external field. In this respect the theta pinch is related to the classical pinch, with which it also has high-voltage, high-current technology in common. They differ in the direction of the plasma current; the 'azimuthal pinch' was named theta-pinch in reference to the cylindrical coordinate system r, θ, z, and the classical pinch became known as the Z-pinch. The proponents were aware of the high amounts of circulating energy that would be required—as in the case of the Z-pinch—in a reactor based on the theta pinch, and most of them saw their devices mainly as simple means to produce a thermonuclear plasma and to study its dynamic behaviour.

At NRL, Kolb had investigated two versions of the theta pinch, one using colliding axial shocks launched by two single-turn coils placed on either side of a pair of mirror-confinement coils, and another using a radial shock wave generated by the single-turn mirror-field coil itself. In the second of these systems, a fast preheater capacitor bank was connected to the coil, producing a 200 kA current ringing at 1 MHz and being damped in 2 μs, after which the main capacitor bank was switched on to

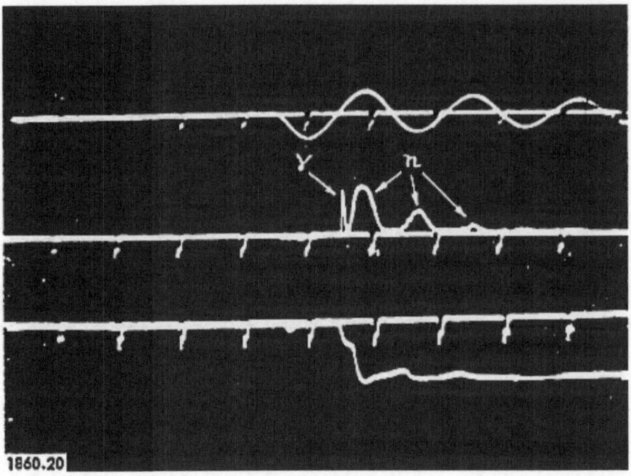

Figure 2.7. Oscillograms showing neutron production in the Scylla theta pinch experiment. From top to bottom: oscillograms of coil current, neutron and gamma (X-ray) intensity, and general light intensity; timing marks at 2.7 μs.

further shock-heat and compress the plasma in the first half-cycle of the main discharge. Alternatively, without the pre-heater, the first half-cycle of the 1.5 MA main discharge would serve as such and shock heating would occur in the second half-cycle. The implosion velocity was about 10^5 m/s. In some discharges, a relatively quiescent plasma was maintained in the compressed state for a good part of an 8 μs half-cycle, but in others, radial oscillations, later identified as a rotating $m = 2$ instability, set in about 2 μs after maximum compression. Estimated plasma parameters were $n \approx 10^{23}\,\mathrm{m}^{-3}$, $T \approx 0.7\,\mathrm{keV}$, but no direct measurements were available. In some cases neutrons were observed which, however, could as well be ascribed to instabilities as to high ion temperatures.

A similar device was Scylla in Los Alamos. It had produced interesting neutron yields just before the Conference which were reported in an addendum to the original paper (figure 2.7). Optimization of the single-turn coil for Scylla had led to a mirror ratio of 1.33. The current rose to a maximum of about 800 kA in a 1.25 μs quarter period. Reproducible neutron yields of 10^6 per pulse were reported, with occasional yields of 2×10^7. Neutron energy spectrum measurements provided a strong indication of reactions originating in a $T_\mathrm{i} \approx 1.3\,\mathrm{keV}$, $n_\mathrm{i} \approx 6 \times 10^{22}\,\mathrm{m}^{-3}$ deuterium plasma.

Both the Scylla paper and Tuck's overview of the work in Los Alamos were careful in their wording, leaving open a margin for a non-thermonuclear origin, and Teller's review of the US programme was even more reticent about the Los Alamos neutrons. Although sceptical voices were heard already at the Conference (we come back to the controversial aspects in the next chapter), Scylla was recognized as the first experiment to show strong evidence of thermonuclear neutrons. The equipment was on display at the exhibition on the grounds of the Geneva conference centre and its simplicity and performance stimulated a host of theta pinch experiments to spring up after 1958 all over the world.

2.4 Stellarators

The stellarator concept was proposed in 1951 by Spitzer from Princeton [8]. This is a toroidal field configuration with rotational transform—the property that field lines, while revolving in the toroidal direction, twist around a magnetic axis (box 2.2). The concept of magnetic surfaces of toroidal geometry originated with Tamm [20], but while his early work with Sakharov had gone in the direction of combining the toroidal field with the poloidal field of a toroidal current—either in a metal ring or in the plasma—Spitzer conceived the stellarator as a steady-state field without an induced current. An axially symmetric vacuum field has no rotational transform because $\nabla \times \mathbf{B} = 0$, but in three-dimensional fields the vacuum requirement is compatible with a net twisting around the magnetic axis. The first method to produce rotational transform was to deform the torus into a 'figure-eight'. Later, Spitzer found that in a circular or racetrack shaped toroid with solenoidal coils providing a toroidal field, the transform can also be obtained by additional helical windings, which make field lines describe double helices. These helical windings impart a non-circular, helically periodic shape upon the magnetic surfaces (box 2.3, figure 2.8).

Rotational transform served to circumvent the loss of particles through gradient and curvature drift that occurs in a simple toroidal field (box 1.3). This drift, which for ions and electrons is in opposite vertical directions (we think of the torus as lying on a horizontal plane), causes them to separate and so to produce a vertical electrostatic field that sweeps out the plasma in the direction of the major radius by $\mathbf{E} \times \mathbf{B}$ drift (figure 2.9). In a macroscopic picture, this radial expansion results from a lack of MHD equilibrium, the magnetic pressure being greater on the inside of the bend than on the outside. Spitzer argued that, in the figure-eight torus, charges accumulating in one bend may flow along field lines to neutralize those accumulated in the other one. Similarly, in the helical field geometry the gradient and curvature drifts alternate in direction as one follows a field line. Hence, space charges developing in one sector can neutralize those in other sectors by flowing along field lines, so that in a highly conducting stellarator plasma the electrostatic fields and the concomitant $\mathbf{E} \times \mathbf{B}$ drifts are greatly reduced.

From the point of view of 'ideal', i.e. non-resistive, MHD, Spitzer demonstrated that if there is rotational transform the current required for pressure equilibrium has no divergence, hence does not cause charge to accumulate on field lines. Early estimates for the equilibrium beta-values allowable in a figure-eight stellarator ran to a few per cent and, for one

Box 2.2 Magnetic surfaces

For good confinement in a closed system, it is required that the field lines form a set of nested toroidal surfaces; the degenerate innermost one of these is called the magnetic axis. The rotational-transform angle, ι, introduced by Spitzer, defines the poloidal angle through which a field line rotates in the course of one complete transit in the toroidal direction. In work on tokamaks and toroidal pinches, one encounters the inverse quantity $q = 2\pi/\iota$, called the safety factor because of its role in stability analysis. (See, however, box 2.5.) It is a topological property of the surface and cannot change in time unless field lines are cut and reconnected.

The surfaces may be labelled by the minor radius r or more elegantly by a flux coordinate, X, which denotes the toroidal flux contained within the surface. When q varies as a function of X, the field is said to have magnetic shear. On a 'rational surface' with $q = m/n$, where m and n are integers, a field line closes upon itself after m toroidal and n poloidal revolutions. In regions of low shear, rational surfaces, especially those with $n = 1$ combined with low values of m, are vulnerable sites for the growth of MHD instabilities with toroidal mode number n and poloidal mode number m. In the absence of shear, magnetic surfaces may develop flute distortions aligned with the field lines without any magnetic energy being expended. With shear, such fluting costs energy because it forces field lines in neighbouring surfaces to wobble over or under the flutes. These instabilities can cause a change in the magnetic topology; field lines break and reconnect to form a secondary set of closed surfaces—a magnetic island—nested about a local, helical magnetic axis. This magnetic island as a whole is then characterized by a particular q-value. Islands can be caused also by stray magnetic fields from external or internal currents and similar effects can result in regions where field lines stochastically fill a volume.

The field configuration can be visualized by following a number of field lines and marking each point where they intersect a given cross-section of the torus. Inasmuch as the field lines lie in closed surfaces, this procedure generates closed loops. In stochastic regions no closed surfaces exist and each field line generates only scattered points. Since particularly the electrons may travel many revolutions along the torus between collisions, both magnetic islands and stochastic regions will lead to enhanced transport.

The condition for pressure equilibrium, $\mathbf{j} \times \mathbf{B} = -\nabla p$, implies that magnetic surfaces coincide with the surfaces of constant pressure and with those described by the current lines. If $\mathbf{j} \parallel \mathbf{B}$ it follows that $\nabla p = 0$ and the configuration is called a force-free field.

Box 2.3 Stellarator fields

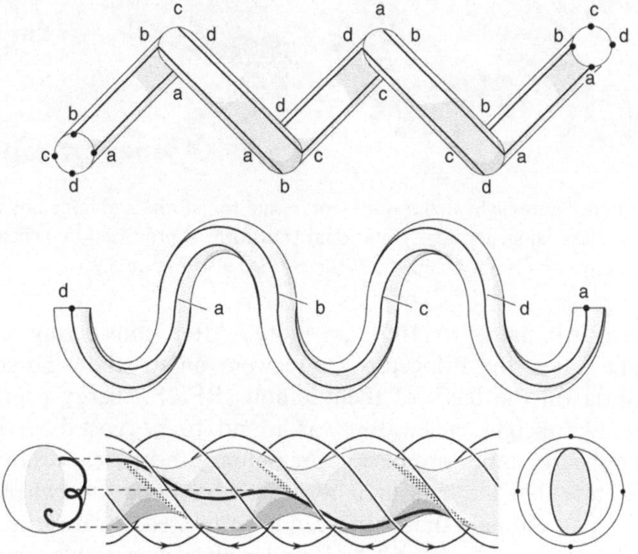

In a stellarator, rotational transform is produced by external coils, that is without recourse to currents in the plasma. One way to do this is to deform a cylinder with an axial field into a helix. This is illustrated by a helical tube built up of 180° bends lying in perpendicular planes. The upper part of the figure shows two projections of a tubular magnetic surface on which are drawn four field lines, marked a, b, c and d. The rotational transform reaches 360° after two full periods of the helix. The principle of twisting the magnetic axis was employed in the early Princeton stellarators like B64. It is further elucidated in box 2.4.

The second way, illustrated in the lower part of the figure, is to wrap pairs of helical windings carrying opposite currents around a tube in which a longitudinal field has been provided by a solenoid (not shown). The poloidal component of the current in one of the helical windings strengthens the axial field, that in the other weakens it. Thus a magnetic surface is pushed inwards under the former and pulled outwards under the latter. Now the longitudinal components of the helical currents cause the field lines to make alternating excursions in poloidal directions as they pass under the conductors. Because this effect is stronger where the longitudinal field is weaker and moreover the field line is closer to the conductor, the result is a net rotational transform. The Princeton C-stellarator was followed by many other devices based on this principle. Sophisticated modern designs employ both a twisted tube and twisted or warped windings in one form or another.

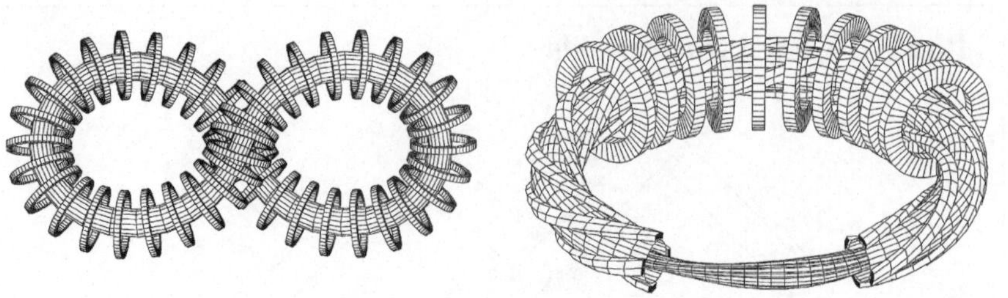

Figure 2.8. Left: figure-eight stellarator—rotational transform is obtained by twisting the torus. Right: classical stellarator—rotational transform is produced by helical windings. The figures display coils and magnetic surfaces. (See also Figure 5.9.)

with helical windings, to 10% or more. After considering the MHD equilibrium states, the Princeton group went on to study the stability of the equilibria on the basis of their famous BFKK energy principle [21]. While the figure-eight stellarator was found to be unstable, the helical windings appeared more promising; preliminary theoretical results even led Spitzer to state that the beta limit would probably lie somewhere between 10 and 40%, an estimate that Princeton soon revised downwards.

The Princeton laboratory had undertaken a sizeable experimental effort (table 2.2). In a series of devices with either figure-eight shape or helical windings, some of the basic experimental techniques of high-temperature plasma physics had been developed and applied, such as RF pre-ionization, Ohmic heating with induced currents, optical and X-ray spectroscopy and microwave interferometry for density measurements. Both the limiter (a

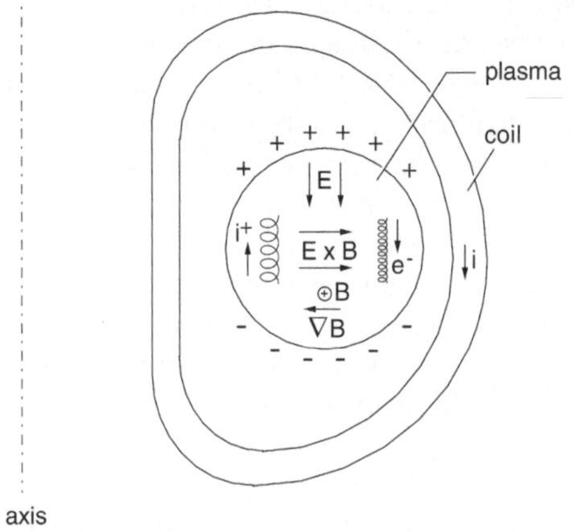

Figure 2.9. Radial expansion of a plasma in a toroidal field. The curvature of the field lines and the field gradient associated with the curvature both cause the ions to drift upward and the electrons downward. The resulting electric field gives rise to an outward $\mathbf{E} \times \mathbf{B}$ drift of the plasma as a whole. The magnetic field points into the page.

Table 2.2 Early Princeton stellarators

Name	Year	L (m)	a^* (cm)	B_t (T)	Rotational transform	Remarks
A	1953–58	3.5	2.5	0.1	Figure-eight	RF breakdown
B-1	1954–58	4.5	2.5	5	Figure-eight	Ohmic heating
B-64/65	1955–67	6	2.6	2	Figure-eight†	Divertor/ICRH
B-2‡	1956–58	6	2.5	1.8	Figure-eight	Magnetic pumping
A-2 (Etude)	1957–61	3.2	2.5	0.85	$l = 3$	Helical windings
A-3	1961–70					
B-3	1958–66	6.4	2.5	5	Figure-eight with $l = 3$	High vacuum/ICRH

* In some cases the vessel radius rather than the plasma radius is given.
† Later versions were racetrack and the final version had $l = 3$ helical windings.
‡ B-2 was exhibited at the Geneva conference.

diaphragm to restrict the cross-section of the plasma column) and the divertor (a set of coils to divert the outer magnetic surfaces on to a collector plate removed from the main vessel) proved their effectiveness in reducing contamination of the plasma.

Although the stellarator field, in principle, was thought of as being in a steady state, the heating methods that could maintain a hot plasma were only beginning to become available, and most experiments had to rely on pulsed heating by induced currents as in pinches. This led to studies of kink instability and to confirmation of the current limit, found independently [22] by Kruskal and Shafranov, which requires that $q > 1$ for stability against an $m = 1$ kink-mode.

One favourable experimental result in relation to single-particle behaviour was long-time confinement of energetic runaway electrons [23], confirming the existence of at least a class of closed or nearly closed particle-drift orbits. The plasma, however, was rapidly lost by an undetermined cause; all that could be said about this 'pump-out' was that the plasma was perhaps not in a quiescent state. Inevitably, the study of 'anomalous' losses and their possible connection with Bohm diffusion was to be the main area of concern for stellarator research in the years to come.

Kadomtsev and Braginskii in Moscow proposed a different toroidal system with stellarator-like properties. Instead of making field lines twist, they arranged for a rotational particle drift around the magnetic axis, which should provide for similar single-particle confinement. This would be achieved by modulating the field strength in the straight sections of a racetrack-type toroid, so that the field there would be like a series of magnetic mirrors. While passing through these mirrors, a particle would undergo an alternating curvature drift in the azimuthal direction, but because its longitudinal velocity would also be modulated it would spend more time in field

Box 2.4 Rotational transform of the figure-eight

The rotational transform of the figure-eight geometry may be illustrated by Spitzer's figure from 1958, which shows how a field line twists around the axis. The left-hand side (a) shows top and end views of a figure-eight stellarator: the 180° bends NK and LM are in planes tilted at angles $\pm\alpha$ with respect to the parallel planes containing LK and MN. The right-hand side (b) shows the rotational transform: in the cross-sections at K, L, M and N, the magnetic axis is represented by the points O, a particular field line by crosses. The cross-section at K is translated to L, then reflected to M, translated to N and reflected to K. As a result, after one passage along the toroid the field line P_1 at K returns as P_2 and the rotational transform angle equals 4α.

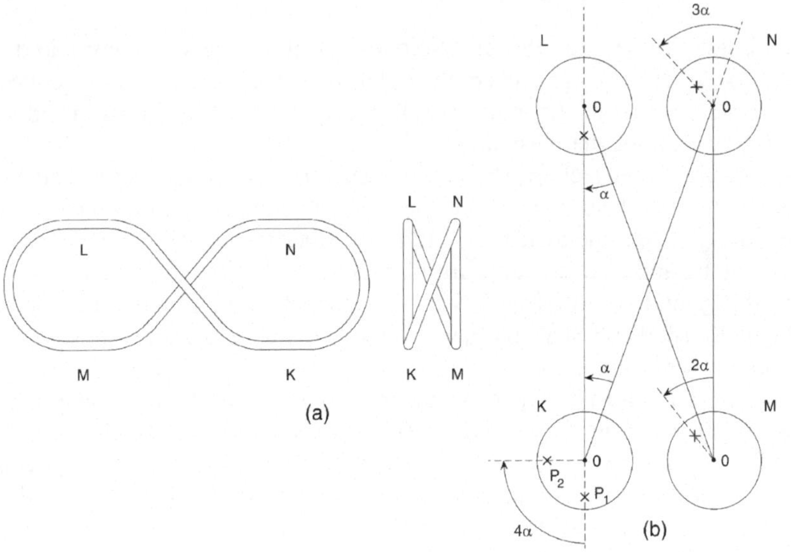

maxima than in minima and as a second-order effect would trace out a helical orbit. (Spitzer had considered similar 'scallops' in the curved sections of a racetrack, but had opted for the helical windings.)

In his book, Artsimovich credits Spitzer for having originated the concept of rotational transform, but reports to have understood already in 1955 that, in the figure-eight torus, particles with sufficient longitudinal speed will not suffer a net torus drift because the contributions from the two curved sections cancel. (The plane figure-eight is indeed, as discussed in box 2.5 [24], a case where one can understand drift cancellation without

Box 2.5 Rotational transform and helicity

The concept of rotational transform has evolved in a significant way since its inception in the early fifties. Spitzer and contemporary authors like Glasstone and Lovberg, Rose and Clark and Artsimovich make no difference between transform angles of 0 and 2π. Spitzer's much quoted formula $\iota = 4\alpha$ (box 2.4) extrapolates to $\iota = 0$ for a plane figure-eight (plane except for a deformation at the cross point) and to $\iota = 2\pi$ for a plane racetrack-shaped toroid—which would have $\alpha = \pi/2$. It would be more appropriate, however, to assign $\iota = 2\pi$ to the former and $\iota = 0$ to the latter, because in the figure-eight field the particle drifts in the opposite curved sections do cancel and accumulated charge can be neutralized by flow along field lines. Indeed, the plane figure-eight field is topologically equivalent to a $q = 1$ field in a torus, each field line once interlinking each other.

The parameter q was introduced by Shafranov in 1956 (in early papers it is named K) and takes the place of ι in tokamak literature. It is a topological constant characterizing a magnetic surface. Invariably it is stated that $q = 2\pi/\iota$, although this is at odds with the way Spitzer defined ι for the figure-eight torus. The ambiguity in the definition of ι disappears if one thinks in terms of helicity instead of transform angle, but this was not appreciated until much later (box 4.1).

invoking rotational transform.) Artsimovich alluded to this by inserting a sketch of a figure-eight torus in his preprint for the Geneva Conference, but after a confrontation between the US and Soviet delegations, this figure and the paragraph relating to it were deleted in the Conference Proceedings, leaving their trace only in a question from Spitzer and in a parallel publication of the Geneva Proceedings [25].

2.5 Toroidal pinches

Although most work on toroidal pinches had been revealed already in prior publications, Geneva was the first full exchange of information on this subject; there were contributions from Britain, the US, the Soviet Union, France and Sweden (see table 2.3). As the publicity storm around ZETA (figure 2.10) had subsided and the neutrons had been demonstrated to be of non-thermonuclear origin already in the spring of 1958, the discussion focused on other items, like the current and field profiles and the ion and electron temperatures.

Table 2.3 Geneva 1958 toroidal pinches

Location	Name	R (cm)	a (cm)	I_{max} (kA)	Liner
Los Alamos	Perhapsatron S-4	35	7	320	Quartz
Livermore	Gamma	25	5	250	Ceramic
Kurchatov Institute, Moscow	Torus-1	15	5	25	Quartz
	Torus-2	20	8		Cu*
	Torus-3	20	8		Cu*
	Torus-4	50	13	200	CrNiFe*
	Torus-5	62.5	24	200	CrNiFe†
Harwell	ZETA	160	50	200	Al‡
AEI, Aldermaston	64-Section torus	50	15	20	Al*
	Racetrack torus	52.5	15	30	Al*
	Sceptre III	56	15	200	Al*
Fontenay	—	39	4	60	Pyrex
	TA-2000§	100	15		Al
Saclay	Equator 1	40	3.5	100	Al*
Uppsala	—	30	4	200	Al*

* With insulating gaps.

† Continuous.

‡ ZETA's first liner had 48 segments of aluminium alloy insulated from each other. This was replaced in 1959 with a continuous liner of thin corrugated stainless steel.

§ In construction at the time of the conference.

In ten years, the toroidal pinches had evolved from small glass tori wound with cables to well engineered constructions. In some cases, where it was attempted to set up a thin current sheath, the torus was surrounded with a thick conducting shell that could serve both as a stabilizing shell and as the primary winding. Mostly, however, glass or ceramic tubes were replaced by sectioned aluminium tori. The continuous metal liner made its first appearance in the slow discharges that would later be recognized as a separate class—the tokamaks. Iron cores often served to improve the coupling with the primary toroidal current, which was fed by capacitor banks.

The measurements on ZETA reported in Geneva showed evidence for spontaneous reversal of the toroidal magnetic field, but the paper made no point of this; the reversed-field pinch did not emerge until ten years later. Aldermaston noted the phenomenon, Los Alamos had no evidence for it, but Livermore found spontaneous reversal in both linear and toroidal pinches and ascribed it to an unstable $m = 1$ helical distortion of the current channel. We shall return to toroidal field reversal in section 4.2.

The early theoretical work on pinch stability had assumed that the pinch current would be induced in a thin skin separating the stabilizing field (the

Figure 2.10. The ZETA experiment at Harwell soon after its first operation in 1957. The transformer core consists of two 4 m diameter rings surrounding the 1 m minor diameter torus.

axial B_z in the linear pinch or the toroidal B_φ in the toroidal pinch) from the confining field (the azimuthal or poloidal B_θ).* Rosenbluth's 'snowplow' model described how during the formation of the pinch the initial plasma—formed by some kind of pre-ionizing discharge—would be swept

* In early literature, B_z is also used for a toroidal field. And in modern literature the φ and θ components are often marked with t and p for toroidal and poloidal. But the toroidal plasma current is sometimes confusingly referred to as as I_p (where the subscript denotes plasma rather than poloidal); we shall use I_t or simply I.

up by $\mathbf{E} \times \mathbf{B}$ drift in the inward-moving current sheath. Rapid heating would produce a plasma of sufficiently high conductivity to maintain this configuration and, hopefully, of thermonuclear temperature. Stability would be ensured by a combination of a trapped field for the shorter wavelengths and a conducting wall for the longer ones. But work by Rosenbluth [26] on finite-thickness skins and by Suydam on diffuse profiles [27] indicated that, in addition to the trapped axial field, an external field of reversed direction would increase the shear and thus would have a stabilizing effect.

Experiments succeeded in producing the expected skin currents for only short times [28] or not at all [29]. Apparently, the diffusion of magnetic field relative to the plasma column proceeded much faster than was predicted, although the energy dissipated in the process did not result in efficient heating. Instead of finding high temperatures, high conductivities and slow diffusion rates, everyone observed rapid diffusion and energy dissipation, yet low electron temperature and low conductivity.

The Kurchatov Institute [30] reported on toroidal pinches with either weak or strong toroidal magnetic field. This work focused on the interaction of the plasma with the wall of the discharge chamber and the ensuing contamination of the plasma, and on MHD instability. Most experiments were done in a torus with major radius 62.5 cm, minor radius 24 cm, toroidal field 0.5 T, stainless steel vacuum liner, and copper shell. Discharge currents reached up to 400 kA, well above the Kruskal–Shafranov limit. Rising, oscillating, and steady toroidal fields were tried, but there was a strong plasma–wall interaction and macroscopic oscillations of the plasma column indicated MHD instability, irrespective of the Kruskal–Shafranov condition being satisfied. The electron temperature derived from the electrical conductivity of the plasma reached 15–30 eV. When operating with a strong stabilizing field—with a current below the Kruskal–Shafranov limit—this type of device was later to be named the *tokamak*.

2.6 RF fields and other subjects

Aside from questions relating to specific magnetic confinement systems, a variety of basic plasma physics issues were touched upon at the Conference. Some of these had been discussed in the open literature, but in the political climate of the time direct contacts between Western and Soviet scientists had been rare. Geneva, therefore, was not only a place where secrets were revealed, but also a meeting ground where Western and Soviet plasma physicists could summarize and assess the state of affairs on either side of the Iron Curtain.

Theoreticians compared their notes on subjects like kinetic theory of plasmas, shock waves and MHD instabilities, often to find that they had covered much the same ground. Radio-frequency waves were discussed in the context of plasma confinement by radiation pressure or time-averaged forces on oscillating electrons (ponderomotive forces), in papers from the Kurchatov Institute, the Argonne National Laboratory and the Ramo–Wooldridge Corporation at Los Angeles [31]. Heating by damping of low-frequency oscillation or absorption of high-frequency waves had been studied in Princeton, as well as in Moscow and Kharkov [32]. Princeton had successfully applied ion cyclotron resonance heating in a stellarator and this method had also been considered in the Soviet Union, although they had no comparable results. On the other hand, Sinelnikov's group in Kharkov was ahead with electron cyclotron resonance heating. Transmission and reflection of microwaves and detection of RF noise were used for plasma diagnostics in all major fusion laboratories [33], notably in Livermore and Harwell, and in institutes like MIT and Stanford that had been involved in radar development [34]. Finally, Trubnikov from Moscow had an unpleasant surprise for the Conference: synchrotron radiation* was to be taken seriously as a loss mechanism in hot-electron plasmas.

Anomalous diffusion across a magnetic field, sometimes called 'drain diffusion' before the term Bohm diffusion stuck, was a central issue in discussions about plasma losses. The problem was addressed specifically in contributions by Simon, who maintained that no anomaly was required to explain Bohm's loss, and by Lehnert from Stockholm, who found a transition from classical to anomalous loss when an axial magnetic field was applied to a discharge tube long enough to reduce end effects. The nature of the loss was not fully understood, however, until two years later, when Kadomtsev and Nedospasov [35] described a current-driven resistive instability that could explain it.

We finally note the concept of 4π-focusing of ion beams in combined magnetic and electrostatic fields [36] which has reappeared from time to time in the thermonuclear literature. We return to this in section 3.5.

2.7 Looking back at Geneva

When, at Geneva, the cards were laid on the table, a picture emerged in which four broad categories of confinement systems could be distinguished: these had closed or open magnetic surfaces and their operation was intrinsically in a pulsed mode or potentially in a steady state. There were no conclusive scientific arguments for favouring one approach above the other as a potential reactor, but for the international research effort that began to take shape in 1958, that choice was not on the top of the agenda; the most urgent question was what could contribute most to the physics of magnetically confined plasma. Yet,

* In later years this is referred to as cyclotron radiation.

the wealth of new ideas that came forward raised great excitement, both among scientists and the public. In the opening session, where the newly declassified programmes of the nuclear powers were reviewed, Thonemann said:

> *I think that the papers to be presented at this Conference, and the discussions which follow them, will show that it is still impossible to answer the question, 'Can electrical power be generated using the light elements as fuels by themselves?' I believe that this question will be answered in the next decade. If the answer is yes, a further ten years will be required to answer the next question, 'Is such a power source economically viable?'*

But Thonemann's reservations were hardly noticed; one heard what one wanted to hear and the general impression was that fusion power would be there within twenty years. In a climate of boundless confidence in the future, how could anyone who had seen what nuclear science had achieved in merely two decades have any doubt that soon this would culminate in the ultimate nuclear reactor, the deuterium-burning man-made sun? The prevailing mood was euphoric, but it is wrong to say, as has been said over and over again, that the fusion scientists at Geneva promised abundant, cheap, safe and clean fusion energy within two decades. Edward Teller, who spoke for the USAEC, said about fusion energy production:

> *I think that it [thermonuclear energy production] can be done, but do not believe that in this century it will be a thing of practical importance.... It is likely that we shall be dealing with an intricate machine which is inaccessible to human hands because of radiation and on which all control and maintenance must proceed by remote control. The irradiation of materials by neutrons and gamma rays will cause the properties of these materials to change.... These and other difficulties are likely to make the released energy so costly that an economic exploitation of controlled thermonuclear reactions may not turn out to be possible before the end of the twentieth century.*

And in his overview of the Soviet effort Artsimovich wrote:

> *... there begins to emerge a rough outline of the scientific foundation on which the methods of solving the problem of controlled fusion reactions will probably rest.... We do not wish to be pessimistic in appraising the future of our work, yet we must not underestimate the difficulties which will have to be overcome before we master thermonuclear fusion.... The solution of the problem of thermonuclear fusion will require a maximum concentration of intellectual effort and the mobilization of very appreciable material facilities and complex apparatus. This problem seems to have been created*

> *especially for the purpose of developing close cooperation between the scientists and engineers of various countries.*

The most prominent sceptic was Spitzer, who saw clearly where the main problems could be expected. The equilibrium of a plasma in a magnetic field and its MHD stability, as well as the particle and energy losses through binary collisions* in simple magnetic-field configurations, were reasonably well understood, but the theory dealt with a quiescent plasma, free of microscopic turbulence, and Spitzer was concerned about Bohm's warning that such turbulence might well be an intrinsic property of the plasma.

In fact, the Conference made clear that the optimism of the early 1950s was built on loose ground and that plasma physics was not ready for the challenge posed by the fusion reactor. One may wonder why; Maxwell's theory of the electromagnetic field, Boltzmann's fundamental equation of statistical mechanics and the principles of fluid dynamics all date from the nineteenth century. The only thing that had kept physicists from combining these disciplines, it seems, was lack of interest. In the first half of the twentieth century, quantum mechanics, relativistic theory, atomic structure, nuclear physics and, finally, solid state physics drew more attention. Electrical discharges in gases with usually a low degree of ionization were studied in the context of atomic physics and the few astronomers who ventured into the realm of ionized matter received little support from mainstream physicists, who mostly saw this as dirty physics. So, when the first wave of excitement subsided, it became clear that the principal outcome of the Conference had been the insight that first the forgotten chapter of plasma physics needed to be written, before the question of the scientific feasibility of controlled fusion could be addressed.

* This term is used to distinguish encounters between individual particles from interactions through collectively generated fields.

Chapter 3

Open systems

After 1958, research on open-ended systems developed mainly in two directions. The first was quasi-steady-state, moderate-field, mostly low-beta mirror confinement, where the plasma is produced and maintained in a pre-existing magnetic field. The second was fast high-field, high-beta theta pinches where the currents induced in the plasma by the rise of the magnetic field are essential, both in setting up the desired configuration and in producing the hot plasma. Thinking of a reactor, one would approach the Lawson criterion, $n\tau \approx 10^{20}\,\mathrm{m}^{-3}\,\mathrm{s}$, with parameters like $n \approx 10^{20}\,\mathrm{m}^{-3}$, $\tau_E \approx 1\,\mathrm{s}$ in a steady-state reactor or with $n \approx 10^{22}$–$10^{23}\,\mathrm{m}^{-3}$, $\tau_{\mathrm{pulse}} \approx 10^{-3}$–$10^{-2}\,\mathrm{s}$ in the pulsed case. We first address steady-state systems in sections 3.1 and 3.2, and return to Z-pinches and theta pinches in sections 3.3 and 3.4.

3.1 Simple mirror machines

Already, at the Geneva Conference, it was apparent that the main problems confronting mirror confinement were

- ion escape and electron heat loss through the mirrors
- charge-exchange loss and burnout
- MHD instabilities
- microscopic or velocity-space instabilities.

We shall, in separate sections, trace these issues from 1958 up to 1976, when two events occurred which changed the face of mirror research. First, Coensgen's 2XIIB ('two X two B') experiment broke through long-existing barriers in the areas of burnout and stability and next, Dimov's tandem-mirror proposal placed the ambipolar loss in a different perspective.

3.1.1 Mirror loss

The fundamental loss mechanism in mirror systems is escape of ions through the mirrors as a result of diffusion in velocity space by ion–ion collisions; the Coulomb-scattering cross section scales as T_i^{-2}. In contrast, empirical scalings of the loss rate in toroidal systems show a weak dependence on the temperature. Thus, mirror reactors must operate at higher ion temperatures or beam energies than toroidal systems—around 100 keV for D–T, depending on the mirror ratio, against some 10–20 keV for D–T in closed systems. The ratio, Q, between fusion energy output and external energy supplied to the plasma* would lie in the range 5–10, according to early mirror-loss computations [1]. This was thought to leave a sufficient margin for other plasma energy losses and for the power requirements of the plant. On this basis, Post carried out an elaborate study of mirror-reactor economics, in which MHD instabilities were supposed to be taken care of by low β, and Q was assumed to be close to five.

The first objection raised against this model was that the ion loss would be enhanced by the escape of electrons. At equal temperature, electrons have a shorter collision time than ions, so they are scattered faster into the loss cone, leaving behind a positive space-charge potential—the ambipolar potential—that drives out some of the ions. The ambipolar potential adjusts itself so as to equalize the two loss rates; its magnitude depends on the temperatures of the two species. Even for stable and quiescent plasmas, American and Soviet authors, when taking into account both the effect of the ambipolar potential on ion escape and additional energy losses via the electron channel, arrived at Q-estimates hardly exceeding one, making net power gain entirely dependent on efficient recovery of the energy of escaping ions [2].

With regard to the electrostatic field, George Kelley [3] from Oak Ridge proposed to add a short mirror section on either side of a long central mirror cell. The potential in these three sections would be equalized by electron flow, so that only the ions in the outer sections would be affected by the ambipolar field. Other schemes to reduce mirror loss [4] were electrostatic plugging and radio-frequency plugging (section 3.5). Moreover, much thought was given to improving the efficiency of mirror systems by converting kinetic energy of charged particles into electric or magnetic energy [5]. In view of the end losses it was clear that mirror economics favoured a very long central cell with a high mirror ratio, but as understanding of macroscopic and microscopic instabilities grew, it became difficult to maintain these assumptions. The minimum-B principle (section 3.1.3) forced mirror systems to develop

* In some early work, Q was defined as fusion energy produced over total energy escaping from the plasma. The difference lies in the fusion energy deposited in the plasma. In either definition, Q strictly includes not only the dominant mirror loss, but also radiation and transport. In pulsed systems, the energy deposited into the plasma in the course of its formation enters into the definition of Q.

in the opposite direction, so that once again reactor studies had to cope with marginal values like $Q \approx 1$–1.5. The solution was sought at first in energy multiplication in the blanket via (n, 2n) and (n, γ) reactions and direct conversion of mirror losses. In 1974, Livermore presented a conceptual design study of a 170 MW(e) mirror reactor [6], which differed from Post's earlier proposal in that the length between mirrors had decreased from 100 to 7 m and the β value had increased from 3 to 85%. Very high efficiencies had to be assumed for the injector and the direct energy converter and, even then, the circulating power in the system was high. We return to reactor concepts in chapter 9.

3.1.2 The quest for burnout

After the Kurchatov Institute and Oak Ridge had presented their beam injection experiments, OGRA and DCX, in Geneva, they were soon joined by other institutes in their attempts to build up hot-ion plasmas by injection of high-energy beams of neutral particles, molecular ions or hydrogenic ions [7]. In the UK a group under Sweetman (first in the AWRE at Aldermaston, later in the new Culham Laboratory) injected atomic hydrogen beams in the Phoenix device. In Livermore, Post's group likewise started with atomic hydrogen beams in Alice and in France, Prévôt led a group in Fontenay-aux-Roses (originally in Saclay) that studied injection of molecular ions. The first concern of these groups was to achieve burnout. The idea [8] is illustrated by figure 3.1, which shows the characteristic dependence

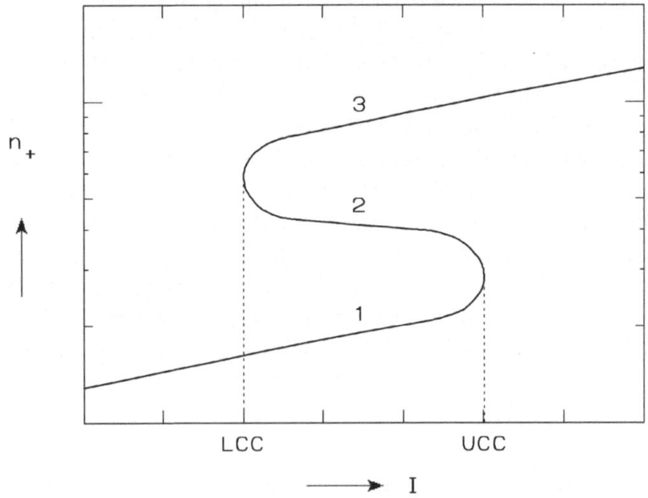

Figure 3.1. Characteristic dependence of hot-ion density on injected beam current in a mirror experiment, indicating stable (1 and 3) and unstable (2) branches (see text) as well as lower and upper critical current limit (LCC and UCC).

of the hot-ion density on the injected beam current. On branch 1, the density remains low because only a small fraction of the beam is dissociated or ionized, either by neutral gas or by a seed plasma or an arc, and the trapped ions are rapidly lost by charge exchange. On branch 3 the neutral gas is burned out, so the beam is trapped by collisions with plasma ions or electrons and the loss is by pitch-angle scattering into the mirror-loss cone. On the intermediate branch, all these processes contribute, but the equilibrium is unstable. Hence, if the injection current rises above the upper critical current, UCC, where trapping on plasma equals charge exchange loss, the density will rise—at first exponentially—to the saturation value on branch 3.

Exponentiation of the trapped-ion density, as a first step toward burnout, was eagerly looked for in all beam-injection experiments. With neutral beam energies of 30 keV and currents of order 100 mA, injected into plasma volumes of order one litre that were common in the 1960s, an interesting fusion plasma with ion density of order 10^{20} m^{-3} and temperature of $\sim$20 keV seemed to be within reach. To reach burnout with such conditions, the background neutral gas pressure would have to be low enough for trapping by collisions with ions (which for the prevailing ion energies and electron temperatures exceeds trapping by electrons) to outweigh plasma loss by charge exchange. This required a neutral density $n_0 < 10^{14}$ m^{-3}, equivalent to a neutral pressure of 4×10^{-7} Pa (3×10^{-9} Torr).

The prospects for achieving burnout improved when Sweetman proposed Lorentz-trapping (ionization of highly excited H or D atoms by the induced field, $\mathbf{E} = \mathbf{v} \times \mathbf{B}$, felt by the injected beam in the mirror field) as an additional means for building up the plasma. On the other hand, he realized that Franck–Condon neutrals (H^0 or D^0 formed by ionization and subsequent dissociation of H_2 or D_2 coming from the wall), with a kinetic energy of 2 eV, would penetrate much deeper into the plasma than thermal molecules or atoms with typically 0.01–0.1 eV. And all groups sooner or later ran into anomalous losses of the hot plasma by some form of instability.

Burnout was pursued by a variety of means. Oak Ridge [9] had started with DCX-1, which had a carbon arc to dissociate a 600 keV H_2^+ or D_2^+ beam making one single passage through the system. A disadvantage of the arc was that incompletely stripped carbon ions also acted as centres for charge-exchange loss of trapped H^+ or D^+ ions. Therefore, completely ionized Li or D arcs were employed in DCX-2, which also had the injector so arranged that the beam could make several passages before being lost [10]. The arc would be turned off after the trapped plasma could take over its function as target for the beam. Later, Oak Ridge pioneered in producing hot-electron target plasmas with electron cyclotron resonance heating (ECRH) in their EPA and ELMO facilities [11]. By 1965, EPA had a density approaching 10^{18} m^{-3}, with $T_e \approx 100$ keV in a 50 litre volume, capable of effectively screening the plasma from room temperature neutrals and to some extent decoupling the problems of burnout and exponentiation. The plasma, however, remained transparent for Franck–Condon neutrals, which still limited the confinement time to some 10–20 ms. The Kurchatov Institute followed a line that went via OGRA with 160 keV H_2^+ to OGRA-2 with 75 keV H^0 injection in a volume of 50 litres. The highest trapped-ion density reported for this experiment [12] was $3 \times 10^{13}\,m^{-3}$. In a much smaller companion device, Ogrenok [13], 10 keV H^0 was injected and trapped-ion densities reached up to $10^{13}\,m^{-3}$.

Fontenay first opted for gas trapping of 80 keV H_2^+ ions, injected radially from an annular source located in the midplane of their MMII mirror machine [14] but, after reaching $2 \times 10^{14}\,m^{-3}$ in 0.7 litres, turned towards injection of plasmoids to fill their mirror machines. Culham chose 20 keV H^0, making use of a strong magnetic field to enhance Lorentz trapping. After the simple mirror machine Phoenix 1, Culham built Phoenix 1a and Phoenix II with magnetic wells (section 3.1.3). When the trapped-ion density reached $3 \times 10^{14}\,m^{-3}$ in Phoenix 1a, they observed the onset of exponential increase, and with an 8 keV beam in Phoenix II they could bring the density up to $4 \times 10^{15}\,m^{-3}$ [15]. Lorentz-trapping in this device was measurably enhanced by the beam atoms being in part in high excited states.

Livermore finally succeeded in pushing neutral injection to the point where they could produce really dense hot-ion plasmas [16]. Their first attempt along this line was Alice, which had 20 keV H^0 injection into a simple mirror system, later transformed into a magnetic well configuration. The modified Alice–Baseball reached a peak hot-ion density of $8 \times 10^{14}\,m^{-3}$. Baseball-II had a superconducting coil and used a variable-energy injector for 1–20 keV D^0 beams. When the density was built up initially with low beam energy, it was possible to maintain a compromise between collisional energy broadening of the trapped-deuteron spectrum (desirable for stability) and collisional scattering loss. With a 0.5 A_{eq} injector, it was expected that n_i would reach close to $10^{18}\,m^{-3}$. The experimental programme, however, ran into problems with instabilities already at a few times $10^{15}\,m^{-3}$ at $E_D = 2$ keV. By that time, most other laboratories had given up mirror research; only the Kurchatov Institute [17] and Grenoble [18] joined Livermore in presenting mirror work at the 1974 Tokyo Conference and of these, only the OGRA-3 group used neutral injection. We shall come to the Grenoble work in section 3.5; OGRA-3, the last in its series, achieved a measure of success with feed-back stabilization. Injecting 50 mA_{eq} of 20 keV H^0, the group built up an ion density of $1\text{–}2 \times 10^{14}\,m^{-3}$, limited by flute modes. When the $m = 1$ mode was stabilized, this went up to $4 \times 10^{14}\,m^{-3}$ and when also the $m = 2$ mode was stabilized, to $9 \times 10^{14}\,m^{-3}$, at which density the $m = 3$ mode set in.

It was evident by the early 1970s that the ion or neutral beam sources then available were not capable of pushing the trapped plasma parameters through the prevailing instability thresholds, and that a heroic effort would be needed to save mirror research from getting totally submerged under the rising tide of tokamak results. The challenge to make injectors capable of maintaining a close-to-reactor-grade plasma in a mirror device was faced by Berkeley, which together with Livermore took on the design first of a 10 A_{eq} source and next of a 50 A_{eq} source. With 12 sources supplying a total of 370 A_{eq} of deuterium atoms, Livermore [19] was finally able to produce a high-beta plasma by beam trapping and exponentation on a low-density cold target plasma. The plasma was maintained for 1–2 ms, so one could say that, transiently, burnout conditions were achieved. It turned out, however, that a flow of cold gas injected along the axis was required for stability, so that 2XIIB still did not reach the goal of a plasma shielding its interior from interaction with external gas. Nevertheless, the emergence of powerful neutral beams, together with improved vacuum techniques such as discharge cleaning of the plasma vessel, laid the issue of burnout to rest. But the beams did not only change the face of mirror research, they also strengthened the competition by providing an invaluable tool for heating plasmas in toroidal devices like tokamaks and stellarators.

3.1.3 MHD stability

By 1958, theory had predicted MHD instability of mirror devices, particularly against flute distortions, but the early experiments in Livermore [20] had shown apparently stable confinement, in some cases for a thousand expected instability growth times, and the group was more concerned with microinstabilities. M S Ioffe and co-workers in Moscow [21], however, took flute instabilities seriously. In the PR-2 experiment they had found anomalous loss and had unequivocal evidence that this was caused by flutes. They came to the 1961 Salzburg Conference with post-deadline results which clearly established both their existence and the method to cure them.

Previously, simple mirrors and cusped geometries were thought of as contrasting schemes. The mirror was expected to be flute unstable but had the advantage that single-particle confinement, based on conservation of magnetic moment and destroyed only by collisional scattering, was good. Cusps (figure 3.2), on the other hand, promised to be MHD stable, but intrinsically had a region of zero or weak magnetic field where the magnetic moment was not conserved. Theoreticians had suggested already in the 1950s that the two features had to be combined to achieve both adiabatic particle orbits and MHD stability [22], but these ideas had not been pursued by experimentalists until Ioffe applied the scheme and proved its effectiveness. By superimposing on the simple mirror field of PR-2 a transverse hexapole field produced by currents in six nearly straight wires (figure 3.3), Ioffe obtained a 'magnetic well' or minimum-B configuration (box 3.1) in which the field strength increased not only towards the mirrors, but also in the radial direction. By switching the currents in the additional windings—which came to be known as *Ioffe bars*—off and on, his group could make

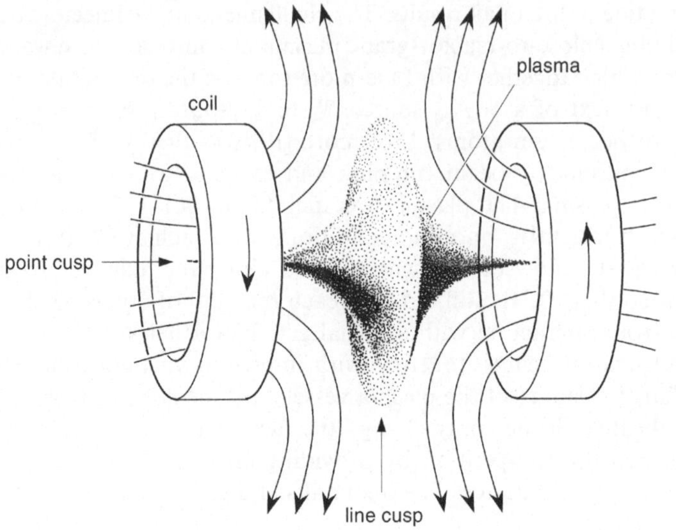

Figure 3.2. Cusped magnetic field, produced by two coils carrying currents in opposite directions.

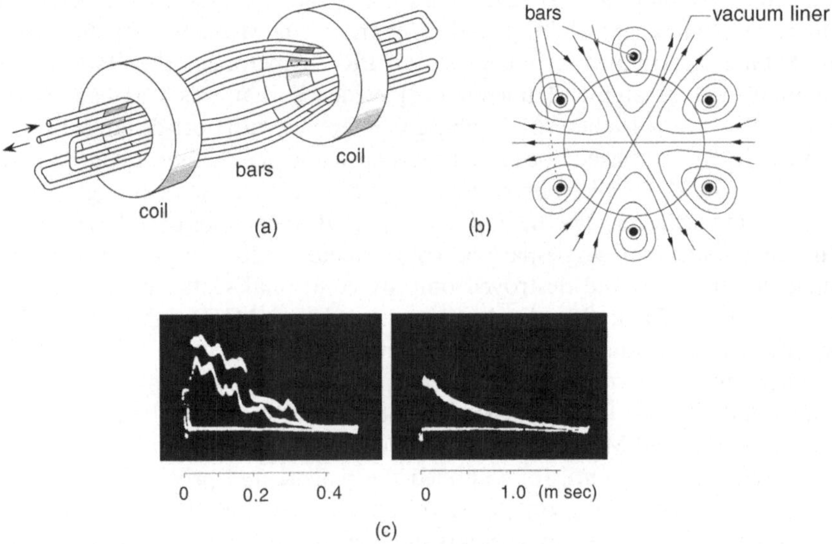

Figure 3.3. Ioffe's experiment to stabilize a mirror plasma. (a) Mirror coils with Ioffe bars; (b) equatorial field plot; (c) containment without (on left) and with (on right) Ioffe bars energized (note different time scales). Two overlapping traces give an indication of the reproducibility of the discharges.

Box 3.1 Flute instabilities in mirror machines

For mirror machines, the most important instability is that of the 'interchange' or 'flute' type. If a magnetic flux tube with higher-pressure plasma is interchanged with one carrying equal flux, $d\Phi$, but lower-pressure plasma, the magnetic energy is unchanged but thermal energy is released if the higher-pressure plasma expands and the lower-pressure plasma is compressed. Therefore a volume element, $dA\,dl = d\Phi\,dl/B$, where A stands for area and l for length, contributes to instability if dl/B increases in the direction of lower pressure. If the field lines are curved and the higher-pressure plasma lies on their concave side, dl increases and B decreases outwards everywhere, so that the equilibrium is certainly unstable. It can be shown that in the opposite case, in which the higher-pressure plasma lies on the convex side, the plasma is stable against these interchanges as long as the pressure gradient stays below a critical value. This is called a minimum-B configuration. Generally, the integral over dl/B has contributions from regions of 'good' and 'bad' curvature (concave or convex side facing higher plasma pressure, respectively), so that an 'average minimum-B' configuration is necessary for stability.

The mechanism that drives this and other drift-type instabilities is the same as that which sweeps the plasma out when there is no equilibrium, as in a simple torus. Particles drifting in opposite directions give rise to charge separation and the resulting electric field causes the plasma as a whole to move in the $\mathbf{E} \times \mathbf{B}$ direction. In the present case, the responsible agent is a net flow of charges caused by a combination of field gradient and curvature drift (box 1.3) and diamagnetic current (box 1.4). These add up in regions of bad, and compete in regions of good, curvature. When the pressure is anisotropic, as is usually the case in mirror devices, this affects the drift, so that the treatment needs refining. The theory of mirror confinement was reviewed by Post [30].

fluctuations and concomitant particle losses appear or disappear, and the confinement time vary fivefold. In the larger PR-5 machine, 4 m long with a central field of 0.5 T, they succeeded with this technique in producing an $n_i = 10^{16}\,\mathrm{m}^{-3}$, $T_i = 5\,\mathrm{keV}$ plasma, that was confined for about 100 ms [23].

Until the Salzburg Conference, other groups had been inclined to believe that the experimental evidence against the occurrence of flutes in conditions where they were expected theoretically, might extrapolate into the future. Plausible explanations were low values of beta, line-tying by electrons flowing along field lines between the plasma and external conductors, and finite

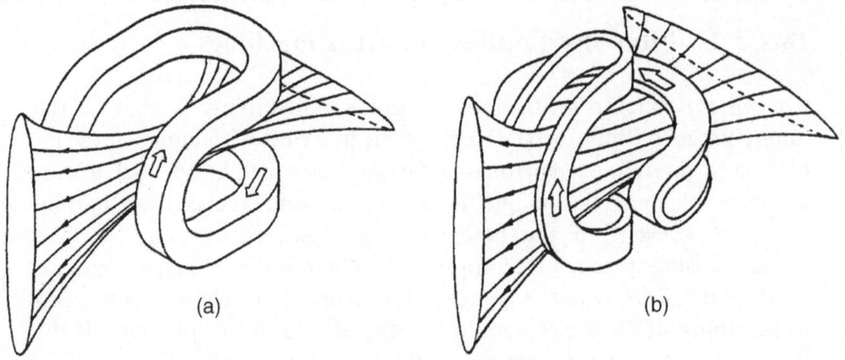

Figure 3.4. Coil arrangements to produce quadrupole minimum-B fields. (a) Baseball mirror coil; (b) yin-yang coil set.

ion gyro-radius effects. These mechanisms were not appropriate for reactors, though; low beta because it would affect efficiency, line-tying because it would be accompanied by electron heat loss, and finite gyro-radius effects because they would restrict the ratio between plasma radius and gyro-orbit size. Yet it was hoped that progress could be made even with MHD-unstable configurations [24].

After Livermore had observed flute instabilities in their own experiments [25], however, interest in the 'simple mirror' waned and minimum-B systems became standard. An elegant coil configuration, first named 'tennis-ball' by the British inventors [26], but now generally known as a 'baseball' coil (figure 3.4a), was used in several laboratories, until Livermore transformed the single baseball into a set of two 'yin-yang' coils (figure 3.4b) for better efficiency and greater access to the plasma centre.

These so-called quadrupole mirrors achieved their greatest success when 2XIIB in Livermore produced MHD-stable plasmas with high beta values. Under these conditions, the plasma altered the vacuum magnetic field in such a way as to decrease the central field and, hence, to increase the mirror ratio. This had a favourable impact on the Q-value expected for a reactor, which depends logarithmically on the mirror ratio. Moreover, this experiment proved the minimum-B principle to be valid for suppressing flute instabilities in plasmas with high mirror ratios and with beta values limited only by equilibrium considerations. In fact, beta—when defined so as to have the central vacuum field strength in the denominator—even exceeded one in the 2XIIB experiments (figure 3.5).

3.1.4 Velocity-space instabilities in mirror machines

The first velocity-space instability (box 3.2) encountered in mirror research was predicted theoretically by Harris [27]. It can be understood as a process

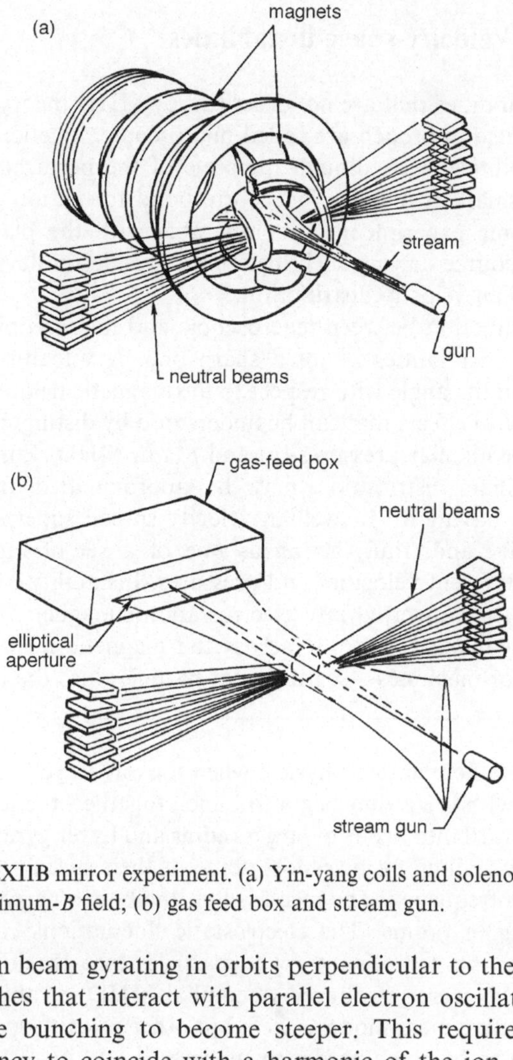

Figure 3.5. The 2XIIB mirror experiment. (a) Yin-yang coils and solenoids producing the quadrupolar minimum-B field; (b) gas feed box and stream gun.

in which an ion beam gyrating in orbits perpendicular to the magnetic field develops bunches that interact with parallel electron oscillations, which in turn cause the bunching to become steeper. This requires the electron plasma frequency to coincide with a harmonic of the ion gyrofrequency, $\omega_{pe} = n\omega_{ci}$, which implies a density threshold. The instability can be suppressed by various mechanisms, such as Landau damping by hot electrons.

The first experimental evidence for emission at harmonics of the ion cyclotron frequency $n\omega_{ci}$ was reported [28] by the OGRA team and similar observations were reported in Salzburg by Oak Ridge. In neither case was the unstable mode positively identified, however. In fact, the Kurchatov group thought of an explanation not in terms of an instability but of a Cerenkov effect, while the Americans considered, and later confirmed, a 'negative mass' instability as an alternative mechanism [29]. This instability

Box 3.2 Velocity-space instabilities

Plasma instabilities that are not described by MHD theory but require a more detailed approach are called microscopic, kinetic, or velocity-space instabilities, even though the scale of the perturbations is not necessarily small. It may, in fact, correspond to the ion gyro-radius, which in some experiments is comparable with the plasma radius. The energy source driving such instabilities resides in deviations from the Maxwellian velocity distribution.

The distinction between macroscopic and microscopic, or MHD and kinetic, instabilities is not a sharp one. A velocity distribution depending on the angle with respect to the magnetic field, for instance, may give rise to effects that can be understood by distinguishing parallel and perpendicular pressures, $p_{\parallel}$ and $p_{\perp}$, in MHD. Furthermore, a non-Maxwellian distribution may be approximated by a set of beams, each having a Maxwellian velocity spread superposed on the mean velocity and, thus, acting as one of a set of interstreaming fluids with different velocities and pressures. Instability then occurs if beam energy is converted into electrostatic or magnetic fluctuations. These may enhance transport across the magnetic field, or—in the case of mirror machines—scattering of particles into the loss cone.

was known from accelerator physics; when ions in a cyclotron with a field that decreases with the radius begin to bunch together, the leading particles are pushed forward into orbits of larger radius and lower gyration frequency. Conversely, the trailing particles lose some of their speed and so acquire a higher gyration frequency. The result is that the bunch narrows in azimuth, while spreading in radius. The electrostatic fluctuations caused by these bunches tend to scatter other trapped particles into the loss cone. When minimum-B fields were introduced to combat MHD instabilities, this took care of the negative-mass mode as well, but other velocity-space instabilities would turn out to be much harder to suppress.

The catalogue of velocity-space instabilities grew rapidly in the early 1960s, making it difficult to assign the proper mode to whatever high-frequency oscillations were observed in injection experiments. The velocity distribution typical for particles in mirror configurations deviates from a Maxwellian, firstly because particles whose velocities have directions within the loss cone escape at once, secondly because particles with smaller velocities are scattered more rapidly into the loss cone, and thirdly because a positive ambipolar potential drives out low-energy ions altogether. Oscillations in this non-Maxwellian distribution with an over-population of high-energy states may couple to various electrostatic, acoustic, drift, and hybrid waves in the plasma and stimulate their growth in a manner analogous to the action of a maser or laser. Details of the geometry are important, in particular when the parallel component of the wave vector, $k_{\|}$, is comparable with the inverse distance between mirrors. Particularly dangerous are the modes that cause rapid velocity-space diffusion of ions into the loss cone, although a moderate amount of enhanced scattering, both in velocity and in direction, may have a damping effect. Other stabilizing effects such as finite gyro-radius and Landau damping come into play, as do reflection or absorption of the wave at the ends of the plasma column and non-uniformity of the magnetic field. Some modes respond to injection of a warm plasma stream partly filling the loss cone [30].

While the Harris instability and the negative-mass instability required beamlike velocity distributions and, therefore, were not dangerous for reactor-like plasmas, theoreticians proposed more troublesome modes, like Kadomtsev and Pogutze's 'modified negative-mass instability' and Post and Rosenbluth's 'drift cone instability' [31]. Yet, there were theoretical predictions that with $\beta \leq 0.5$ and with the right ion- and electron-velocity distributions, a mirror system might hold stable reactor-grade plasmas if it had a minimum-B configuration and if its length, L, and plasma radius, a, satisfied $L \leq 200\rho_{\rm i} \leq a$, where $\rho_{\rm i}$ is the thermal-ion gyro-radius. This forced mirrors into a very short, fat configuration—the opposite of the long cell favoured by economic considerations. The discussion in the late 1960s was focused on the DCLC (drift-cyclotron loss-cone) mode, and in order to make experimental plasmas resemble more closely the reactor plasma with collisional velocity distributions and high dielectric constant,* injection energies were lowered to, typically, 20 keV, while beam currents were pushed to their technical limits. Still, all attempts to break through the density thresholds into the high-density regime by means of beams failed, so that other schemes like Livermore's plasma injection or the Kurchatov Institute's turbulent heating remained indispensable.

The results of many years of mirror research, in which theory and experiment had gradually come together in understanding velocity-space instabilities, distinguishing dangerous from harmless ones and finding conditions for their suppression, culminated in the Livermore 2XIIB experiment, the successor of 2X and 2XII, which as we saw had been designed to accommodate neutral-beam injection. In Berchtesgaden in 1976, Fred Coensgen reported successful trapping of 370 $\rm A_{eq}$ of deuterium atoms (produced by accelerating a mixed $\rm D^+ + D_2^+ + D_3^+$ beam to 15–20 keV, so that the resulting neutralized beam had full, half, and one-third energy components) on a gun-injected target plasma. Using either a pulsed plasma generator or neutral-gas feed to stabilize the DCLC mode, they achieved peak densities above $10^{20}\,\rm m^{-3}$ at mean ion energies around

* Alfvén waves propagate as if the plasma had a dielectric constant $(\omega_{\rm pi}/\omega_{\rm ci})^2$.

10 keV, and with volume averaged $n\tau_E = 2.4 \times 10^{16}\,m^{-3}\,s$. The ion energy loss was through classical ion–electron collisional transfer [32]; the central β-value was in excess of unity. Their method for stabilization had earlier been demonstrated in the PR-6 and PR-7 machines in the Kurchatov Institute [33] but 2XIIB was the first to produce high-energy, high-density, stable mirror plasmas.

3.2 Tandem mirrors

Notwithstanding the impressive performance of 2XIIB, mirrors would not have survived as a viable approach to a reactor if it had not been for the tandem mirror. Their principal weakness was insufficient energy multiplication and in the mid-1970s American mirror research was under great pressure from the Energy Research and Development Administration, ERDA (successor to the Atomic Energy Commission, AEC, and in turn succeeded by the Department of Energy, DOE) to come forward with some form of Q-enhancement scheme. Three broad categories of proposals were being considered [34]: linking of mirror machines, field reversal, and end plugging with DC or RF electric fields, but none of these was entirely convincing for reactor application. Just in time, Dimov [35] from Novosibirsk presented his revolutionary 'open trap with ambipolar mirrors' at the 1976 Berchtesgaden meeting. Similar work [36] from Livermore was described in a preprint circulated at that meeting and was recognized by the Soviet colleagues as an independent discovery of the ambipolar mirror principle. Contributions on 'open traps' at this conference amounted to only four Livermore papers, including one on a $Q = 1.1$ mirror-reactor design, in addition to the Dimov paper. The tandem mirror proposal restored the hope that Q-values of five or more would be possible.

The tandem mirror as originally proposed consisted of a long central solenoid terminated at each end by two mirror coils forming a short mirror system, the end cell or plug (figure 3.6). That much was already contained in Kelley's paper, mentioned above, from 1967; the new idea was to make a plasma with a high ambipolar potential in the end cells, so that the ions in the central section would be confined electrostatically in the axial direction. In the central section, where the confinement would be beta limited, the ion temperatures or energies would be in the region of 10–20 keV to optimize the reaction rate, but they would be as high as 1 MeV in the end cells to minimize scattering loss of the barrier population. In the original tandem mirror, the electron temperature was assumed to be the same in the end cells as in the centre section. So, a high potential, which by Boltzmann's law would follow the electron density according to $e\Phi = kT_e \ln n_e$, called for a high plasma density in the end cells. The proposal led to a new wave of experimental activity [37]. Livermore set up the Tandem

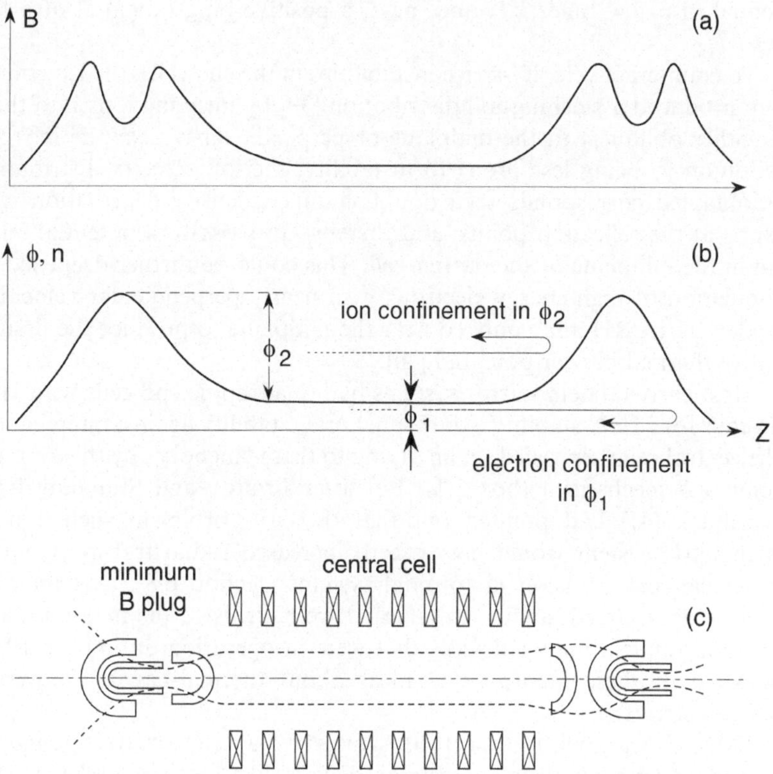

Figure 3.6. Principle of tandem mirror confinement. From top to bottom: (a) magnetic field on axis; (b) plasma density and potential on axis; (c) coils for axisymmetric central cell and quadrupole plugs.

Mirror Experiment, TMX, and Novosibirsk started construction of Ambal. But the Japanese were the first to succeed in putting a tandem-mirror system into operation; GAMMA-6 in Tsukuba reported clear signs of ambipolar potential barriers already at the 1978 Innsbruck Conference.

The original version of the tandem mirror, however, still suffered from a high energy transfer from the energetic plug ions to the much cooler electrons. Thus, there would be an advantage in creating the desired plug potential with a lower-n_e, higher-T_e plasma, if this could be thermally isolated from the central cell. For this purpose, Baldwin and Logan [38] from Livermore proposed the thermal barrier, a region of depressed electron density and negative electrostatic potential between the plug and the central cell. To maintain the barrier plasma at a low density, it was necessary to pump out ions by some means such as charge exchange on a neutral beam directed so as to convert an ion trapped in the barrier into an untrapped one. The central plasma would then see at both ends first a negative thermal barrier potential of a few times kT_e and, next, a positive plug potential of a few times kT_i.

To counteract velocity-space instabilities in the end cells, the Livermore group produced a sloshing-ion distribution [39] by injecting ions into these cells either obliquely in the midplane or perpendicularly near the mirrors. In addition to being less prone to instability, the ion velocity distribution thus obtained corresponds to a double-humped density distribution with maxima at the reflection points, and thereby gives rise to a potential minimum in the midplane of the barrier cell. This could be further deepened by a population of high-energy electrons with nearly perpendicular velocities, provided by ECRH and confined near the midplane to produce the desired negative thermal barrier potential [40].

Most early tandem mirror systems had quadrupole end cells with yin–yang coils for MHD stability (MHD anchors). Stability against interchanges in the central cell then relied on line-tying to these 'anchors', by the average-minimum-B mechanism (box 3.1). But after Ryutov and Stupakov from Novosibirsk [41] had pointed out that the drift orbits in such a non-axisymmetric system would give rise to increased radial transport, much like neoclassical diffusion in toroidal systems (section 6.8), axisymmetric mirror fields regained favour. In table 3.1 we have listed the major second-generation tandem mirror devices that were in operation during the 1980s [42]. Most of these incorporated ideas about thermal-barrier formation and symmetrization.

TMX-U was, with respect to symmetry, a conservative design; it consisted of two quadrupolar mirror devices like the successful 2XIIB, linked by a central solenoid. It succeeded in demonstrating the formation of the desired plugging and thermal barrier potentials, but the barrier tended to fill up with trapped ions when the central density reached a few times $10^{18}\,\mathrm{m}^{-3}$, so that in the end there was no improvement in the $n_i\tau_E$ value of $4 \times 10^{16}\,\mathrm{m}^{-3}\,\mathrm{s}$ obtained in normal tandem-mirror operation [43].

GAMMA 10 has separate cells for plugging and stabilization; the MHD anchors are two sets of quadrupole mirrors, located between the central cell and the plugging cells, all of which are axisymmetric. It achieved a barrier potential up to 2 keV and provided good axial as well as radial confinement, with $\tau_E = 10$–20 ms, of a plasma with electron density of up to $5 \times 10^{18}\,\mathrm{m}^{-3}$ and a perpendicular ion temperature reaching more than 5 keV. At first it ran into a problem that all potential barriers have in common—that they are potential wells for opposite charges and so tend to be filled up with trapped low-energy particles, The energy loss associated with this barrier decay [44], however, scales favourably with the ion temperature. With 10 keV in the central cell, the plug and barrier potentials in GAMMA 10 were sustained for 50 ms (limited by the ECRH power supply) with an energy confinement time of 8 ms [45]. Radial potential control was found to be effective in suppressing drift wave fluctuations and attaining good radial confinement.

Table 3.1 Major second-generation tandem-mirror devices

			Central cell			Plugs			
Location	Name	Year	L (m)	B_0 (T)	Heating	B_{max}	Heating	Anchors	Remarks
Tsukuba	GAMMA-10	1983	6	0.5	NBI/ICRH	3	NBI/ECRH	4-poles inside plugs	Effectively axisymmetric
Livermore	TMX-U	1982	8	0.3	NBI/ICRH	2	NBI/ECRH	4-poles as plug	Potential distribution control
MIT	TARA	1984	5	0.2	ICRH	3	NBI/ECRH	4-poles outside plugs	Stabilization by divertor/ponderomotive RF
Nagoya	RFC-XX-M		3	0.35	ICRH	2.1	ICRH	Cusps as plugs	ICRF-sustained tandem mode
Madison	Phaedrus-B	1987	3	0.6	ICRH	1	ICRH	4-poles inside plugs	Ponderomotive stabilization; floating rings
Kyoto	HIEI		1.6	0.18	ICRH	0.8	ICRH	None	Pondromotive stabilization
Novosibirsk	AMBAL-M	1962	13	0.45		6	NBI/ECRH	Cusps	Fully axisymmetric
Livermore	MFTF-B	1985	20	1.0	ICRH/ICRH	12	ICRH/ECRH		Not put into operation

The inverse arrangement, with a pair of axially symmetric plug-and-barrier cells (axicells) located between the central cell and the quadrupole anchors, was tried out in the TARA mirror device at MIT [46]. When the axicell coils in this arrangement were fully energized, plugging was entirely separated from anchoring. Alternatively, the axicells could act as intermediate plugs which still allowed a fraction of the central-cell ions to reach the end plugs, thus providing a connection between these and the central cell while reducing the asymmetry-driven neoclassical transport. The axicells were filled with sloshing ions by NBI and the electrons in the central planes of the cells were heated by ECRH to raise the barrier potential. Both these plug cells and the central cell could be heated by ICRH. A high-β plasma was maintained in the quadrupole cells to provide MHD stability.

One of the questions addressed by TARA was whether the connection between the anchors and the central cell would be sufficient for stability [47]. It turned out that an $m = 1$ flute mode could be stabilized, both by these anchors and by RF ponderomotive forces, but that trapped-particle instabilities persisted in the bad-curvature axicells. As a consequence, the build-up of the central-cell ion density stagnated at $4 \times 10^{17}\,\mathrm{m}^{-3}$, reaching the intended $4 \times 10^{18}\,\mathrm{m}^{-3}$ only when a novel stabilizing system was energized. This consisted of auxiliary coils, which diverted field lines to the metal wall (as in an axisymmetric divertor) and acted to stabilize the plasma by providing a short circuit across field lines [48].

As an alternative to quadrupole stabilizers, various axisymmetric field configurations with minimum-average-B properties or with other stabilizing features have been considered [49]. Some of these will be discussed in section 3.5; here we mention those that have actually been employed in tandem mirror experiments.

Phaedrus-B at the University of Wisconsin in Madison [50] originally was a five-cell tandem mirror system with quadrupole end cells, but was also used as an axially symmetric three-cell or five-cell tandem, in which case stable confinement was achieved with the radial ponderomotive force of ICRH waves. The central-cell plasma reached $n_\mathrm{e} = 9 \times 10^{18}\,\mathrm{m}^{-3}$ and $T_\mathrm{i} = 80\,\mathrm{eV}$ with an energy confinement time $\tau_\mathrm{E} = 200\,\mu\mathrm{s}$. The energy confinement times in all tandem mirror devices are brief when calculated not on the basis of particular loss channels, but considering all the NBI or RF power supplied to produce or heat the plasma and to sustain the potential distribution.

Ponderomotive confinement (box 3.3) was studied also in the RFC-XX-M device in Nagoya [51] (figure 3.7). This was a central mirror system with cusped end cells plugged with ICRH waves. Although not originally designed as a tandem mirror, it could sustain the potential distribution characteristic for such a system entirely by means of ICRH. It was fully axisymmetric to avoid neoclassical transport loss and derived its stability from the favourable curvature of the cusp fields. The RF plugging potential, Ψ, which reached 1 kV for $(\omega_\mathrm{pi}/\omega_\mathrm{ci})^2 = 10$, decreased however with increasing plasma density:

Box 3.3 The ponderomotive force

The ponderomotive force on a charged particle (sections 1.1.3, 1.3.6) is the time-averaged sum of the electric and magnetic forces experienced by a charged particle oscillating in a non-uniform electromagnetic field. This force can be written as the gradient of a pseudo-potential, which—if the oscillation is perpendicular to a constant magnetic field—takes the form

$$\Psi = \frac{e^2 E^2}{4m(\omega^2 - \omega_c^2)}$$

where ω is the frequency of the wave and ω_c is the cyclotron frequency in the constant field. The resonant denominator allows Ψ to have a significant value, even for ions.

$\Psi \propto (\omega_{pi}/\omega_{ci})^{-1.6}$. Because $(\omega_{pi}/\omega_{ci})^2 \propto n_i/B^2$, this implies a rather severe limitation on the density in the spindle cusps. In 1986, Sato could report simultaneously achieved central-cell parameters $n_i = 6 \times 10^{18}\,\mathrm{m}^{-3}$, $T_i = 0.4\,\mathrm{keV}$, $\tau_E = 1\,\mathrm{ms}$. A parametric study revealed that the RF plugging field can be enhanced by a plasma resonance (a Bernstein mode) and that, at typical operating plasma parameters [52] of $n_i = 2.4 \times 10^{18}\,\mathrm{m}^{-3}$, $T_i = 100\,\mathrm{eV}$, the central plasma potential indeed assumes a positive value governed by the ponderomotive force on the ions in the plugs. A supporting study of RF heating and ponderomotive stabilization was carried out in the HIEI device in Kyoto [53].

Ambal-M [54] in Novosibirsk was also entirely axisymmetric. It had so-called semi-cusp stabilizers outside the ambipolar-plugging cells. The semi-cusp was one of a variety of axisymmetric configurations with

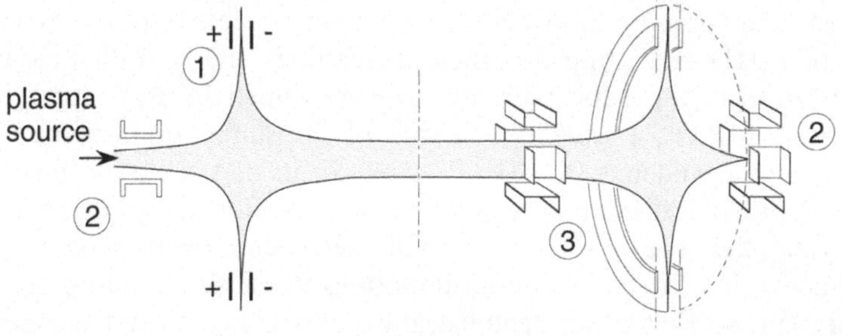

Figure 3.7. RFC-XX-M. 1, line cusp plugging electrode; 2, point cusp plugging coil; 3, ICRH coil.

Figure 3.8. MFTF-B. One of the two 6 m high yin-yang coils sets for the end plugs being delivered at the Livermore Laboratory.

minimum-*B* properties, the first of which had been proposed already in the early 1960s by Andreoletti and Furth [55]. We refer to Ryutov [56] for a review of the problem of stability in axisymmetric open-ended systems, including multiple-mirror devices, the gas-dynamic trap and a rotating plasma, all of which have been studied extensively in Novosibirsk.

The largest tandem mirror device, MFTF-B (figure 3.8), at Livermore [57], had quadrupole end plugs and axicells to reduce the passage of ions between these and the central cells. The design of tandem mirror systems involved many compromises between the requirements imposed by various forms of MHD and trapped-particle instabilities and by radial as well as axial transport [58]. Moreover, it was constrained by its history, which began with MFTF, a scaled-up version of the simple quadrupole 2XIIB, then became a tandem with MFTF as one of its end cells, and finally was equipped with axicells to reduce radial ion loss. Although the basic principles underlying the MFTF-B design had been confirmed separately, all experiments thus far had run into limitations in density build-up or energy confinement, so that a full demonstration of a consistent set of operating conditions for such a high-performance tandem-mirror experiment was still lacking. In fact, the performance expected from MFTF-B was one order of magnitude higher in temperature and two orders in $n\tau$ than had been achieved in any tandem mirror experiment.

This large extrapolation from the existing knowledge base sadly turned out to be a bridge too far. In the face of budget restrictions, the facility was closed down in February 1986, immediately after its main components had been tested and the assembly officially dedicated. The large tokamaks TFTR, JET and JT-60 (chapter 8) had already produced excellent results and could, according to empirical scaling relations, be extrapolated to reactors with confidence, whereas TMX-U had run into problems. Mirrors never had any margin for deviation from the highest theoretical expectations and particularly the threat posed to MFTF-B by its lack of axial symmetry made it a higher-risk venture than the tokamak programme that was competing for the same funds.

As compared with simple mirror machines, tandem mirrors have a lower β limit; early theoretical estimates [59] nevertheless yielded acceptable β values and provided a basis for expecting reactor Q values of order 25. This encouraged both a number of reactor studies and an upswing of small-scale supporting experiments, but the closure of MFTF-B in effect stopped the US mirror programme, because with it also the supporting studies at Livermore, Madison, MIT and other places came to an end. Since Europe had already abandoned the subject in the early 1970s, mirror research is now carried forward only by Russia (Novosibirsk and Moscow) and Japan (Tsukuba).

3.3 Z-pinch and plasma focus

The classical linear pinch emerged from the Geneva Conference with little credit. It had been openly discussed already since 1956 and all it produced seemed to be 'false' neutrons coming from an unstable plasma. Yet, these relatively simple experiments yielded the highest energy density and X-ray brightness then available in laboratory plasmas, so they were interesting not only from a purely scientific point of view but also because one could envisage various applications outside the area of magnetic-confinement fusion. Moreover, there were signs that the compressed plasma broke up not quite as rapidly as MHD-instability theory would have it and with some optimism one could imagine scaled-up versions which, even in the short life-time of the discharge, might satisfy the Lawson criterion because of the high plasma density. So, interest in the Z-pinch was kept alive in universities as well as in military and industrial laboratories and the subject re-emerged from time to time throughout the history of fusion research.

In the Kurchatov Institute, early studies of linear pinches in symmetrical, metal-walled discharge tubes had revealed strong cathode-to-anode asymmetry of the discharges. Filippov concentrated subsequent

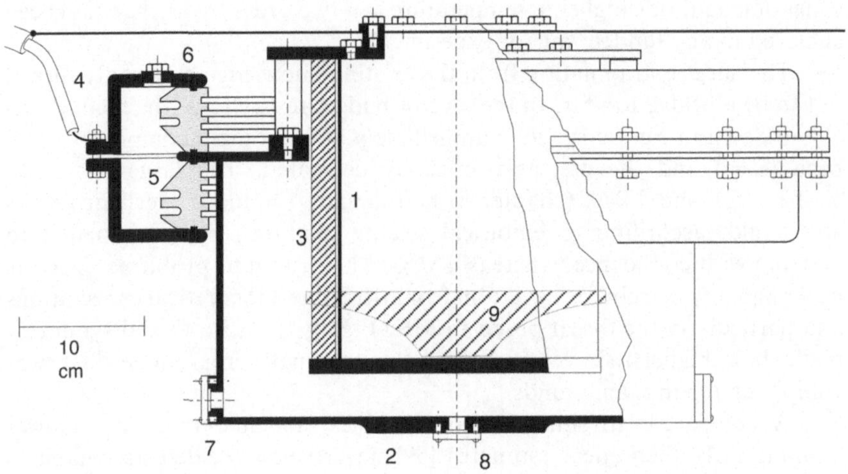

Figure 3.9. Filippov's plasma focus. 1, anode; 2, copper housing (cathode); 3, porcelain insulator; 4, coaxial current feed; 5, vacuum spark gap; 6, trigger electrode; 7, 8, observation windows; 9, conical version of anode.

investigations on the spot near the anode, which he called the focus, where the plasma first contracted on the axis and where both neutrons and X-rays were generated. His discharge chamber became asymmetric (figure 3.9), the cathode being located in what would have been the mid-plane of a traditional pinch device. In their paper at the Salzburg Conference [60], Filipov and his co-authors give evidence, mainly from soft X-ray and pressure measurements, for the existence for 0.2 μs of a plasma blob with $n_e \approx 10^{25}\,\mathrm{m}^{-3}$, $T_e \approx 1\,\mathrm{keV}$. With the assumption that $T_i \approx T_e$, this could explain the observed emission of 10^7–10^8 neutrons per pulse. No such claim was made for the up to 5×10^9 neutrons observed in some high-energy discharges, but in some cases the spectrum of reaction products (protons from the $\mathrm{d} + \mathrm{d} \rightarrow \mathrm{t} + \mathrm{p}$ reaction, rather than neutrons from $\mathrm{d} + \mathrm{d} \rightarrow {}^3\mathrm{He} + \mathrm{n}$) seemed to suggest a thermonuclear origin as well. Although Artsimovich in his summary lecture to the conference said '*that serious consideration could hardly be given to this method as a practical means of obtaining intensive thermonuclear reactions on a scale which was of technical interest*', not everyone was so sceptical. So, while the Salzburg Conference still featured a variety of straight, hollow and hard-core linear pinches, the groups that chose to continue along this line all turned towards the plasma focus.

In Los Alamos, there had long been an interest in guns for injecting plasma blobs (plasmoids) into some kind of magnetic confinement system. Marshall developed first an electrodeless gun using a travelling wave in theta-pinch geometry and then a coaxial gun in which a plasma ring was

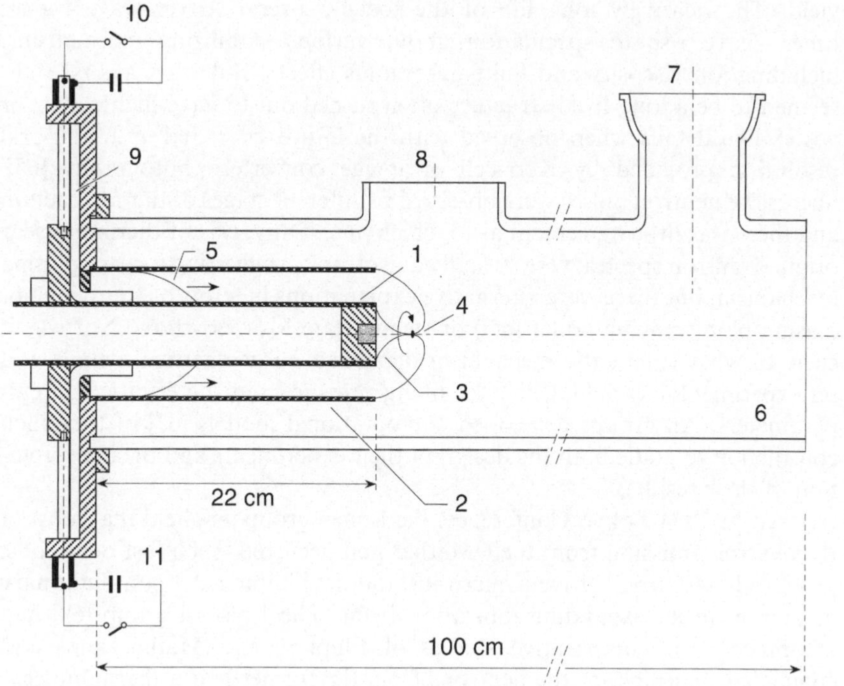

Figure 3.10. Mather-type plasma gun. 1, inner electrode (5 cm diameter); 2, outer electrode (10 cm diameter); 3, tungsten insert; 4, plasma focus; 5, current sheath before the focus has formed; 6, Pyrex glass vacuum vessel; 7, vacuum outlet; 8, viewing port for soft X-ray spectrometer; 9, insulator; 10, vacuum spark gap; 11, capacitor bank.

accelerated axially by a radial current and an azimuthal magnetic field [61]. The same geometry was chosen by Mather *et al.*, also from Los Alamos, who noted that the plasma at the end of the barrel was subject to radial Z-pinch forces. When it turned out that this pinch (figure 3.10) had characteristics very similar to those of Filippov's plasma focus, this transient phenomenon became more important than plasma propulsion [62]. A third route was explored by Morozov [63] from the Kurchatov Institute, who studied steady-state acceleration of a plasma stream in a coaxial, conical configuration and found a similar high-density focus in front of his central electrode. Although a comparative study in Frascati [64] did not yield a conclusive answer to the question whether and, if so, why one scheme offered better evidence for thermonuclear reactions taking place, most laboratories had some preference for the Mather-type plasma focus.

The activities in this area reached their peak in the early 1970s when more than a dozen laboratories all over the world did experiments on the dense plasma focus. The discussion centred on the stability of the focus, the nature of the neutron emission and the scaling laws for the neutron yield. The relatively long life of the focus—several Alfvén-wave transit times—gave rise to speculation about various stabilizing mechanisms, including ion viscosity and finite-gyroradius effects. But what at first sight seemed to be a long-lived structure often turned out to have filamentary or vortex-like details when observed with the improved radial and temporal resolution provided by Kerr-cell or image converter photography [65]. Successive neutron pulses were observed in different stages of the contraction and there was little agreement as to which one, if any, was of thermonuclear origin. Neutron spectra were strongly anisotropic, suggesting a beam-plasma mechanism, but there were alternative explanations in terms of a moving hot plasma blob (moving boiler [66]) or steady-state Morozov flow. Nor was it clear to what extent the mechanism depended on geometrical details and on experimental variables like the filling pressure and the electrical circuit parameters. All groups developed computational models [67] of the pinch contraction to guide both the design of their experiments and the interpretation of their results.

At the 1974 Tokyo Conference, the Limeil group [68] held that most of the neutron emission from their Mather gun occurred when fast deuterons, produced by current-driven micro-instabilities, impinged upon stationary deuterium in an expanding ionization front. The Frascati group [69] had continued their comparative studies of Filippov and Mather guns and concluded that most of the neutrons from the former had a thermonuclear origin, explained by turbulent heating due to fast-electron beams, while the evidence concerning the latter remained ambiguous. The Kurchatov group [70] confirmed that the neutron yield, if taken to be of thermonuclear origin, was consistent with their estimate of the temperature in the second contraction stage. Willard Bostick from the Stevens Institute [71], finally, kept stressing the spatial, vortex-like, and temporal fine-structure of the focus.

Until then, it was generally held that the neutron yield, Y_n, in optimized focus discharges scaled with the stored energy, W_0, or with the discharge current, I, at the time of neutron emission, according to $Y_n \propto W_0^2$ or $Y_n \propto I^4$, approximately. A scaling [72] of the type:

$$Y_n(\mathrm{d{-}d}) = 10^{13} W_0^2 \quad [W_0 \text{ in MJ}]$$

would imply a conversion efficiency from capacitor bank to nuclear reactions of $3.5 \times 10^{-8} W_0$, if the reaction rate were extrapolated to that expected from deuterium–tritium fuel, which in the relevant temperature range reacts 50–100 times faster than D–D. A useful reactor would then operate at a level of a few hundred MJ per pulse, not unacceptable from an engineering point of view. The outcome is, however, sensitive to the exponent in the scaling. This assumes that the energy of the capacitor bank is converted into magnetic energy right at the moment of the neutron burst; the design criteria for this were reviewed by the Frascati group [73]. This group was the first to bring the energy up to the 1 MJ level, and they found that the energy scaling started to degrade around $W_0 = 0.5$ MJ because of imperfect impedance matching and parasitic currents flowing along the insulator, hence outside the focus. At the same time, the Limeil group [74] disposed of the notion that the Filippov focus was inherently different from the Mather geometry. Although Filippov's group [75] insisted that they had a thermonuclear plasma, other groups had by that time given up hope of finding ways to scale up the device to an energy-producing reactor. Studies to clear up the physics and to exploit the focus as a source of neutron and X-ray flashes, however, continue both in universities and in mission-oriented fusion laboratories.

Before the Frascati group turned towards the plasma focus, they had employed Z-pinch techniques for rapid compression of hollow plasma cylinders [76]. The idea was to implode a heavy-ion plasma layer upon a deuterium plasma to compress the latter and to retard its subsequent expansion. For the same purpose, Frascati experimented with metal liners driven inwards by chemical explosives, a technique to which we shall briefly return in section 3.5. Here, we cross the line between magnetic and inertial confinement and enter a domain that is beyond the scope of this review.

The development of high-voltage ultra-short-pulse power supplies for inertial confinement with electron or ion beams, made it possible for the original, symmetric Z-pinch to stage a come-back in the magnetic-confinement fusion arena. To avoid the current maximum coming long after the compression, further increase of I had to be accompanied by an increase of $\mathrm{d}I/\mathrm{d}t$, hence by higher voltage, not by merely switching more capacitors in parallel. Malcolm Haines [77] argued that the Bennett relation predicts thermonuclear temperatures in a Z-pinch with a current of 1 MA and a line density of about $5 \times 10^{18}\ \mathrm{m}^{-1}$ and that under those conditions the ion gyro-radius is not much smaller than the pinch radius, giving rise to enhanced stability against particularly the $m = 0$ mode. His group at Imperial College had built a 600 kV transmission line [78], with which they studied compressional, gas-embedded, and gas-puff pinches; a compressional pinch remained stable for 4 MHD growth times. They went on to push the technology to higher voltages; the 2.4 MV MAGPIE generator can produce a 1.9 MA discharge of 150 μs duration in carbon or aluminium fibres, which are thereby compressed to higher than solid-state densities but rapidly develop hot spots suggestive of $m = 0$ instabilities [79]. In the Kurchatov Institute, which has a collaboration on this subject with the TRINITI Laboratory at Troitsk (formerly a branch of the Kurchatov Institute), the S-300 generator can deliver 4 MA with a rise-time of 60 ns. Discharges in fibres of organic material again display $m = 0$ instabilities, but this group suggests that the hot and dense plasma in the constriction might launch nuclear burn waves into the undisturbed sections of the column [80].

These and other institutes, for example the Sandia Laboratories in Albuquerque in the USA, continue with similar studies, mostly inspired by non-fusion applications or directed primarily towards inertial-confinement fusion; this work is reported in a series of conferences on dense Z-pinches [81].

3.4 Theta pinches

Although, as we already observed in section 2.3, the results of Scylla obtained in the spring and summer of 1958 were presented at Geneva in a very cautious manner, particularly with respect to the origin of the neutrons, the Los Alamos group was confident that it saw thermonuclear reactions in a plasma with an ion temperature of 1.3 keV. Strong arguments in support of this contention were the time dependence of the neutron emission from the $d + d \rightarrow {}^3He + n$ reaction and the energy spectrum of protons from $d + d \rightarrow t + p$ (figure 3.11). But they admitted that the plasma lived long enough for only one thermalizing ion–ion collision, not enough to establish a fully Maxwellian velocity distribution. Already during the Geneva Conference it was noted [82], however, that the occurrence of neutrons in the second half-period suggested that a reverse field trapped in the plasma during the first half-period played an important role. This was elaborated by the group at NRL who challenged the Los Alamos picture [83] of energetic deuterons being produced by Ohmic and shock preheating followed by adiabatic compression in the rising field. Kolb [84] suggested that rapid annihilation of the trapped reversed field due to enhanced resistivity would induce an azimuthal electric field strong enough to accelerate deuterons to the energies required to explain the neutron emission.

The importance of the trapped reversed field was indeed confirmed in subsequent experiments in Scylla which showed that the neutron pulse would occur already in the first half-period if a reversed bias-field was applied [85]. The annihilation of this field gave rise to preferential ion heating which—after adiabatic compression—finally resulted in the observed 1.3 keV ion temperature. There was lively debate about the putative thermonuclear nature of the neutron emission both at the Uppsala meeting in 1959 and at the 1961 IAEA conference in Salzburg [86]. Although Artsimovich in his summary lecture in Salzburg characterized the theta pinch as '*not a stable plasma process . . . obscure in many respects*' and declined to deal with the question of the nature of the neutron production, it was broadly accepted that the ion temperatures were indeed in the keV range and this was confirmed by subsequent studies [87].

In all early theta pinches, the plasma was lost through the ends with approximately the ion thermal velocity, and it was clear that simply lengthening the devices would lead to an unacceptably large energy per pulse.

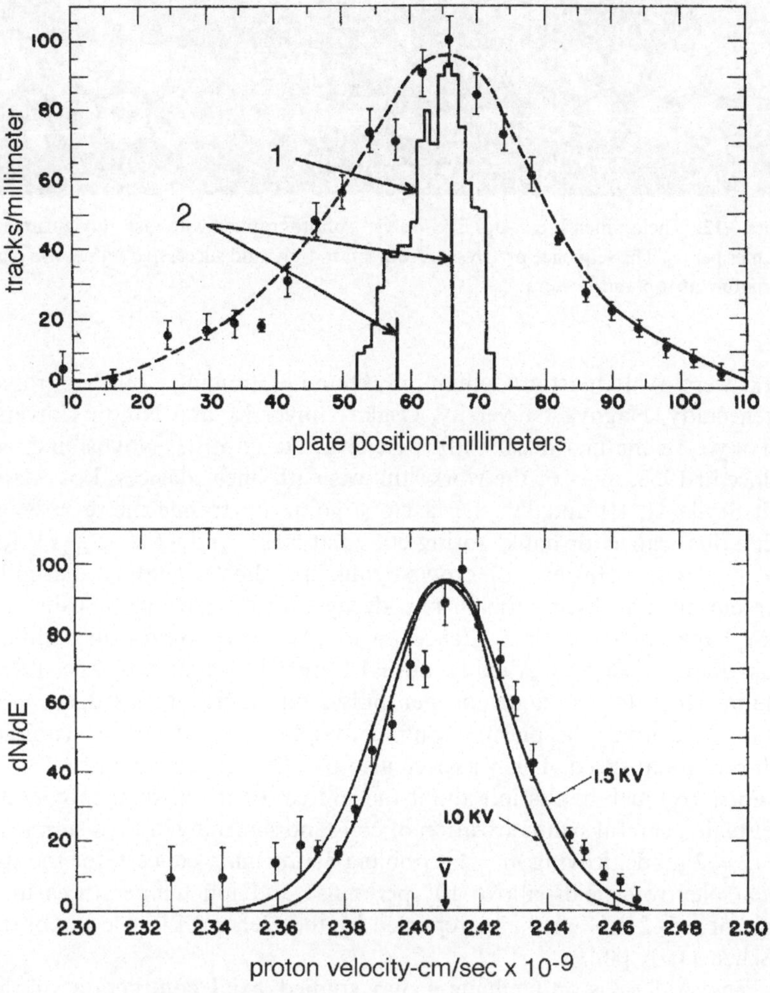

Figure 3.11. Protons from d–d reactions in Scylla. Top: proton spectrum and calibration run (1) with ^{240}Pu α-particle lines (2); bottom: proton spectrum compared with computed curves for 1.0 and 1.5 keV ion temperatures.

Hence, finding other ways to extend the life of the plasma and to study confinement on a correspondingly longer time scale was on the top of the agenda. Eventually, this path would lead away from the theta pinch as an open system; either the straight tube was bent into a torus, or a reversed field was trapped in the plasma to form closed field lines when linking up with the external theta-pinch field.

In the 1960s and early 1970s, the theta pinch and its modifications were in the forefront of fusion research. Among the laboratories taking part in this

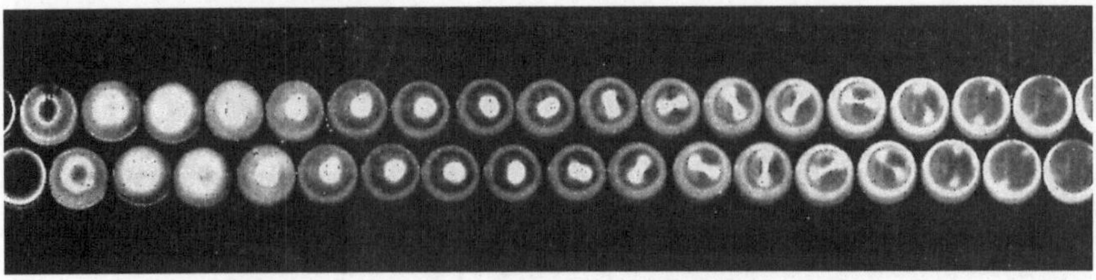

Figure 3.12. Theta pinch break-up. End-on view photographed with fast movie-camera at 8 frames per μs. The sequence progresses from left to right and successive frames are placed alternately at top and bottom.

work were Aldermaston/Culham, Sukhumi, Garching, Jülich, Frascati, Schenectady, Nagoya University, Osaka University and Nihon University in Tokyo. In the first decade after Geneva, i.e. until the Novosibirsk conference in 1968, most of the work still was with linear devices. Los Alamos built Scylla II, III and IV, the last one going up to one metre in length, with a dual capacitor bank storing 600 kJ at 50 kV and 3 MJ at 20 kV for a power-crowbar. (Instead of crowbarring, i.e. short-circuiting the coil at peak current and letting the current decay with its L/R time constant, the 20 kV bank could also be switched on to power the coil for an additional 20 μs quarter period.) Already at Salzburg, they reported $T_i \approx 1.8$ keV yielding close to 10^8 neutrons per pulse, but there appeared an $m = 2$ instability causing the plasma column to split in two when it acquired a high rotational speed. This was seen also by other groups (figure 3.12), and gave rise to much discussion about the sources of the angular momentum. Eventually, careful symmetrization of coils and screening of gaps suppressed the $m = 2$ mode, leaving $m = 1$ wobbling as the main concern for the theta pinch. Neutron yields above 10^7 per pulse and ion temperatures in the range of 1.5–2 keV were also reported by the General Electric Laboratory in Schenectady [88].

The Aldermaston/Culham group studied axial contraction of theta pinch plasma columns in conditions with or without reversed bias field. In the former case, a closed configuration developed, which turned out to be unstable. The length of the column was increased to 8 m to allow a study of radial loss of plasma in the central region on a time scale such that it would not be affected by end effects. At the Novosibirsk Conference, Hugh Bodin [89] from Culham reported a stable period of 10 μs with $T_e \approx 240$ eV or 25 μs with $T_e = 120$ eV, and radial confinement at least twenty times better than predicted by Bohm's formula. Thus, the theta pinch joined tokamaks, stellarators, internal-ring devices and reversed-field pinches in defying Bohm diffusion.

In retrospect, one can conclude that Scylla 1, as reported in Geneva 1958, was indeed the first laboratory-produced thermonuclear plasma, albeit by a fortuitous combination of circumstances. The Los Alamos group had optimized the pulse length without being aware of the pitfalls on either side: axial contraction and anomalous heating had done their work before the moment of maximum compression, and the wobbling instability did not develop until after this moment.

The US NRL team, which had first seen the $m = 2$ break-up of the rotating theta pinch, turned to a hard-core version in which a current-carrying rod was placed along the axis, and so produced a ring-shaped plasma with a tokamak-like field topology. This scheme was later also investigated in Jülich, after reversed bias field and cusped end plugs had been experimented with. Linhart and Knoepfel in Frascati, rather than trying to modify the basic theta-pinch scheme, chose to push the original concept to its limits by going to ultra-high magnetic field strengths. A part of the volume within the single-turn coil was surrounded by explosives, which were fired at the instant of current maximum to squeeze the flux into the remaining theta-pinch volume and so to raise the magnetic field and the plasma to extreme values of the energy density.

The strongest effort in the theta-pinch area, except for that in Los Alamos, developed in Garching. At the Culham Conference in 1965, Andelfinger [90] reported neutron emission during 5 μs, yielding up to 5×10^9 neutrons per pulse; the ion temperature reached 3.7 keV. Various aspects of shock heating and confinement of theta-pinch plasmas were studied also in the other places mentioned above. One of the main concerns remained the $m = 1$ instability, which transported the plasma column as a whole to the wall, often in about 10 μs, i.e. on the time scale of the magnetic-field pulse. But in those cases where the $m = 1$ amplitude would saturate before the plasma touched the wall or would be absent altogether, sufficient evidence of good radial plasma confinement was acquired to justify making the next large step forward: a toroidal version of the theta pinch. We come to this in the next chapter.

Long after the mainstream of theta pinch research in Los Alamos and elsewhere had turned towards toroidal systems, however, the Scylla IV group in a final attempt to reduce end losses in linear theta pinches tried 'end stoppering' with solid, low-Z material plugs [91]. The experiment yielded a threefold increase in the energy containment time—from 9 to 29 μs—but this was not sufficient to open up a new path towards a reactor for the theta pinch. In fact, after Scylla achieved an $nT\tau \approx 10^{17}\ \mathrm{m}^{-3}\ \mathrm{keVs}$* plasma in 1958, linear theta pinches could hardly improve on this performance two decades later, whereas tokamaks by that time were operating already in the 10^{17}–10^{18} range.

* The significance of this 'fusion triple product' is discussed in section 8.2.3 and box 8.2.

3.5 Unconventional schemes

Without some special provision to reduce mirror loss, a simple straight mirror machine or theta pinch would not meet the requirements of efficient energy multiplication. By making the machine longer than the mean free path of the ions, one could reduce the effect of the end losses, but this would lead to a kilometres long reactor with a power output in the multi-gigawatt range [92].

Oleg Lavrentiev, who had shown his passion for fusion already in 1949, had subsequently taken courses in physics and had found a position in the Physical–Technical Institute in Kharkov. His electrostatic confinement scheme [93] had taken the form of a $\beta = 1$ plasma, separated from a magnetic cusp field by a thin boundary layer. A positive electrostatic potential was imposed in the ring cusp and the two point cusps by means of external electrodes. To avoid Debye screening of these electrodes, electrons had to be prevented from accumulating in the regions of positive potential. Although Lavrentiev achieved some positive results, work on the Atoll facility in the Kurchatov Institute revealed anomalous transport in the boundary layer [94]. While the idea of electrostatic confinement re-emerged from time to time [95], there has not been a large experiment along this line.

In section 3.2, we discussed the use of RF ponderomotive forces for stabilization of interchange modes as well as for plugging magnetic mirrors. The second of these schemes is called adiabatic RF confinement, to distinguish it from a non-adiabatic change of the magnetic moment induced by the RF, which affects the reflection of particles in the DC mirror field. The effect can be to increase or to decrease the magnetic moment, but on average it is positive, leading to enhanced confinement. This was studied in the 1960s, notably by Consoli [96] in Saclay/Grenoble. His group developed a two-stage electron cyclotron plasma source, in which first the gas was ionized, then the plasma was accelerated down a magnetic slope (from strong to weak field) to separate it from the residual gas and finally it was further heated as it passed through one mirror of the confinement region. This was a simple mirror stabilized with six Ioffe bars; the group succeeded in accumulating an $n_e = 5 \times 10^{17}\,\mathrm{m}^{-3}$, $T_e = 85\,\mathrm{keV}$, $T_i = 270\,\mathrm{eV}$, $n_e/n_0 > 100$ plasma with a lifetime of 4 ms, which was meant to serve as a target plasma for neutral injection. As interest in open systems waned, however, this line of research was not pursued any further.

Ryutov in Novosibirsk proposed the gas-dynamic trap, in which the mirror ratio is made very high and the plasma is nearly isotropic because the—very small—loss cone is filled by collisions, so that the loss is like that of a gas escaping through a small orifice. Experiments with low-temperature plasmas show unexpected flute instabilities [97]. Another scheme investigated at Novosibirsk is to increase confinement in a simple mirror by centrifugal forces (figure 3.13). This idea already stems from the

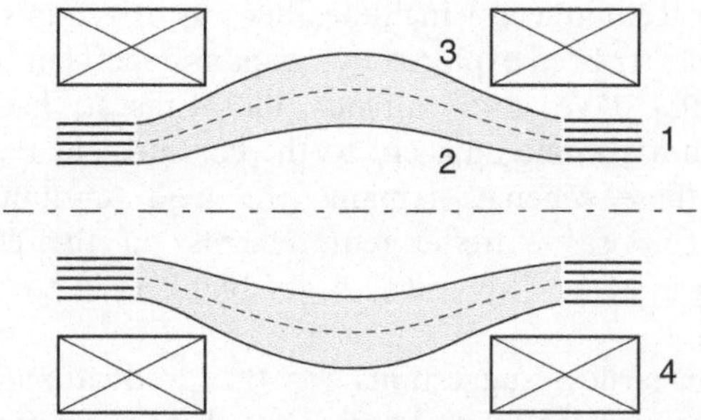

Figure 3.13. Rotating plasma in mirror machine. The coaxial electrodes (1) are connected with a voltage divider which ensures that the potential difference between the inner and outer liners (2 and 3) is evenly distributed over the magnetic surfaces. The radial electric field, combined with the mirror field produced by the coils (4) causes the plasma to rotate in the azimuthal direction.

1950s; what was new here is a system of concentric high-voltage electrodes behind the mirror which generates the radial electric field desired for $\mathbf{E}_r \times \mathbf{B}_z$ rotation and provides line-tying against flute instability [98].

Mirror losses can also be reduced by an arrangement of multiple mirrors in a long, straight series [99]. When the collision frequency of ions is high enough, particles escaping through one mirror may be scattered and trapped again before reaching the next one—through which they would pass when their magnetic moments were constant. The scheme would significantly reduce the size of an open system, but in one reactor study [100] would still extrapolate to a 400 m long reactor with a power output of 3 GW_e. Several laboratories considered pulsed high-density reactors in which, as suggested by Sakharov [101] already in 1951, the plasma would be surrounded by neutral gas and the magnetic field would serve only to reduce radial heat loss. Combining these ideas, Budker's group at Novosibirsk investigated a $\beta > 1$ multiple-mirror system in which wall confinement balances the radial plasma pressure through an intermediate gas blanket [102]. This is investigated in the GOL devices, which would extrapolate to a reactor with an estimate length of 200 m [103]. Multiple-mirror systems with or without wall confinement were reviewed by Mirnov and Lichtenberg [104], who show that generally one arrives at 100–1000 m.

Gross's group at Columbia University in New York [105] studied a wall-confined plasma generated in a coaxial-electrode shock tube. Oliphant at Los Alamos [106] considered quenching a theta-pinch plasma in a cold gas flow, after the burning phase. Finally, groups in Moscow, Frascati, and Los Alamos studied compression of plasma by an imploding liner [107]. The plasma could be produced *in situ* by an axial current or electron beam or be injected into the confinement volume as a toroidal plasmoid, and the liner was a length of thin-walled metal tubing compressed by Z-pinch forces. A similar technique in which the liner was driven inward by chemical explosives had earlier been explored by weapons experts in the Soviet Union and the US [108]. (We noted already that Frascati had compressed a magnetic field in a volume adjacent to the confinement region.) A serious problem with these schemes remains the high amount of circulating energy, which imposes extreme requirements on the efficiency of the energy-handling systems. We come to gas blankets in toroidal systems in the next chapter.

Among the earliest suggestions for the production of energy from nuclear reactions were colliding beams, but these were soon dismissed on elementary considerations [109] of power density and dispersal of the beams by Coulomb collisions. A possible answer to these objections was to direct a beam, or two beams if reactions between unlike particles were desired, into a small volume in a focusing field configuration. Particles emerging from this volume would return to it so that they could make a large number of passages. The first proposal to this effect was a Japanese one, dating from 1957 [110]; it involved an axisymmetric electrostatic quadrupole field for axial confinement combined with an axial magnetic field for radial confinement—4π focusing. This was not presented at the 1958 Geneva Conference by the Japanese, but the idea was discussed in an Italian paper [111]. Without space-charge neutralization, power production is limited to insignificant values (the power generated in a sphere with radius r m and ion density n_i m^{-3} is at most of order $10^{-3}r^{-1}$ or $10^{-12}n_i^{1/2}$ watts if the Coulomb energy is to stay below 10 keV per ion) (box 3.4). With electrons compensating the ion space charge, competition between energy production in ion–ion collisions and energy loss in ion–electron collisions requires $kT_e > 3.5$ keV in the case of colliding d and t beams. Noting further that the power density in a colliding-beam system can be a factor 2 higher than in a thermal plasma of equal energy density, one sees that the scheme has two cutting edges: higher gain and lower loss through the electron channel than in the case of a thermal plasma with $T_e \approx T_i$. But it poses questions with respect to both the efficiency of ion acceleration and the stability of these non-Maxwellian energy distributions against disturbances in velocity space.

Injection of ions into a mirror machine by means of an annular ion source enclosing the return flux of the mirror field produces particles with zero generalized angular momentum which, if given sufficient radial momentum to penetrate through the reverse field, must pass through the axis (figure 3.14) [112]. This was used by the Fontenay team [14] to make a peaked density distribution of thermalized ions. In the USA, Maglich [113] chose the field index of the mirror field—its dependence on the radius—so as to focus the particles also in the axial direction and introduced the name Migma for the neutralized, self-focused ion beam so obtained. His privately funded Aneutronic Energy Laboratory in the Princeton area studied the Migma,

Box 3.4 The need for charge neutrality

For a given ion density, the highest possible reaction rate per unit volume is achieved when deuterium and tritium beams with equal densities collide with a relative velocity of 4×10^6 m/s, where the reaction cross section has its maximum of 5×10^{-28} m^2. The resulting $\sigma v = 2 \times 10^{-21}$ m^3 s^{-1} yields as upper limit for the fusion power density

$$\tfrac{1}{4} n_i^2 \sigma v E_{DT} = 1.4 \times 10^{-33} n_i^2 \text{ W m}^{-3}$$

and for the power generated in the sphere:

$$P_{fus} < 5.8 \times 10^{-33} n_i^2 r^3 \text{ W.}$$

The mean potential energy of ions with density n_i in a sphere with radius r amounts to

$$E_{pot} = n_i e r^2 / 5\varepsilon_0 = 2.25 \times 10^{10} n_i r^2 \text{ eV.}$$

From the last two equations, one may derive scaling laws for the radius and the density

$$r < 1.2 \times 10^{-47} E_{pot}^2 / P \text{ m} \quad [E_{pot} \text{ in keV}]$$

and

$$n_i > 3 \times 10^{86} P^2 / E_{pot}^3 \text{ m}^{-3} \quad [E_{pot} \text{ in keV}]$$

which show that to generate as little as 1 W, one arrives at utterly unrealistic parameters. Evidently, the beams must be neutralized, which leads one to beam–plasma interactions; these are discussed in box 8.1.

and did achieve some success, building up ion densities two orders of magnitude above the space-charge limit. He reported [114] $n_i = 3 \times 10^{16}$ m^{-3}, with $E_i = 700$ keV and $\tau_{Ei} = 20$–30 s, and compared the triple product of these numbers with what is achieved in tokamaks. One should note, however, that for $n_i T_i \tau$ to be a proper figure of merit for fusion energy efficiency, T_i must be in the range $10 < T_i < 20$ keV (box 8.2), or $15 < E_i < 30$ keV when beams are considered, and τ must represent the time actually spent in the reaction volume, not in the external-storage portions of the orbits. Moreover, the volume in which these parameters are reached must be capable of being extrapolated to one of economic interest.

The original Japanese 4π-focusing system was reconsidered in 1993 by a group [115] that included the authors of the 1957 paper. But some of the advantages of beam–plasma or beam–beam systems are not restricted to

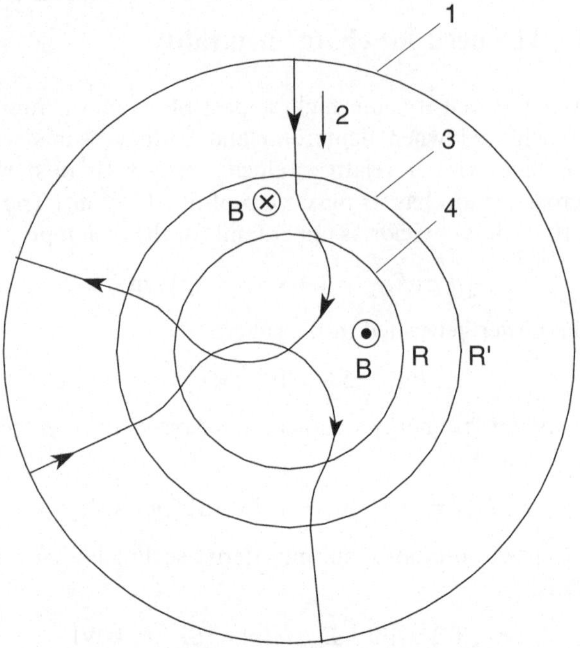

Figure 3.14. Radial injection in mirror machine. When ions from an annular source (1) are injected radially (2) through the return flux (3) into the main mirror field (4), they will pass through the axis, provided they have sufficient momentum to penetrate into the main field.

self-focusing systems. In fact, any fusion device that uses particle beams for fuelling or for additional heating will take advantage of the collisions undergone by the unthermalized beam. For example, the two-component—*beam–plasma*—tokamak proposed by Dawson [116] is a viable option to operate toroidal systems as energy amplifiers in a sub-ignited regime and ideas that were developed at Princeton along these lines in the 1970s found application in the TFTR experiment to which we return in chapter 8. Advocates of aneutronic fusion [117] favour tandem-mirrors and FRCs for their potential to serve as colliders for d and ^{3}He or for p and ^{11}B beams. The need to confine a d–^{3}He plasma with a 15–20 times higher $nT\tau$ than required for d–t, not to speak of the even more difficult requirements for p–^{11}B, seems to put such schemes beyond the present horizons for fusion.*

* The main problem with the p–^{11}B reaction lies in the electrons that are required to neutralize the beams (see box 3.4). If the electron temperature is too low, the beams are stopped before having had a high enough reaction probability; if it is too high, the loss through bremsstrahlung and synchrotron radiation outweighs the reaction power yield. Whether there really exists a workable operating window remains to be demonstrated; it will depend both on the quality of the confinement and on the efficiencies with which ion and electron energy losses can be reconverted into electrical energy.

3.6 The status of open systems

Rapidly pulsed magnetic confinement systems have played an important role in fusion research; the first hot plasmas were produced in linear Z-pinches and the first thermonuclear neutrons came from a linear theta-pinch. Yet, they have all but vanished from the scene. Even if problems of energy confinement and stability had been overcome, reactors based on these schemes would carry a high fraction of circulating energy and be subjected to great cyclic stresses. With the possible exception of the dense Z-pinch, which is still adhered to by some as a potential reactor, the fast-implosion systems, after having helped to elucidate rapidly developing MHD instabilities, had to yield to systems capable of demonstrating plasma behaviour on longer time scales.

The situation is less clear when it comes to steady-state open systems. MFTF-B was to have been a proof-of-principle experiment for a tandem mirror reactor with central cell, ion plugs, thermal barriers and quadrupole MHD anchors. Its premature closure has precluded a test of all these systems in consonance, so that questions about radial and axial losses in such a system remain unanswered. Also the potential of axially symmetric tandem mirrors has not been sufficiently explored to demonstrate their future as reactors. It is clear, however, that mirror machines need extremely efficient energy recovery and recycling systems and leave hardly any margin for anomalous energy loss. If a D–T burning mirror reactor, tandem or otherwise, ever proves to be feasible, the step from there to a D–^{3}He reactor is probably smaller than from a D–T burning tokamak or stellarator. In fact, whilst the closed systems are better suited for D–T, open systems appear to be the only way to exploit the aneutronic fuels. Sadly, the institute in Novosibirsk, which has generated so many new ideas for mirror research, is now lacking the means to pursue these lines with sufficient vigour, so that the main burden rests on the GAMMA-10 experiment at the University of Tsukuba.

Chapter 4

Pulsed toroidal systems and alternative lines

After the open systems, we now come to devices with closed magnetic surfaces. The main representative of this class is the tokamak, to which gradually the greater part of fusion research has been directed and which will be covered in the later chapters. There is, however, also a category of closed systems that were never regarded as serious candidates for reactor development, but were useful for studies of toroidal confinement in general. And between those extremes, there are those that may or may not have a potential of becoming competitive with the tokamak as energy-producing reactors. Precisely where the line between reactor-oriented and supporting or alternative studies should be drawn is a matter of judgement. We draw this line between the stellarators, which together with the tokamaks will be discussed later on, and the reversed field pinch, which has been given a place in the present chapter.

4.1 High-beta stellarators

So easy as it had been to produce dense and hot plasmas in linear theta pinches, so difficult it was to make them last for any length of time. Simply bending the tube into a torus would have meant loss of equilibrium and the plasma would have expanded in the direction of the major radius within a microsecond. But already in 1958, Meyer and Schmidt [1] in Garching had proposed a toroidal high-β equilibrium with corrugated flux surfaces (figure 4.1). In a field-free plasma of uniform pressure, separated from a confining field by an infinitely thin boundary layer, pressure equilibrium requires the surface current density as well as the magnetic field strength to be constant on the surface. If there is no rotational transform, and field lines close upon themselves, the fact that all lines on a surface enclose the same currents (running in external coils), so that $\oint B\,\mathrm{d}l$ is constant in the surface, then requires all field lines on the surface to be of equal length.

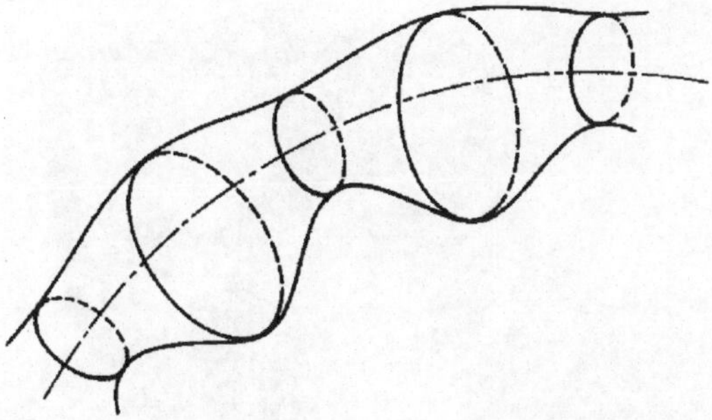

Figure 4.1. M&S scheme for toroidal confinement. Magnetohydrodynamic pressure equilibrium in a torus requires all field lines on a magnetic surface to be of equal length. In the Meyer–Schmitt field, this is achieved by a corrugation that is deeper on the inside than on the outside.

Hence, the corrugation must be deepest on the inside of the torus. An alternative to this 'M&S' principle is to introduce rotational transform as in a stellarator. In fact, the toroidal theta pinches studied in Garching and Los Alamos had various combinations of M&S corrugation and helical distortion of the magnetic surfaces, and were therefore also known as high-β stellarators. The magnetic field was shaped by a thick single-turn coil, the inner surface of which was machined to match the outermost magnetic surface. (It takes milliseconds, long compared to the life of these discharges, for the field to penetrate these ~1 cm deep corrugations.) The required shape was derived from MHD-stability computations [2], particularly for the $m = 1$ mode, in which the plasma column moves about as a whole. To minimize the curvature—to approximate an infinitely long column—these toroids had aspect ratios in the order of 100. The Scyllac toroidal device, which was put into operation in 1974, was designed along these lines, but the theta-pinch plasma was destroyed within 6–10 μs by an $m = 1$ instability.

One method to stabilize these 'wobbling' modes that was considered in several institutes was dynamic stabilization with auxiliary RF fields. The time-averaged force on a plasma resulting as a second-order effect when a non-uniform alternating field is applied (section 2.6), had been abandoned long before as the principal means for confinement because of excessive power requirements, but now returned as a possible way to stabilize a neutral or weakly unstable equilibrium [3]. But since $m = 1$ was the only mode that could not be stabilized otherwise, e.g. by the finite gyro-radius effect, it seemed a more promising approach to apply the auxiliary field through a feedback system consisting of sensors, power amplifiers and correction coils.

Los Alamos took the lead in studying feedback stabilization of the toroidal theta-pinch plasma. Their original 15 m circumference helical

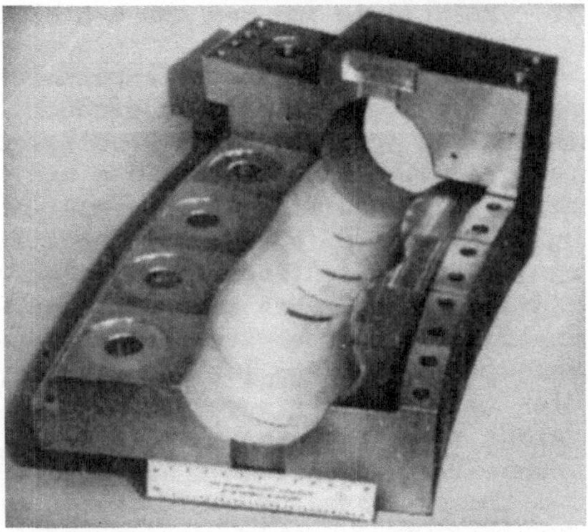

Figure 4.2. Section of the 5 m radius Scyllac sector coil. The coil is machined so as to produce a combination of $l = 0$ M&S field and $l = 1$ helical field.

torus had been rebuilt with 25 m circumference to reduce the stabilization requirement. They did preliminary experiments on a 3 m long linear model and a 5 m long 120° sector (figure 4.2), which did confirm both the equilibrium and the $m = 1$ instability that were expected theoretically [4]. In view of the change to the 25 m torus, the length of the 120° sector was increased to 8 m. The power rating of the feedback system, however, proved to be inadequate to cope with the $m = 1$ growth rates [5], so that feedback stabilization was never brought to a full-scale test. Los Alamos could demonstrate partial success on a downgraded version of the 8 m sector with reduced $m = 1$ growth rate [6], and then abandoned this scheme. Both in Los Alamos and Garching, attention briefly turned to wall stabilization, despite the fact that this set limits to the compression ratio which, in turn, limited the heating. Garching considered a 50 m circumference device relying on wall stabilization [7], but this was never built. Meanwhile, however, the outlook for the conventional stellarator to reach acceptable β values in steady state had improved considerably, causing support for the fast-implosion high-β approach to evaporate. The theta-pinch technique, however, did survive in the field-reversed configuration (FRC), which will be discussed in section 4.4.

4.2 Stabilized and reversed-field pinches

In early Russian papers, a distinction was made between toroidal pinches with a weak or a strong stabilizing field. Later, the name tokamak was assigned to the latter, while (stabilized) toroidal pinches became understood to be systems with a weak toroidal field and a current above the Kruskal–Shafranov limit. In these pinches, the toroidal magnetic field contracts with the plasma into a peaked radial profile upon application of the toroidal plasma current, and remains so for the duration of the discharge, whereas in tokamaks it is allowed to relax towards its undisturbed configuration.

The year 1958 had been a bad one for toroidal pinches. All suffered from a rapid energy 'drain' and, even worse, the neutrons from ZETA had proved to be 'false'. The disrepute into which the pinch had fallen, however, had more to do with the relatively advanced stage of research on it than with any proven advantages of its competitors. But less charted lands seemed more attractive and most laboratories abandoned the field. Some turned towards theta pinches, mirror machines or stellarators, while others sought to learn about toroidal, current-carrying plasmas in hard-core pinches and multipole devices. Also the British diversified their programme, but a small team managed to keep ZETA running in Harwell long after the UKAEA was to have moved all their fusion work from Harwell and Aldermaston to the new laboratory in Culham. One of the motivations for this move had been to acquire space for a large, fast-implosion toroidal pinch, named ICSE (Intermediate-Current Stability Experiment), which, according to prevailing stability theory, was to have the toroidal current concentrated in a thin skin between the compressed toroidal field in the plasma and the poloidal field in the vacuum outside. ICSE was never built, but ZETA continued to dominate the toroidal-pinch scene for a full decade, from 1958 until the Novosibirsk Conference in 1968.

At the Salzburg Conference in 1961, the Culham–Harwell group reported far-infrared spectroscopy and bolometry on ZETA, along with more conventional diagnostics like Doppler spectroscopy, and it was clear that in most discharge regimes there was a rapid circulation of plasma to the wall and neutral gas into the plasma, associated with strong impurity radiation. Electron temperatures were low, not much in excess of 10 eV; ion energies were in the 100 eV range and appeared to be only partly thermalized. Although the authors called their discharges 'grossly stable', these clearly suffered from violent small-scale fluctuations which were held responsible for the rapid loss, on the one hand, and for the ion heating, on the other. The nature of the instabilities could not be established unambiguously, but most theoretical thought turned towards current-driven resistive modes. At the same conference, Afrosimov from the Ioffe Institute in Leningrad presented extensive diagnostic work on Alpha, including measurements of neutral particles emitted by the plasma. This machine had quickly been assembled, as a response to ZETA, at the Efremov Institute for Electrophysical Apparatus, also in Leningrad. It showed plasma behaviour similar to ZETA and the Russians left it at that; it did not become a lasting part of their programme [8].

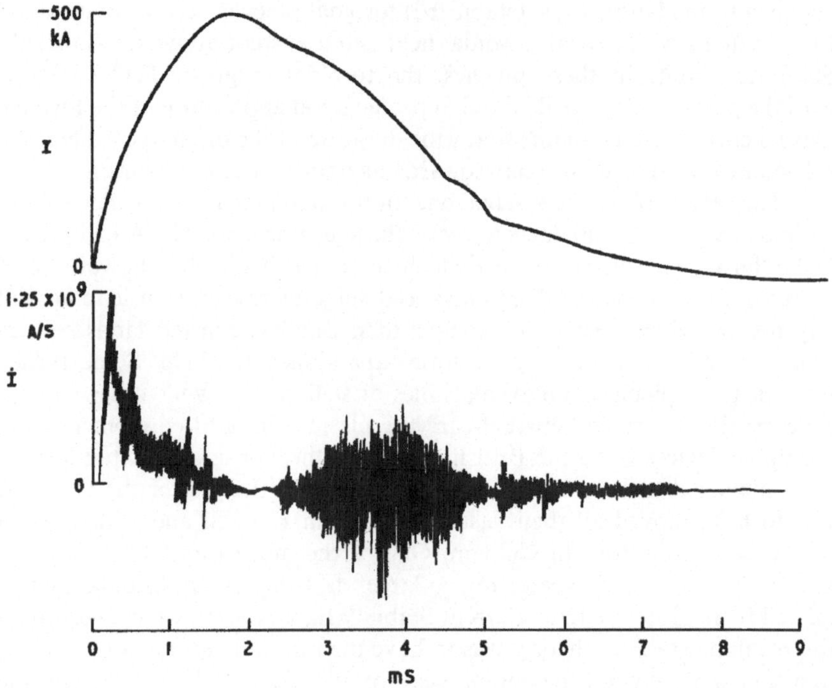

Figure 4.3. The quiescent period in ZETA. The upper trace shows the plasma current which peaks at about 500 kA. Quiescence manifests itself here for a period of about 0.5 ms shortly after the peak current in the oscillogram of the toroidal induced voltage (lower trace), which is a measure for dI/dt.

By the time of the Culham Conference in 1965, two papers on ZETA were all that remained of the toroidal pinches. They were scheduled at the end of a session on tokamaks and stellarators and Spitzer did not even mention them in his summary of experimental work presented at that conference. Yet, the ZETA team [9] had found in some discharges a quiescent period, appearing just after the current maximum and lasting about 1 ms. At $n_e = 2 \times 10^{20}\,\mathrm{m}^{-3}$, $T_e = T_i = 50\,\mathrm{eV}$, the energy confinement time during the quiescent period rose to 1 ms (figure 4.3), but the explanation—spontaneous reversal of the toroidal field outside the plasma column—was not at once apparent.

The first theoretical hints at the stability of reversed magnetic fields had been given already in Geneva in papers by Rosenbluth from the General Atomics (GA) laboratory at San Diego and Suydam from Los Alamos. They had been picked up by the experimentalists working on the 'perhapsatrons' at Los Alamos, Gamma at Livermore and unnamed toroidal pinches at San Diego. In Britain, ZETA at Harwell and SCEPTRE at Aldermaston showed signs of spontaneous field reversal without the experimentalists, at this stage, paying much attention to the phenomenon. Later, however, those at Harwell took note and searched for an explanation, but all that could be said was that the spontaneous reversal could not be explained by a helical distortion of the pinched column. Reversed-field configurations were also seen in Alpha in Leningrad, and were discussed by Kadomtsev as arising from a turbulent relaxation process in which a Suydam-unstable configuration loses its excess pressure by small-scale interchanges of neighbouring flux tubes. This could explain why the fields observed in these toroidal pinches closely resembled a family of force-free (i.e. with $\mathbf{j}||\mathbf{B}$) fields, first described by the astrophysicist L Woltjer, in which field reversal develops at high values of the toroidal current. Ohkawa at San Diego performed systematic studies of various toroidal pinch configurations, among which a field-reversed theta pinch gave the highest temperatures and confinement times. Reversal of B_{t} outside a toroidal pinch gave poor results, unless also the toroidal electric field* E_{t} was reversed after initial formation of the pinch [10].

The 1965 paper that first described the quiescent period in ZETA only mentioned reversal of the toroidal electric field as a condition for stability. The matter remained unclear until, three years later at Novosibirsk, Robinson and King pulled all experimental and theoretical strings together. Recognizing the importance of the spontaneous reversal of the toroidal magnetic field outside the pinched column with which the quiescent period turned out to be associated, they signalled the death of the traditional stabilized toroidal pinch and the birth of the reversed-field pinch (RFP) [11]. But it would take six more years before Taylor grasped the theoretical significance of the relaxation process that produced the B-profiles found in reversed-field pinches and tokamaks.

Robinson's Novosibirsk paper suggested that energy confinement during quiescence in ZETA exceeded the Bohm-diffusion limit by about a factor of 3.5. In this respect, ZETA was not alone; tokamaks, stellarators and theta pinches had performed the same feat. Yet, the new evidence for stability, combined with the long recognized intrinsically high Ohmic heating power in pinches, encouraged several groups to start new experimental RFP programmes. Culham, Los Alamos, Padua, Tokyo and Nagoya built HBTX1, ZT-1, ETA-BETA-1, TPE-1 and STP-1, respectively. The latter two started as screw pinches (section 4.3) but were later converted to study RFP configurations. As their names indicated, all were intended to become the first in a series; all used high-voltage capacitor banks and glass or ceramic vacuum vessels to allow the desired field configuration to be established by rapid programming. After the toroidal field was trapped in the plasma, the current in the external windings was reversed to produce the reversed field

* Many papers of this period use B_z and E_z to denote toroidal components of the magnetic and electric fields.

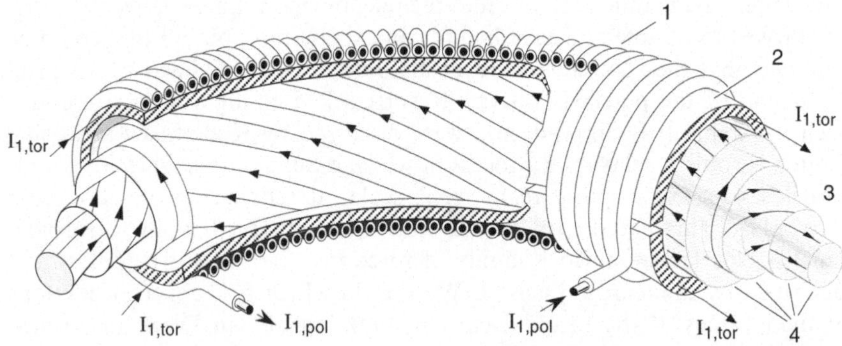

Figure 4.4. Schematic of a reversed-field pinch. 1, toroidal-field coils; 2, conducting shell; 3, plasma; 4, magnetic surfaces. Arrows indicate the direction of the magnetic field; the plasma current is in the opposite direction.

outside the plasma. A conducting shell served both for stabilization and as primary winding for inducing the toroidal plasma current (figure 4.4). These experiments succeeded in transiently producing high-β plasmas, but the temperatures stayed low (order of 10 eV) because of line radiation from impurities, mainly oxygen, which had caused so many problems in non-metallic tori. Resistive decay of the current profile caused these discharges to develop violent MHD instabilities within 10–20 μs [12].

A major step towards understanding not just the RFP, but a wide range of phenomena in laboratory and astrophysical plasmas as well, was made when J B Taylor [13] proposed his stability theory, based on the hypothesis that small-amplitude instabilities or turbulence cause the plasma to relax towards a minimum-magnetic-energy state under the constraint—the 'Taylor conjecture'—that the helicity is conserved. He showed that in the low-β limit the ratio, θ, of toroidal current times radius to toroidal flux—the 'normalized current'—uniquely determines the current and field profiles, leading to field reversal for $\theta > 1.2$ and to a helical equilibrium at $\theta > 1.6$. The degree of reversal is expressed by the parameter $F = B_\phi(a)/\langle B_\phi \rangle$ so that the theory prescribes F as a function of θ (figure 4.5). Helicity [14] is discussed in box 4.1.

In an ideally conducting plasma, the topology of flux lines is conserved. With finite resistivity, field lines can break and reconnect, magnetic energy can be dissipated or transferred between a poloidal and a toroidal field, and helicity can be exchanged between different parts of the plasma. But the helicity integral changes on a longer time scale than the energy content. So, the plasma can relax quickly to a state of lower energy, while the total helicity stays nearly constant. Specifically in the RFP, this process causes the initially applied toroidal flux to be driven towards the axis and to reverse in the outer part of the column. (When a trapped axial field is compressed by

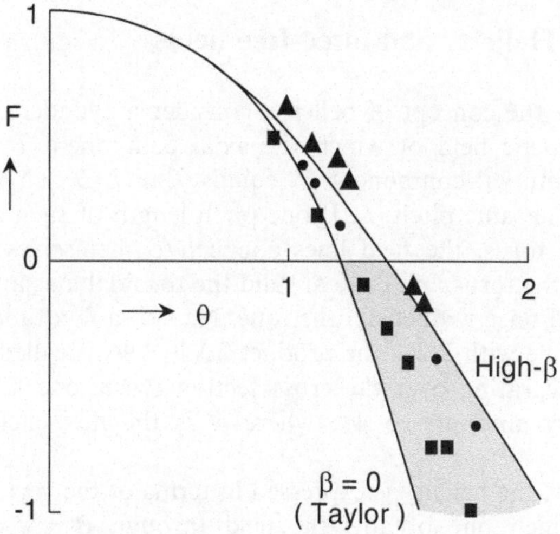

Figure 4.5. Reversed-field pinch F–θ plot. Here, $F = B_z(a)/\langle B_z \rangle$ and $\theta = B_\theta(a)/\langle B_\theta \rangle$. After the initial formation of the plasma, the field configuration has quickly relaxed towards Taylor's minimum-energy state.

the pinch effect, the plasma is said to be paramagnetic, but this applies to the axial field only; with respect to the total field the plasma is always diamagnetic.)

The fast RFP experiments in operation during the 1970s [15] confirmed that plasmas in a wide range of toroidal currents and fields did indeed relax towards states not far removed from the predicted $F(\theta)$ curve and stayed near that curve during build-up and decay of current and flux. In the centre of the plasma the profiles corresponded well to the predicted $\mathbf{j} = \lambda \mathbf{B}$, with λ constant over half the radius, although the current density fell to zero towards the wall. But neither in terms of the electron temperature, nor of the confinement time, did these fast experiments come near the old ZETA results. Only when RFP behaviour was re-interpreted in the light of Taylor's relaxation theory did it become clear that the way to move forward was to set up the discharge on a much longer time scale—with the side benefit that one could use a metal-wall chamber—and to let the fields relax towards and stay at the minimum-energy state; in other words, to go back to where ZETA had stopped.

From a reactor point of view, arguments in favour of the RFP were the demonstrated relatively high values of β and the expected possibility to reach thermonuclear ignition with Ohmic heating only. Reactor design studies [16] were carried out in Culham, Padua and Los Alamos to validate these considerations. What was assumed in all these studies, but remained to be demonstrated, was that relaxation would be a benign process, not involving excessive energy drain or plasma–wall interaction, that no destructive MHD

Box 4.1 Helicity and force-free fields

To visualize the concept of helicity, consider a cylinder containing a helical magnetic field of which the axial component B_z is uniform and the azimuthal component B_θ equals $(2\pi r/L)B_z$, so that all field lines have the same pitch, L. If one pitch length of such a cylinder is bent into a torus, the field lines connect to themselves after going once along the torus and once around the toroidal axis and each field line is linked once with each other one. Now, if a flux tube containing $\Delta\Psi_1$ links one with $\Delta\Psi_2$, the product $2\Delta\Psi_1\Delta\Psi_2$ is called the mutual helicity. Integrating over the cross section twice, one finds that the total helicity amounts to Ψ^2, where Ψ is the flux enclosed in the cylinder.

Usually, the helicity is expressed in terms of the vector potential, **A**, from which one obtains the field through $\mathbf{B} = \nabla \times \mathbf{A}$. In the cylindrical model this is satisfied by $A_\theta = rB_z/2$, $A_z = \pi r^2 B_z/L$. With these values of **A** and **B**, one finds that $\mathbf{A}\cdot\mathbf{B} = 2\pi r^2 B_z^2/L$, so that the volume integral of $\mathbf{A}\cdot\mathbf{B}$ over our cylinder with radius r and length L equals

$$\int 2\pi r^2 B_z^2 2\pi r \, \mathrm{d}r = \pi^2 r^4 B_z^2 = \Psi^2.$$

This example illustrates that—in agreement with a general theorem—the flux linkage equals the volume integral of $\mathbf{A}\cdot\mathbf{B}$. The toroidal field considered here in a cylindrical approximation is characterized by $q = 1$, but the treatment can be generalized to arbitrary field configurations.

In this example there is supposed to be a plasma that carries the axial current. Away from the axis, this exerts an inward $j_\theta B_z$ (the r-component of $\mathbf{j} \times \mathbf{B}$) force on the plasma. If the plasma pressure is negligible, **j** is everywhere parallel to **B**; this field configuration is called force free. A special case is $\mathbf{j} = \lambda\mathbf{B}$, where λ is uniform throughout the volume under consideration. In cylindrical symmetry, the equation $\mu_0\mathbf{j} = \nabla \times \mathbf{B}$ with $\mathbf{j} = \lambda\mathbf{B}$ is satisfied by a 'Bessel function profile', $B_z = B_0 J_0(r/\rho)$, where the characteristic length, $\rho = 1/\mu_0\lambda$. Within a certain range of λ-values, this turns out to be the profile that minimizes the magnetic energy for given axial flux and helicity. In these Bessel function profiles, the pitch decreases away from the axis until the axial field and the current density reverse their direction at $r/\rho = 2.4$, where the Bessel function J_0 falls to zero.

Table 4.1 Reversed-field pinch devices

Location	Name	Operation*	Major radius R (m)	Minor radius a (m)	Toroidal current I_p (MA)
First generation					
Culham	HBTX-1	1970–78	1.0	0.06	0.3
Padua	ETA-BETA-1	1974–77	0.40	0.05	0.1
Los Alamos	ZT-1	1970–74	0.38	0.05	0.25
ETL, Tokyo	TPE-1R, RM	1975–79	0.5	0.1	0.15
Nagoya	STP-1(M)		0.12	0.04	0.11
Representative second-generation devices					
Culham	HBTX-1B	1981	0.80	0.25	0.5
Padua	ETA-BETA-II	1979–90	0.65	0.12	0.28
Los Alamos	ZT-40, 40M	1981	1.14	0.20	0.44
GA, San Diego	OHTE	1981–88	1.24	0.19	0.5
Univ. of Tokyo	REPUTE-1	1984	0.82	0.20	0.24
ETL, Tsukuba	TPE-1RM(20)	1992	0.75	0.19	0.2
Nagoya	STP-3(M)	1983	0.50	0.1	0.17
Stockholm	Extrap T1	1988	0.5	0.08	0.06
	Extrap T2†	1994–date	1.24	0.18	0.25
Sichuan	SWIP-RFP	1989	0.48	0.1	0.12
Mega-ampere devices					
Harwell	ZETA	1957–68	1.5	0.5	0.8
Padua	RFX	1991–99‡	2.0	0.45	1.2
Wisconsin	MST	1988–date	1.50	0.52	0.6
Tsukuba	TPE-RX	1998–date	1.72	0.45	0.5 (1)§

* Year of start-up and (where known) end of operation.

† OHTE refurbished.

‡ Stopped by a fire in the power supplies. Operation is due to restart in 2003/4.

§ Design value.

instabilities would develop in high-β plasmas on the larger time scales aimed for and that energy confinement times would show favourable scaling with plasma radius, magnetic field strength and temperature. Moreover, it had to be shown that the close-fitting stabilizing copper shell could be removed—perhaps to be replaced by a feedback system—to satisfy reactor requirements.

Table 4.1 [17] shows both the early RFP machines already mentioned and some of their successors. The first of the new generation to come on line was ETA-BETA-II in Padua, which started operation in the spring of 1979. It was followed later that year by ZT-40 in Los Alamos, then still equipped with a quartz liner, and by TPE-1R(M) (a metal-wall version of TPE-1R) in the Tsukuba ETL (formerly in Tanashi). The Padua group first succeeded in reproducing the quiescent regime found on ZETA, in which the electron temperature increased and the confinement improved. This regime was found in a restricted range of I/N, where I is the toroidal current and N, the line density, stands for N_e, the number of electrons per unit (toroidal) length of the discharge. At the high-density limit, impurity radiation became dominant, while large-amplitude fluctuations developed at low density. The stable period lasted for only half a millisecond in this small device, but could be extended to over 20 ms in ZT-40M, a modified version of ZT-40 in which a metal liner had been installed [18]. Already in ETA-BETA-II, this had proved to be the decisive factor in cleaning up the plasma and breaking through the radiation barrier to reach temperatures in the 100 eV range. This transition from glass or ceramic to metallic discharge chambers, which had been found necessary for tokamaks twenty years earlier and had likewise been successfully applied to pinches already in ZETA, was crucial for pulling the RFP out of the morass of radiation-dominated, cold and rapidly decaying plasmas and for reaching conditions in which studies of long-term stability and transport loss became possible.

Many laboratories took part in RFP research during the eighties and we shall not trace all the modifications of devices and construction of new apparatus through which they went [19]. The basic slow RFP formation process, in which a low level of instability drives the plasma towards Taylor's minimum-energy state, was demonstrated most clearly in ZT-40M. While the toroidal current was ramped up slowly from 70 to 170 kA and the external toroidal field was held negative—i.e. directed against the trapped field—the plasma continuously generated the required amount of positive internal flux through 'dynamo action' [20]. The relaxation was shown to involve instabilities with large m and small n numbers. They were generally thought to be resistive tearing modes, but interchange modes could not yet be excluded. The velocity and field-fluctuations, $\mathbf{v}_\sim$ and $\mathbf{B}_\sim$, give rise to a non-Ohmic contribution to the electric field, $\langle \mathbf{v}_\sim \times \mathbf{B}_\sim \rangle$; the corresponding Poynting flux can sustain the reversed field. The process leads to enhanced ion heating, but also causes anomalous energy loss, so there was an interest in finding less violent methods to establish the configuration. Theoretical work in Culham [21] suggested that auxiliary toroidal and poloidal voltages oscillating out of phase would give rise to a time-averaged electromotive force in the toroidal direction. This was called 'F–θ pumping' and was successfully applied in Los Alamos [22]. In addition to this MHD dynamo effect, a current carried by fast electrons generated in the centre of the discharge and diffusing outwards along tangled magnetic field lines could also play a role in shaping the current distribution [23]. Such fast electrons were found in ZT-40M to carry most of the discharge current and most of the energy flux to the wall. The GA group in San Diego built the OHTE reversed-field pinch, in which helical windings helped to reverse the magnetic field in the outer region of the column [24].

Nearly all RFPs produced plasmas with β_p in the range of 10–20% and all operated at a density within a factor of 2 of the value corresponding to $I/N = 2 \times 10^{-14}$ A m. These observations, combined with the assumption that the resistivity of the plasma follows Spitzer's $Z_{\rm eff} T_{\rm e}^{-3/2}$ dependence, lead to optimistic scaling relations [25] for RFPs. Since $\beta_p \propto nT/B_p^2$ and $B_p \propto I/a$, one has

$$T \propto I(I/N)\beta_p$$

while the power balance, $nT/\tau_E \propto \eta j^2$, yields

$$n\tau_E \propto I^{5/2}(I/N)^{1/2}\beta_p^{5/2}Z_{\rm eff}^{-1}$$

and

$$nT\tau_E \propto I^{7/2}(I/N)^{3/2}\beta_p^{7/2}Z_{\rm eff}^{-1}.$$

Hence, presuming β_p, along with I/N and the impurity content, expressed as $Z_{\rm eff}$ (see box 6.4), all to be constant, one arrived at a promising $I^{7/2}$ scaling for the reactor-relevant triple product (box 8.2). The constant-β_p model received theoretical support [26] and formed the basis for the performance predictions for the high-current experiments RFX in Padua and ZTH in Los Alamos, whose construction was launched in 1984 and 1985, respectively. Until about 1985, constant-β_p scaling was widely held to be the rule, but even before these higher current machines were put into operation there came results [25, 27] from Los Alamos, Padua, Culham and Wisconsin that indicated a drop in β_p and thus a saturation in the proportionality of T_e with I.

RFX was originally proposed around 1974 by Culham as a 1 MA machine, HBTX II. But as the ambition of the protagonists grew and the design was upgraded to 2 MA around 1978, it became more and more difficult for it to compete with the tokamak for support. With JET coming to the Culham site, although as a separate organization with largely European Community funding, the UKAEA had to make a significant commitment and could not afford another large investment in money and manpower. In 1981, they withdrew the RFX proposal, but the Italian Government was willing to save the project for Europe and through the CNR (the Italian organization for basic research) enabled the University of Padua to assume the responsibility for the project. From the iron-core version proposed by Culham, the machine had meanwhile evolved to an air-core transformer of advanced design.

RFX had briefly been envisaged as a tripartite venture in which Culham, Padua and Los Alamos would participate. The machine would have been located in Culham and the three parties would all have participated in the design, construction and operation of it. But the American interest in the project remained at the scientific level and never resulted in a political decision.

When RFX [28] finally made its long-awaited entrance into the current range of 0.5–2 MA for which it was designed, it confirmed the breakdown of the constant-β_p scaling, so that all hopes of a straight ($nT\tau \propto I^{7/2}$) passage towards the reactor regime faded away. The alternative $\beta_p \propto I^{-1/3}$ or $\beta_p \propto I^{-1/3}a^{-1/6}$ scaling relations [29] raise the order of magnitude of the current in a reactor from 10 to a prohibitive 100 MA, so the most urgent task for RFP research was to further clarify the nature of the dynamo fluctuations, to establish their influence on transport and to find ways to mitigate their deleterious effects. RFX in Padua has been in operation since 1992. Among its contributions are studies of the electrostatic fluctuations responsible for particle transport in the outer region and the reduction of these losses by an $\mathbf{E} \times \mathbf{B}$ velocity shear layer associated with a radial electric field at the edge of the plasma. The Padua group also introduced a method for spreading out the plasma–wall interaction caused by localized helical deformations by forcing these structures to rotate in the toroidal direction.

Meanwhile, Los Alamos had involuntarily abandoned ZTH when it lost DOE support for its toroidal pinch programme. But just as RFP research in Europe had shifted from a national laboratory to a university environment, its American counterpart found a refuge in the University of Wisconsin, where MST has been in operation since 1988 as a durable and productive machine. After an analysis of the resistive tearing modes responsible for dynamo bursts and the anomalous transport in the core of the plasma caused by them, the Wisconsin group sought to reduce these fluctuations by active control of the current profile. They introduced pulsed poloidal current drive (PPCD) to regenerate the toroidal field, significantly reducing the magnetic fluctuation level and improving the energy confinement in the core. In Japan, RFPs retained a broader base, with, among others, REPUTE in Tokyo, the STP series in Nagoya and the TPE series in Tsukuba, the latest version of which is a 1 MA machine that produced its first RFP discharges in 1998. The Tsukuba group is committed to continue the divertor studies they performed on their smaller machines on the large TPE-RX [30].

The latter part of table 4.1 lists the new generation of large devices which have 1 m diameter bore and are capable of operating in the 0.5–2 MA range of toroidal plasma current (ZETA is included for comparison).

In spite of much progress in understanding the physics of the RFP, and of the introduction of new methods to alleviate the undesirable consequences of the dynamo mechanism, the crucial question of the scaling of the $nT\tau$ product with the toroidal plasma current has not yet been settled. Whereas early results from MST and RFX could still be fitted to the favourable scaling relations mentioned above, recent work has not confirmed this trend [31]. Not only is the reduction of the magnetic turbulence in the core with increasing current less than expected on the basis of the constant-β model, the electrostatic turbulence in the outer zone also fails to respond favourably to the current.

Clearly, for the RFP to find a course towards a reactor it must deal with the fluctuation-driven transport associated with the dynamo. Recent RFP research does indeed emphasise active methods to control the poloidal current profile, as well as other parameters like plasma rotation. Meanwhile, it continues to serve as a test bed for confinement studies. In a historical perspective, with new understanding in such areas as helicity conservation, marginal stability, anomalous heating and anomalous transport, the RFP has allowed fusion research to pay back much of the debt it owed to astrophysics from the start of the programme.

4.3 Screw pinches

In 1963, van der Laan in Jutphaas,* The Netherlands, studied toroidal pinches with rapid programming of toroidal and poloidal fields and found the screw pinch (figure 4.6), in which the two components were applied in phase, to be the most reproducible. Although the bulk of the plasma was compressed into a relatively narrow column, a dilute, nearly pressureless plasma stayed behind in which flux-conserving currents played an important role. This phenomenon had earlier been recognized by Colgate in the screw-dynamic pinch. In the toroidal screw pinch with external pressureless plasma, field lines maintain their topology when moving inwards, translating their time-history (constant pitch) at the wall into their radial profile (uniform pitch) in the compressed state. As compared with one separated from the wall by a vacuum region, a toroidal plasma column surrounded by a pressureless plasma turns out to be both better centred in the vessel and more stable against MHD disturbances [32].

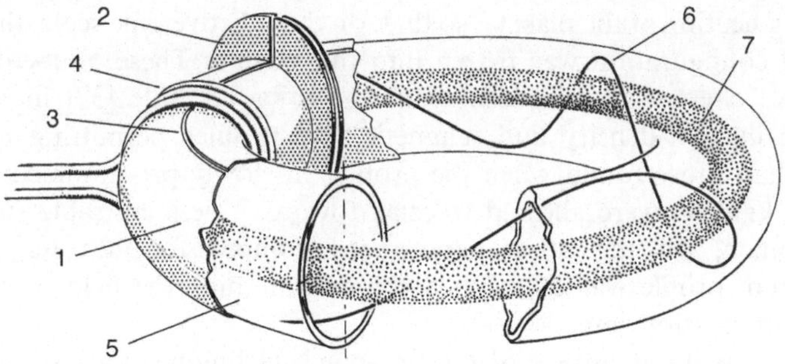

Figure 4.6. Schematic of a screw pinch. 1, copper shield carrying primary toroidal current; 2, toroidal current feed; 3, gap in shield; 4, coil carrying primary poloidal current; 5, quartz glass torus; 6, magnetic field line; 7, plasma.

* The town later became part of Nieuwegein.

The path of these rapidly imploding screw pinches soon converged with that of the toroidal theta pinch. Aside from the high-β stellarators, which occupied the centre of the stage, there were programmes at NRL, Garching, Jülich, Tokyo, Nagoya, Lausanne and Columbia University in New York in which fast-compression theta-pinch techniques were exploited to establish axially symmetric high-temperature plasmas in toroidal geometry [33]. The group at NRL installed a current-carrying hard core in the Pharos theta pinch to produce an elongated toroidal plasma. Garching investigated toroidal pinches with circular and elongated (belt pinch) cross-sections; the confinement of a 20–40 eV, $\beta \approx 0.5$ plasma with a lifetime of 50 μs in Belt Pinch II [34] demonstrated the advantage of an elongated cross-section: by lengthening the poloidal circumference, one can have a higher plasma current for the same value of q, which is beneficial for tokamaks as well as for toroidal screw pinches. The cross-section may be of the racetrack, D-shape, finger ring, doublet or bean type.

The theory of the screw pinch was elaborated mainly in a collaboration between Nieuwegein and Los Alamos [35]. First, these groups developed a sharp-boundary model in which the pinched plasma column was separated from a rigid wall by a pressureless plasma carrying force-free currents (with $\mathbf{j} \| \mathbf{B}$); later they replaced this by a model with diffuse current and pressure profiles. Predicted peak β values were in the range of 20% for screw pinches with elongated (elliptical or D-shaped) cross-sections.

The largest screw pinches were SPICA, with circular cross-section, and the elongated SPICA II, both in Nieuwegein. Although originally intended to operate at $0.5 < q < 1$, these devices performed better at $q > 1$, where they are similar to the high-β tokamaks studied at Columbia University [36]. In the Columbia Torus II, the high-β configuration was established by rapid reversal of the toroidal field outside a stabilized toroidal Z-pinch. This caused turbulent dissipation of the trapped toroidal field accompanied by strong heating of the plasma, so that, on the resistive time scale, the resulting field configuration was frozen into the plasma. These 'flux-conserving tokamaks' were also proposed from the tokamak side [37] in order to generate current-density and magnetic-field profiles permitting higher β values than those found when the profiles in slowly programmed conventional tokamaks were allowed to relax towards a less desirable state. The screw pinch, on the other hand, used compression heating and the equilibrium profile was established by programming the field components in the compression stage.

A major disadvantage of the fast-pinch technique employed in screw pinch and high-β tokamak devices as well as in fast reversed-field pinches proved to be the need for quartz tori to allow rapid penetration of fields. Not only were these tori extremely vulnerable and difficult to produce—in the end, one glassblower in Germany was sustaining the world's pinch programme—they also gave rise to oxygen impurity radiation which made it very difficult to burn through the 50 eV electron temperature barrier (box 6.4).

SPICA had a toroidal current of about 200 kA and maintained a central column with $\beta \approx 15\%$ at $T_e \approx 50$ eV for some 200 μs. Its successor, SPICA II [38], which had a racetrack cross-section and a toroidal plasma current up to 450 kA, did occasionally produce 200 eV plasmas, but no stable operating regime was found and the programme was terminated in 1987 because tokamaks by then showed both better performance and better reactor prospects than screw pinches.

In Japan, the screw pinch was studied in the TPE-1 and TPE-2 devices at ETL in Tokyo (the institute moved to Tsukuba in 1978); TPE-2 was successful in producing high-β plasmas with temperatures up to 400 eV. Both rapidly programmed, shock-heated pinches and slow Ohmically heated discharges were found to relax towards specific q-profiles, independent of the history of the discharge [39]. This behaviour appears very similar to that of reversed-field pinches relaxing towards Taylor's minimum-energy state. The corresponding current and density profiles are nearly flat, as expected from high-β tokamak stability theory.

The high-β tokamak programme at Columbia University [40] explored equilibrium and stability limits by comparing experiments in their Torus II and its successor HBT with numerical codes such as the Princeton PSEC and PEST equilibrium and ballooning instability codes. The HBT-EP device, designed to study passive and active feedback stabilization methods at β values above the Troyon limit, came into operation in 1992. It has shown that a segmented close-fitting shield favourably affects the external kink modes that precede disruptions and that internal tearing modes respond to rotating field perturbations produced by properly phased alternating currents in external saddle coils.

4.4 Field-reversed configurations and spheromaks

A variety of schemes have been investigated to produce configurations in which a toroidal plasma is confined by either a purely poloidal field, or a combination of a toroidal and a poloidal field, embedded in a mirror field. The class is named Compact Toroids or Compact Tori [41] (CTs) and is subdivided into Field Reversed Configurations (FRCs) which have an elongated cross-section, and the truly compact Spheromaks. They have in common that their aim is to confine the plasma in a toroidal field, but without metallic conductors or structures interlinking the plasma. Interest in these devices has not been restricted to fusion research laboratories; like RFPs they also attracted attention in the academic world. Except for early work on reversed-field theta pinches, the European fusion laboratories hardly pursued this line, but it was quite popular in the US and Japan, mainly during the 1980s when rising doubts about the economic perspectives of tokamak reactors stimulated the study of alternatives.

The history of these current-carrying plasma rings goes back to early work on accelerated plasmoids [42] and to attempts to explain the ball lightning phenomenon [43]. The virial theorem [44] forbids self-confinement of plasmoids by the magnetic field of an internal current, but this restriction is removed when the confining field is itself contained by the forces that it exerts on external conductors, as is the case in all magnetic confinement systems, or when the plasmoid is in pressure equilibrium with the atmosphere. The plasma ring model of ball lightning, however, requires a current-drive mechanism to explain the long lifetimes that have been reported, because the L/R (inductance over resistance) time constant of the circuit would be in the order of milliseconds in the kind of plasma found in lightning discharges.

The first attempt to confine a thermonuclear plasma, in what was only later called the field-reversed configuration, was Astron in Livermore (section 2.2) in which a relativistic electron beam was injected into a mirror field. But Christofilos failed to stack up enough beam pulses in his E-layer and Astron was shut down shortly after he died in 1972. A group at Cornell, however, studied a similar field configuration in the RECE facilities and achieved field reversal during short periods with single-pulse electron beams [45]. Later, NRL obtained transient reversal by injecting a proton beam [46]. The 2XIIB experiment at Livermore (section 3.1.4) tried to reach this goal with 20 keV neutral beams (figure 4.7), which did indeed cause a strong diamagnetic effect without, however, quite reversing the

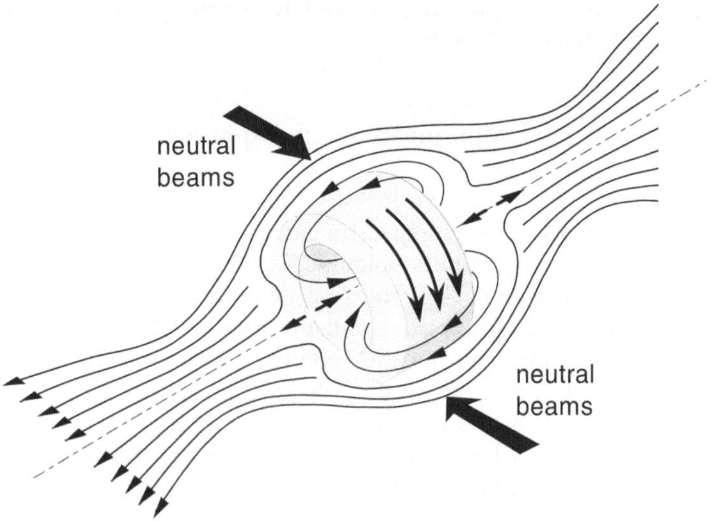

Figure 4.7. Field-reversed mirror configuration produced by neutral beams.

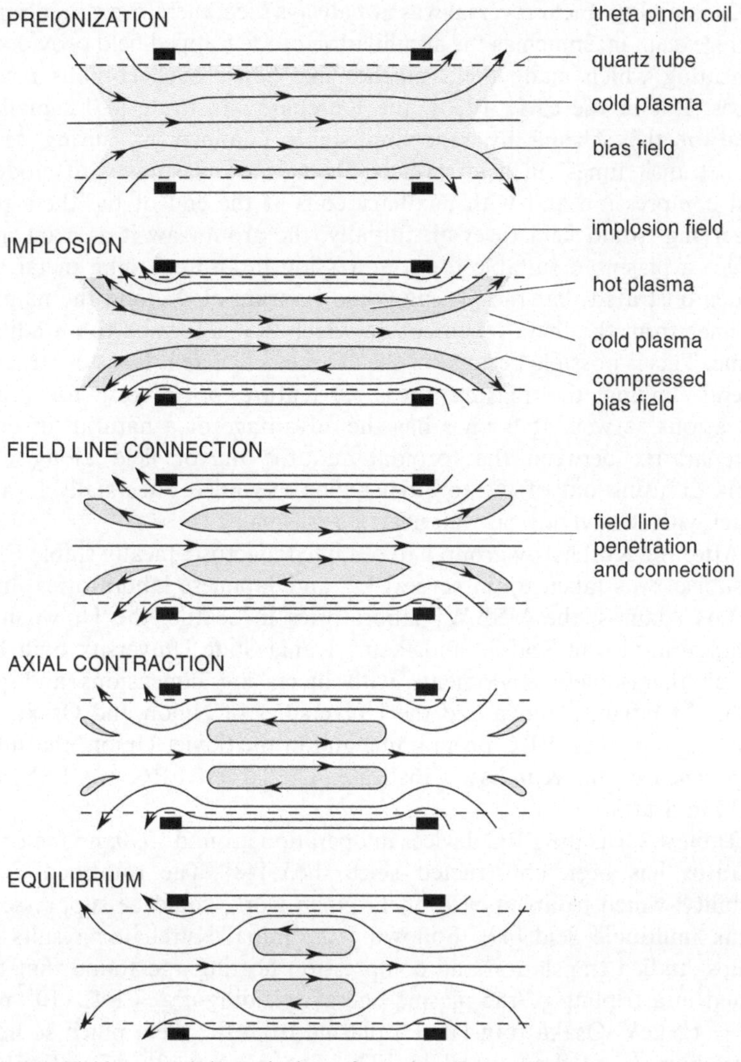

Figure 4.8. Field-reversed theta pinch configuration. Top to bottom: successive stages of formation, in which the plasma is compressed between the bias field and the reversed external field. The field lines connect to form toroidal surfaces.

field on the axis. Rings of relativistic electrons were also used in Nagoya to produce spheromak configurations (see below).

In early theta-pinch work (section 3.4), it had already been found that a trapped reverse axial field could link with the forward theta-pinch field to produce closed field lines lying in toroidal surfaces with elongated cross-section (figure 4.8). The plasma, however, tended to contract axially and to break up or to dissipate within microseconds, or to develop a rotational

$m = 2$ instability. Field reversal was not always seen, therefore, as a desirable feature, except inasmuch as the annihilation of the trapped field provided the ion heating which made theta pinches like Scylla such copious neutron sources. But in the early 1970s, the Kurchatov Institute [47] signalled a revival of the scheme by reporting stable confinement during several Alfvén transit times* in field-reversed theta-pinch plasmas with moderate radial compression and with auxiliary coils at the end of the theta-pinch coil serving to aid field reversal. Initially, the group saw it as a means to produce a plasmoid suitable for compression in an imploding metal liner, and demonstrated that the plasma could be translated along the magnetic field lines from the theta-pinch coil in which it was formed to an adjacent volume. This is possible because of the absence of current leads or structural elements linking the plasma torus, a feature of interest for reactor applications as well. It further has the advantage of a natural 'divertor': the separatrix between the toroidal and the mirror field along which plasma diffusing out of the toroidal region is guided out axially to avoid contact with the cylindrical wall near the plasma.

After the Kurchatov group had obtained macroscopically stable FRCs, this subject was taken up in several US and Japanese laboratories. In the US, Los Alamos, the MSNW† laboratories in Seattle, the University of Washington, also in Seattle, and Pennsylvania State University built field-reversed theta-pinch experiments with increasing dimensions and pulse lengths. In Japan, Nagoya and the Universities of Nihon and Osaka each had an experimental FRC programme and in the Soviet Union, the subject was pursued by the Kurchatov Institute. The largest FRC was LSX, built by STI in Seattle.

Table 4.2 lists the FRC devices in operation around 1990; no major new apparatus has been constructed since then [48]. The rotational $m = 2$ instability, which troubled early theta-pinch work, could be suppressed by a weak multipole field [49]. Following the initial Kurchatov results, Los Alamos studied translation and compression heating and found that these resulted in a tripling of the plasma energy, yielding $n\tau_E = 1.4 \times 10^{17}\,\mathrm{m^{-3}\,s}$ at $T_i = 1.5\,\mathrm{keV}$. Osaka translated a plasma from the theta pinch source to a large-bore ($r = 0.4\,\mathrm{m}$) vessel in FIX. The $n = 8 \times 10^{19}\,\mathrm{m^{-3}}$, $T = 70\,\mathrm{eV}$ plasma, which existed as long as 0.5 ms, was used to study heating by fast compression [50]. The theoretically most dangerous MHD-mode, in which the plasmoid tilts and ultimately turns around in the external field, is not observed in most experiments. This has been ascribed to finite gyro-radius stabilization, whose effectiveness is expected to depend on the dimensionless parameter s, which is the ratio of the radial scale length to the ion gyro-radius ρ_i (in tokamak and stellarator scaling studies, one often uses ρ_i^*, the ratio of ρ_i

* The time for an Alfvén wave to traverse the system.
† Later named STI.

Table 4.2 FRCs in operation around 1990

Location	Name	Length (m)	Diameter (m)
Kurchatov Institute	TOR	1.50	0.30
	T-L	0.83	0.18
Los Alamos	FRX-C	2.00	0.40
	FRX-C/LSM	2.00	0.60
Nihon University	NUCTE-III	1.50	0.28
Osaka University	OCT	0.59	0.184
	PIACE	1.00	0.12
	FIX	1.00	0.40
Spectra Technology	TRX-2	1.00	0.20
	LSX	4.50	0.80
University of Washington	HBQM	2.00	0.20
	CSS	1.00	0.40

to the minor radius). Moreover, theory suggests that stability may be provided by high-energy axis-encircling ions, and it has been suggested that pulsed ion acceleration of the kind developed for inertial confinement might be suitable to supply these ions on the theta-pinch timescale.

Experiments in the Los Alamos FRX-C/LSM (Large-Source Modification) facility showed evidence for tilt instability and degradation of confinement with increasing s/e (here e is the elongation of the plasma cross-section at the separatrix between the regions of open and closed field lines). The Los Alamos finding of poor confinement at $s > 0.5e$ is not in contradiction to results of the large-s experiment, LSX, which had e in the range 5–10 and yielded stable plasmas at s up to 4 or 5, the higher values of s corresponding to higher e. For reactor operation, however, s must be in the range 20–40, where the stabilizing effect should be very weak [51].

Transport processes in FRCs appear to be governed by highly anomalous particle loss. The most elaborate scaling studies [52] find that energy, particle and flux confinement times all scale with nearly $r_s^3 \rho_i^{3/2}$, where r_s is the radius of the separatrix. The energy confinement time is about a factor of 2 lower than the others; such a transport scaling law leads to pulsed reactors of large size.

So far, most experimental evidence has been obtained in a kinetic regime with $s \approx 2$, but LSX has achieved some results at higher values of s. This machine was designed to reach $s = 8$ and produced symmetric, well confined FRCs with s up to 4 or 5, above which no quiescent plasma could be formed [53]. Whether this was a fundamental limitation or one determined by experimental procedures has not been settled definitively; operation of the machine was stopped after one year.

Figure 4.9. Schematic of FRC reactor. 1, formation section; 2, burning section; 3, travelling wave decelerators for direct energy conversion.

Although progress, both in understanding and performance, of FRCs is far behind that of tokamaks, they are advocated as possible candidates for reactor development (figure 4.9), mainly because they are thought to be suitable for burning the 'aneutronic' fuels (chapter 9) D–^{3}He, or even p–^{11}B. A D–^{3}He reactor requires a higher temperature, higher pressure and longer confinement time than a D–T reactor; and would benefit from an efficient scheme for conversion of charged-particle energy to electricity. This applies, *a fortiori*, to a p–^{11}B reactor, which would not achieve a Q-value of interest for a reactor when operating with thermalized ions, but could conceivably operate as a colliding-beam reactor [54]. There are hardly any prospects for burning aneutronic fuels in tokamaks or stellarators, but their advocates observe that, in principle, an FRC reactor might be better suited for the specific requirements of these fuels. There is, however, a long way to go before outstanding problems with respect to β limits and transport at high s, for example, can be satisfactorily resolved.

Around 1980, when the new generation of large tokamaks was under construction and was claiming an increasing share of the available effort, an undercurrent that always had kept running against the trend towards directed research and reactor development came to the surface. Many felt that it was premature to narrow down the field to one or two choices and looked for alternative lines that, on the one hand, could be argued to have at least some (if remote) reactor promise or relevance and, on the other hand, could be significantly advanced by smaller-scale and more broadly oriented research. So, a Princeton paper [55], read by Harold Furth at the Innsbruck Conference in 1978, found an interested audience. It proposed a compact torus (figure 4.10) resembling Hill's spherical vortex [56] and gave it the name 'Spheromak' to emphasize its relationship with the tokamak, although it was in fact more like a stabilized pinch configuration with a trapped toroidal field and purely poloidal external field.

Experiments along this line could be done with modest and sometimes already available equipment, as spin-off from earlier work on linear pinches, theta pinches, plasma guns and plasma foci. So, compact toroids came to rank among the favoured long-shot alternative lines; table 4.3 shows that not only universities and industrial laboratories, but also major fusion

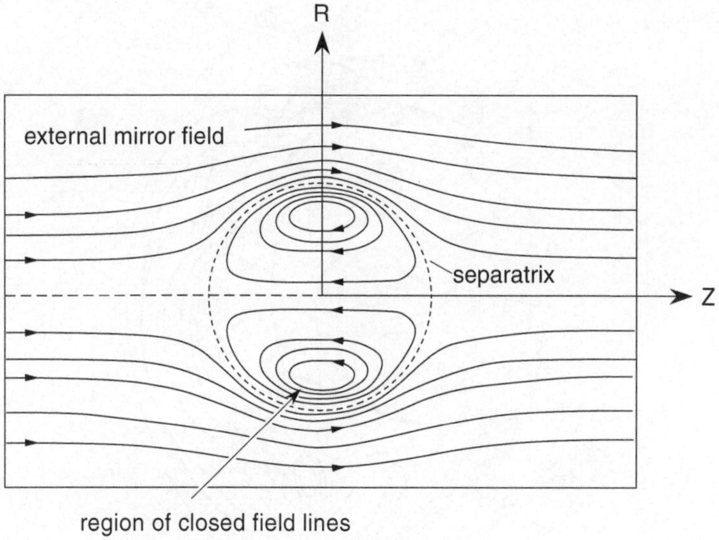

Figure 4.10. Compact torus embedded in a mirror field.

centres placed small bets on them. Theoretical analysis [57] showed the need for a closely fitting conducting shell, the flux container, and for an oblate cross-section of the plasma to stabilize against the $m = 1$ 'tilting' mode which would tend to make the plasma ring turn over in the external guide field. Mercier's criterion suggested stable average β values of a few per cent, but the 'engineering β', which relates the plasma pressure not to the magnetic field at the centre of the plasma but at the coils, came out much higher.

The Princeton laboratory also demonstrated an ingenious way to set up the configuration on a slow time-scale using resistive tearing of field lines to

Table 4.3: Some representative spheromak devices

Location	Name	Year of first operation	Containment diameter (m)	Plasma ring current (kA)	Formation scheme
Princeton	S-1	1983	0.4–0.65	350	Inductive
Los Alamos	CTX	1984	0.6/1.3	300–600	Coaxial source
Maryland	PS-1–PS-3	1983			Theta pinch
Nagoya	SPAC VII*		0.3	20	REB
Osaka	CTCC-I, II	1980	0.75	90	Coaxial source
Himeji	FACT	1987			Current feed
Tokyo	TS-3†		0.45	75	Inductive
Manchester	SPHEX	1988	1.0	200	Coaxial source

* Central conducting rod.

† Central Ohmic-heating transformer coil.

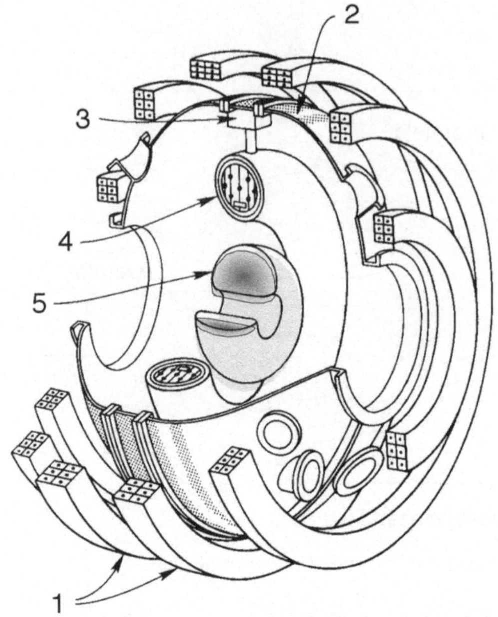

Figure 4.11. Schematic of the Princeton S-1 spheromak. 1, external field coils; 2, vacuum vessel; 3, flux core support arm; 4, flux core; 5, plasma. The formation of the spheromak plasma is explained in figure 4.12.

make the plasma come loose from the flux core: a metal ring containing toroidal and poloidal windings in which currents were programmed so as to induce the required currents in the plasma [58]. This was investigated in the proto-S1 and S-1 devices (figure 4.11). Laboratories with expertise in the pulse techniques developed for theta-pinch and plasma-focus research produced spheromaks by injection of plasmoids into metal-walled containment volumes by means of coaxial guns, as exemplified by CTX in Los Alamos. A third method, practised in TS-3 in Tokyo, was *in situ* formation by combinations of electrodes to produce axial currents and coils to induce toroidal currents. Finally, Nagoya produced spheromak-like configurations—but with central conductors—by means of relativistic electron beams in a series of devices running up to SPAC VII [59]. The highest energy densities were obtained in a 0.3 m radius container in CTX, where an $n_e = 3 \times 10^{20}\,\mathrm{m^{-3}}$, $T_e = 400\,\mathrm{eV}$ plasma was captured. In a 0.6 m radius container, the same gun produced $n_e = 3 \times 10^{19}\,\mathrm{m^{-3}}$, $T_e = 180\,\mathrm{eV}$, with $\tau_E = 0.2\,\mathrm{ms}$. As was seen most clearly in the Princeton S-1, the scaling of the electron pressure with the magnetic field strength tends to follow a constant-β law, much as in RFPs, but it turns out to scale unfavourably with the current density, suggesting an anomalous loss mechanism.

An attractive feature of the spheromak is that the toroidal field in the plasma is either trapped during its formation, or is generated by helicity-conserving processes, anyway without coils interlinking the plasma ring. Intricate systems of external conductors, however, in some cases even including a cylindrical core, have turned out to be necessary to obtain MHD stability and β has been limited to moderate values. It has been suggested, however, that confinement in the core of the plasma is better than the global confinement, which is determined mainly by edge effects,

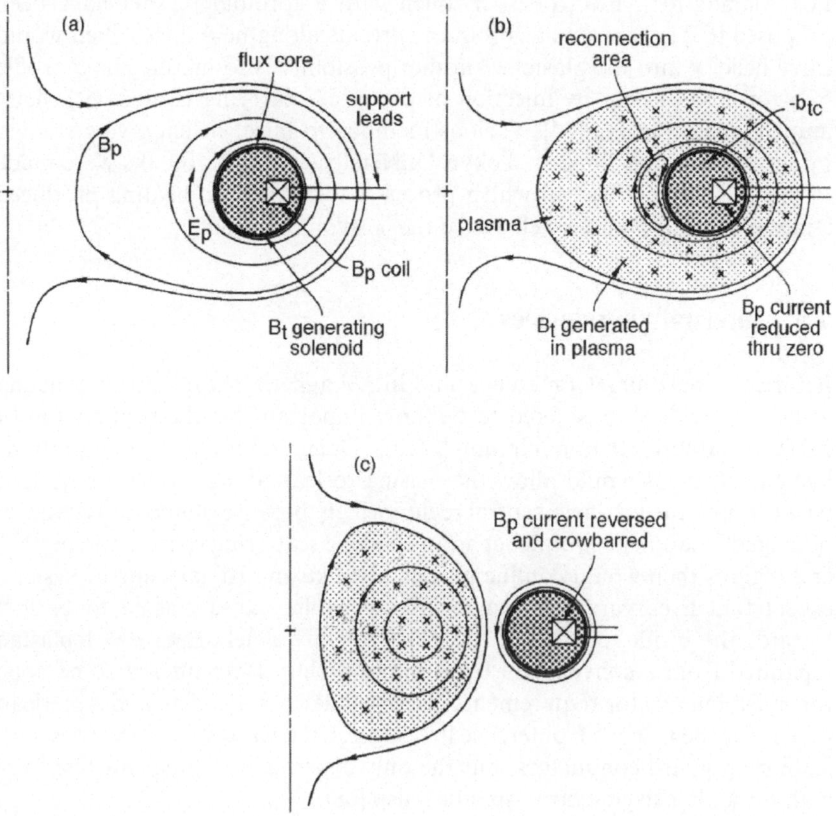

Figure 4.12. Spheromak formation. (a) The vacuum field before the plasma is formed. There is a toroidal field in the flux core, but none outside; the poloidal field is shaped by the B_p coil together with the external field coils shown in figure 4.11. (b) After pre-ionization, both the toroidal field in the flux core and the toroidal current in the B_p coil are reversed, inducing a toroidal field as well as a toroidal current in the plasma. (c) When the poloidal field adjacent to the flux core reverses, field lines reconnect to form a magnetic island, which eventually develops into the detached plasma. The field lines are helical; the figure shows their projection upon the poloidal cross section.

and that the confinement scales favourably with the magnetic Reynolds number* [60].

Perhaps the main significance of compact torus research has been to elucidate the concept of magnetic helicity (box 4.1). This idea was not only applied to the formation of spheromak plasmas, illustrated in figure 4.12, but also to the injection, by a variety of means, of helicity into an already existing toroidal plasma. We saw in section 4.2 that alternating toroidal and poloidal voltages give rise to a dynamo effect in an RFP [61]. Steady state injection of a magnetized plasma stream from a Morozov source (section 3.3) was studied by Jarboe's group in the compact torus CTX at Los Alamos [62]. Electrodes in touch with a toroidal plasma have been proposed [63] as a means to produce currents along field lines which would carry helicity into the plasma. Another possibility is to modify the q profile of a toroidal plasma by injection of plasmoids carrying their own helicity into the main plasma [64]. Various methods to inject helicity were studied by Japanese groups [65]. In Tokyo University, two spheromaks were made to merge into an FRC, in which process anomalous ion heating produced a plasma with a β-value well above the Troyon limit [66].

4.5 Internal-ring devices

Before the Salzburg Conference in 1961, magnetic shear and conducting walls were widely considered to be more important for the suppression of MHD instabilities than minimum-B fields. Open minimum-B configurations like cusped fields would allow the plasma to leak along the field lines that pass through the low-field central region which, because of non-conservation of magnetic moment, would act as an effective scattering centre. The peripheral regions then would confine plasma just like an ordinary mirror system, except that a curvature-driven instability would cause plasma to be lost towards the inside, instead of the outside. Only a field-free, $\beta = 1$ plasma separated from a convex field by a thin boundary layer offered some hope for satisfying reactor requirements in cusped devices. Russian and American papers for the Geneva Conference [67] suggested a variety of geometries with stabilizing guard conductors, but the only experiments along this line were with a simple axisymmetric 'spindle' cusp [68].

A natural modification of the cusps was to close the systems by making the field lines re-enter the main plasma volume after curving around the conductor [69]. Ring-shaped guard conductors then became internal rings, submerged in the plasma. To obtain a large volume of high-β plasma,

* The magnetic Reynolds number is the ratio of the resistive decay time to the Alfvén-wave transit time; it scales as $aBT_e^{3/2}n^{-1/2}$.

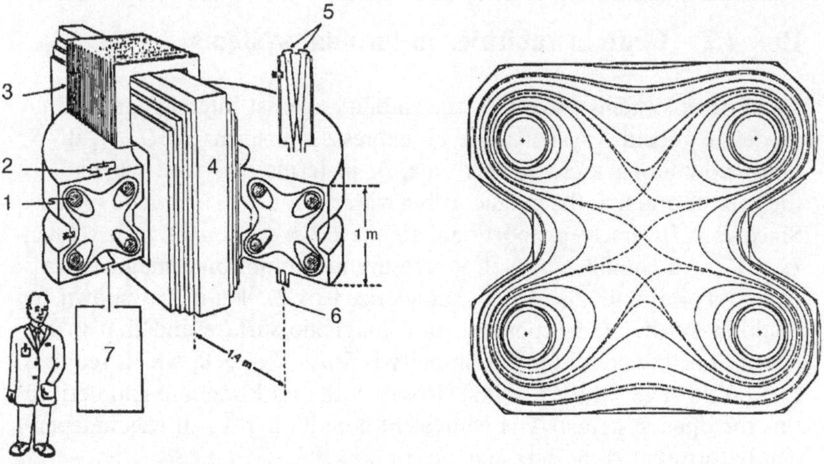

Figure 4.13. Schematic of the Wisconsin Levitated Toroidal Octupole. Left: experimental setup; 1, levitated rings; 2, vacuum outlet; 3, primary transformer windings; 4, iron core; 5, neutral beam sources; 6, plasma guns; 7, ICRH source. Right: section through the magnetic flux surfaces.

Ohkawa and Kerst [70] proposed toroidal multipoles. The octopole* configuration is illustrated by figure 4.13; the four rings carry currents in the same direction and the return current flows in the metal wall. If the pulses are short, the field cannot penetrate into the wall; this is another way of saying that the current distributes itself in such a way that the field lines are parallel to the wall. In the same fashion, one, two or three rings produce dipole, quadrupole or hexapole fields. Quadrupolar and higher-order fields have zero field strength at the toroidal axis. Thus, by 1961 two basic ingredients for the development of toroidal minimum-B systems were at hand: the negative-V'' theory (box 4.2 [71]) and the multipole configuration. But for most reactor-relevant concepts, shear still seemed to be the more appropriate stabilizing agent. It was not until after Ioffe's classical experiment that minimum-B was recognized not only as the standard for open systems, but as something to be incorporated in closed systems as well.

Multipole systems were studied extensively by Ohkawa at GA in San Diego and Kerst at the University of Wisconsin, and on a smaller scale by groups in Garching, Oak Ridge, Culham and JAERI. Already at the Culham Conference in 1965, Garching [72] reported encouraging results with a cool caesium plasma in a toroidal hexapole. The plasma in this device was confined for one to two orders of magnitude longer than expected from the Bohm-like 'pumpout' phenomenon observed in stellarators. Low

* Sometimes written as 'octupole' or even 'octapole'.

Box 4.2 Flute instabilities in toroidal systems

As in open systems (box 3.1), the stability against interchange or flute modes in toroidal systems can be expressed in terms of $U = \oint \mathrm{d}l/B$, taken now along a closed field line, or in terms of $V(\Psi)$, where V is the volume and Ψ is the magnetic flux within a toroidal magnetic surface. Since B is inversely proportional to the cross-section of a flux tube, $U = V'$. An equilibrium with a pressure maximum on a magnetic axis (where $\Psi = 0$ and $V = 0$) is characterized by $U(\Psi)$; it was shown by Kadomtsev that U is constant on a magnetic surface and that in the low-β limit the condition for stability is $\nabla p - \nabla U > 0$, which requires $U' < 0$ throughout the plasma. (Rosenbluth and Longmire had derived this for open systems.) The equivalent condition $V'' < 0$ was independently formulated by Bernstein *et al.*

This negative-V'', or minimum-average-B criterion, shows that U has the character of a potential function like the barometric height. If one assumes an adiabatic equation of state, $\delta(pU^{\gamma}) = 0$, a stable finite-pressure profile $p(U)$ requires $\mathrm{d}\ln p/\mathrm{d}\ln|U| < \gamma$. All this was first derived for ideal (collisionless) plasmas; finite resistivity theory predicted unstable ballooning modes driven by the pressure gradient in the regions of bad curvature, which may be stabilized by magnetic shear, in addition to less dangerous modes.

fluctuation levels, suggesting correspondingly low diffusion losses, were found with gun-injected hydrogen plasmas in the San Diego and Wisconsin octopoles, although losses to the ring supports limited the actual confinement times [73]. The Oak Ridge group [74] obtained $\tau/\tau_{\mathrm{Bohm}} \approx 40$ in their microwave-heated, levitated quadrupole.

Design improvements aimed at reducing both support losses and inductive electric fields associated with the decay of the ring current, were rewarded by further enhanced performance in the GA Octopole, which by the addition of a toroidal field component had a variable rotational transform like a stellarator or a tokamak.* By varying the safety factor, the density and the magnetic field strength and measuring the absolute values of the decay time as well as their parametric dependences, the San Diego group could demonstrate collisional diffusion ranging from the Pfirsch-Schlüter, via the intermediate or plateau, to the 'banana' regime (section 6.8). In these dc octopole experiments [75], the plasma was injected by a coaxial gun and the electron temperature during the measurements was below 1 eV. Other groups applied heating, mostly Ohmic or by microwaves,

* These experiments would lead to the Doublet series of tokamaks (see section 7.1).

to disentangle particle and energy confinement and to study their temperature dependences.

A variety of instabilities has been observed in internal-ring systems. Interchange instability was seen in the outer zone, where it was expected, in the Wisconsin octopole [76]; otherwise, average-minimum-B stabilization of MHD modes was achieved in all these machines. Although the magnetic well was clearly beneficial, low-level density and potential fluctuations were present in many experiments and much effort was devoted to determining their origin. The San Diego group saw ion and electron drift cyclotron instabilities in their octopole in 1968 and the same group [77] a few years later identified thermal fluctuations as the dominant diffusion mechanism in a low-density regime. The earlier paper already hinted at trapped-particle instabilities—for which more evidence came from the Princeton spherator [78]—and the later paper from San Diego introduced the concept of 'poloidal Bohm diffusion', which scales as T_e/B_{pol}, with a coefficient 500–1000 times smaller than the original Bohm diffusion. It was interpreted as being caused by trapped particles and this was confirmed by heat-pulse propagation experiments in the Princeton spherator as well as in later San Diego and Wisconsin experiments [79]. The collisionless trapped-ion mode caused particular concern for the upcoming experiments with neutral-beam heating in tokamaks and for subsequent generations of large experiments. Also the current-driven drift modes identified in the Culham levitron [80] were relevant for explaining the scaling of confinement in tokamaks.

This work on toroidal multipoles culminated in the confirmation of the bootstrap and Pfirsch-Schlüter currents that had been predicted by the 'neoclassical' collisional transport theory (section 6.8) and had long been searched for in tokamaks and stellarators. In the Wisconsin octopole, gun-injected plasmas were heated by ICRH to produce high-β plasma in weakly collisional ('banana') or in plateau regimes, presumably stabilized by finite gyro-radius effects. Such currents have since been reproduced in all kinds of toroidal confinement system, establishing neoclassical behaviour as the standard with which anomalous transport is to be compared.

Internal rings had earlier been in the picture when Colgate and Furth proposed to turn the hard-core pinch, or 'unpinch', into a toroidal 'Levitron' (figure 4.14), the initial purpose of which was to test shear stabilization against resistive tearing and interchange modes. The Livermore sheet pinch (figure 2.2) had evolved into a system in which the hollow plasma cylinder was compressed between an internal, predominantly azimuthal field and an external axial field, the inverse of the traditional stabilized pinch. The configuration of an internal-ring field combined with a toroidal field was chosen because of the high shear resulting from the steep dependence of the poloidal field on the minor radius [81]. The central conductor was a freely floating ring, momentarily suspended in the toroidal discharge chamber after having been levitated to its designated position, and the

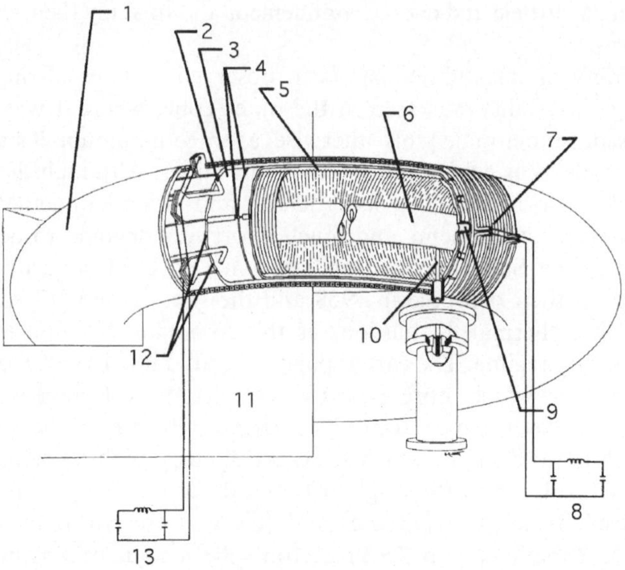

Figure 4.14. Schematic of the Levitron device, seen from below. 1, torus with 60 cm major radius and 15 cm minor radius; 2, cooling for copper shell; 3, copper shell; 4, toroidal and poloidal slots; 5, stainless steel bakeable liner; 6, 10 cm diameter copper core; 7, toroidal-field windings; 8, toroidal-field capacitor bank; 9, diagnostic port; 10, retractable support rods; 11, iron core; 12, poloidal-field winding; 13, poloidal-field capacitor bank.

current in it was excited inductively. A comparable configuration, but with both mechanical supports for the ring and conductors feeding the ring current passing through the plasma, was studied by Lehnert in Stockholm. In fact, among the first things Tamm and Sakharov had considered in 1951, before the matter of stability had come up [82], was to make a poloidal field in a torus by means of a current in an internal toroidal conductor.

Ware and Yoshikawa showed that, with additional poloidal fields, such a single-ring device can be turned into an average-minimum-B torus [83]. Thus, later versions of the levitron in Livermore were equipped with windings to produce an 'axial' or 'vertical' field. Moreover, they made use of flux compression between the ring and the outer wall. All this gave the experimentalists a broad range of control of both the shear and the well depth, and the results confirmed that toroidal confinement had found the right track. Although density fluctuations could not be suppressed completely, gun-injected plasmas were confined during 100 Bohm times, while a hot-electron population produced by electron cyclotron resonance heating even lasted for 10^4 Bohm times [84].

The first levitron was followed by similar experiments in Princeton and Culham. But the early versions were restricted by L/R decay of the ring

Table 4.4 Internal-conductor devices

Location	Name	Year*	$\langle R_{ring} \rangle$ (cm)	Number of rings	Support	Sum of ring currents (kA)	Current feed	B_{tor} (T)
Livermore	Levitron	1960	62	1	Levitated	150	Inductive	—
	SCL	1971	40	1	Levitated	100	Superconducting	0.3
Stockholm	F IV	~1982	16	1	Mechanical		Leads	—
San Diego	Octopole	1965	63	4	Mechanical	400	Inductive	0.01
	Quadrupole	1968	63	2	Mechanical	400	Inductive	0.01
Wisconsin	Octopole	1968	43	4	Mechanical	300	Leads	—
	Octopole	1970	140	4	Levitated	1300	Inductive	0.01
Garching	Octopole W-5	1965	43	4	Mechanical	300	DC	0.02
Oak Ridge	Quadrupole	1968	30	2	Levitated	36	Inductive	
Culham	Climax	1969	80	2	Mechanical	800	Inductive	—
	Sphinx	1968	12	1	Mechanical	45	Leads	0.3
	Levitron	1974	30	1	Levitated	500	Superconducting	0.2
JAERI	JFT-1		43.5	3	Mechanical	250	Leads	0.19
Princeton	Spherator	1968–69	32.5	1	Mechanical	72	Leads	0.1
	SP-3	1969	46	1	Mechanical		Superconducting	
	LSP	1970–71	46	1	Levitated	85	Superconducting	0.1
	FM-1	1971–76	76	1	Levitated	250	Superconducting	0.4
Los Alamos	Quadrupole		65	2	Mechanical		Leads	—
Leningrad	Tornado-X	1990	34 & 48	2 spherical spirals each with 8 turns				1.5

* Year of start-up and (where known) end of operation.

current in the case of floating rings, or by the interception of field lines if the ring was mechanically supported. Moreover, if the current was fed through leads in the support posts, magnetic disturbances appeared to cause convection of the plasma across the main confining field. As in the case of the multipoles, the answer was to use levitated superconducting coils. Princeton went via the supported superconducting ring SP-3 to the levitated LSP, which started operation in 1970 after which Livermore and Culham followed with their superconducting levitrons [85] and Princeton installed the second-generation FM-1 Spherator. Table 4.4 lists parameters of the main internal-ring devices, both single-ring levitrons and multipoles.

In the levitated FM-1, the Princeton group studied the particle confinement time as a function of temperature. Below 3 eV, they found a scaling close to 'classical' (collision induced, hence $\propto T_e^{1/2}$), with an absolute value about one fifth of the classical one. Above 3 eV, the confinement time deteriorated in a manner suggestive of Bohm scaling ($\propto T_e^{-1}$), but it stayed 300 times higher than Bohm's formula predicted. Later, in the same device, they measured the electron thermal conductivity by applying a localized heat pulse and found a similar Bohm-like dependence above 1 eV, at a level of 10–20 times smaller than the Bohm value [86].

Multipoles and levitrons have until recently not been put forward as candidates for a reactor. They have been test beds in which rotational transform, shear and well depth could be adjusted in order to verify theories of stability and transport in toroidal systems [87]. Thus, they contributed much of the evidence against the universal character of Bohm diffusion that was amassed at the Novosibirsk Conference in 1968. Earlier hot-ion and hot-electron experiments in mirror machines had already exceeded Bohm times by large factors [88], but the suspicion kept being nourished, particularly by stellarator experiments, that this was different in closed systems. Along with the tokamak, the levitrons and multipoles played a major role in putting this contention aside. In fact, they confirmed both the rate of cross-field diffusion and the bootstrap current predicted by neoclassical theory. Recently, however, an MIT–Columbia University collaboration has launched the construction of a Levitated Dipole Experiment, LDX, with a levitated superconducting ring. This project was backed up by a reactor study which made a case for a D–^{3}He burning reactor [89].

4.6 Unconventional toroidal schemes

A scheme named Bumpy Torus (figure 4.15) may be regarded either as an attempt to circumvent mirror losses by arranging a series of mirror systems in a toroidal configuration, or as a way to improve single-particle confinement in a torus [90]. With regard to the interchange instability, the curvature of the field lines is favourable in the regions of the mirror coils

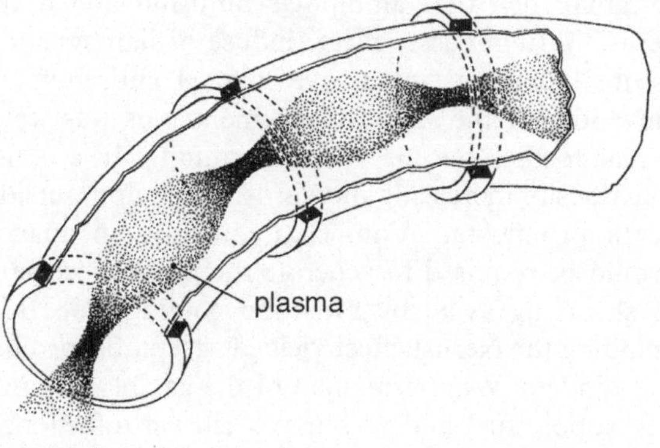

Figure 4.15. Schematic of the Bumpy Torus, a series of multiple mirrors arranged in a toroidal configuration.

and unfavourable in between. This does not preclude average minimum B (section 3.1.3) being achieved, but resistive ballooning remained a problem for which the Oak Ridge group proposed to make use of annular hot-electron plasmas (not axis-encircling beams as in Astron) located between the mirror coils. Their diamagnetism had been shown in simple mirror systems to create a stabilizing local minimum-B configuration. The ELMO Bumpy Torus at first indicated promising results, but the project was abandoned after improved diagnostics led to a downward revision of both the electron temperature and the efficiency of the microwave heating [91]. This line was further pursued in Nagoya, where the β value of the hot electron ring was estimated, by a direct measurement of the magnetic field inside the plasma, to be a few per cent, not enough to create the desired field minimum [92]. In 1988, a review of this work concluded that the hot electrons failed to reverse the local field gradient and that there was rapid convective loss.

As noted in the previous chapter, Sakharov had considered already in 1951 that a magnetic field could reduce heat conduction even if it was not in pressure equilibrium with the plasma. The idea to confine a plasma within a gas blanket was re-introduced in 1960 by Alfvén [93]. His group at the Royal Institute of Technology in Stockholm started experiments with linear discharges centred within a gas vortex, but these failed to test the validity of the scheme because the discharges were produced on such a short time-scale that the plasma contracted and the usual vacuum-pinch instability set in before the gas blanket could follow and provide pressure equilibrium.

The Rijnhuizen institute [94] at Nieuwegein proposed to choose a vacuum configuration for the magnetic field in order to eliminate magnetically driven instabilities; the pressure of the plasma would then be balanced by the neutral gas. Work in Garching [95], however, soon reconfirmed

Sakharov's original idea that ambipolar diffusion and a thermo-electric force, the Nernst–Ettinghausen effect, induce a diamagnetic current such that the pressure balance is taken over almost entirely by the magnetic field. While invalidating the stability argument, this was welcome because of another threat to the concept—bremsstrahlung. In a constant-pressure profile with the density increasing outwards, the region outside the burning core would emit prolific radiation. Consequently, an unacceptably large core radius would be required to generate the power necessary to keep the column from shrinking away by radiation cooling [96]. But a transport model [97] including the Nernst effect yielded acceptable profiles in infinitely long, straight cylinders. What remained of the gas blanket in these models would serve to supply fuel and to remove ash via diffusion, as well as to protect the plasma and the wall against their mutual interaction.

Lehnert [98] formulated criteria for the co-existence of a plasma and a gas blanket in a steady state: a density requirement for the plasma to be impermeable to neutral gas and an energy flow necessary to sustain radiation loss. Experimental work towards steady-state reactors developed in two directions: the Stockholm group [99] set up the series of Intrap experiments, in which a rotating plasma in a poloidal magnetic field was kept away from the inner wall of the discharge vessel by centrifugal forces, and in Nieuwegein the high-density regime was studied in the Ringboog tokamak [100]. Meanwhile, theoretical groups [101] in Nieuwegein and Moscow calculated steady-state profiles with cool plasma in the outer layers of toroidal systems. Experimentally, however, the operating regimes of tokamaks are not consistent with a gas blanket between the plasma and the wall. But the power flow into the divertor must go through a cool plasma in front of the divertor plate, which screens the hot plasma and the plate from each other.

4.7 Status of alternative toroidal systems

The toroidal systems presented at the Geneva Conference were pinches—with either weak or strong stabilizing toroidal field—and stellarators. But each one of these had only limited flexibility for varying the magnetic-field configuration, so that there was a need for facilities specifically designed for investigating stability and confinement. At first, the emphasis was mainly on stabilization by conducting walls and magnetic shear, but when minimum-B stabilization of open systems came forward at the Salzburg Conference in 1961, one of the main topics in fusion research became to incorporate the magnetic well in closed systems as well. Toroidal versions of the hard-core pinch with mechanically supported, magnetically levitated, or freely falling central rings and toroidal multipoles with, typically, four current-carrying toroidal rings immersed in the plasma appeared on the scene as means to vary both shear and well depth. Plans to construct larger machines as successors to ZETA and the C-stellarator were shelved; first priority now became to sort out the physics of the plasma in toroidal fields, rather than to improve the performance of the devices by increasing their scale.

Among the toroidal systems discussed in this chapter, the internal ring devices and the reversed-field pinches stand out as the most productive ones, scientifically. The former were instrumental in confirming predictions of the neoclassical transport theory, as well as in demonstrating MHD stabilization by rotational transform, shear and negative V''. The latter pointed to the conservation of helicity, relaxed MHD equilibria and dynamo mechanisms.

As possible schemes for a fusion reactor, however, their development has been outpaced by the tokamak and the stellarator. Internal conductors are difficult to accommodate in a reactor and field reversal is difficult to sustain in a steady state. But the high-β capability of the RFP is an important advantage, and this chapter is not yet closed. Nor is the field-reversed configuration entirely without reactor potential; if confinement in the FRC meets the highest expectations, it could develop into a D–^{3}He burner.

Chapter 5

Stellarators versus tokamaks

In the previous chapter we discussed the toroidal 'supporting lines'; now we turn to the mainstream toroidal systems—stellarators and tokamaks. We first compare their progress in the decade after Geneva, that is until the tokamak broke through at the Novosibirsk Conference in 1968. Tokamaks came to dominate fusion research and stellarators became scarce. Indeed, after Novosibirsk there would be a remarkable geographic switch, with the Soviet Union relinquishing the lead in tokamaks but maintaining the largest stellarators and the United States, for a time at least, closing down all its stellarators in order to build tokamaks. In this chapter we cover stellarator research from Geneva up to the present time; the tokamak will require more space, so we defer our discussion of its development after Novosibirsk until later chapters. But we include here as an interlude some paragraphs on diagnostics, most of which are not specific for the kind of device to which they are applied.

5.1 Stellarators: Bohm diffusion or not?

The stellarator was the only closed system for which, at the time of the Geneva Conference, steady-state operation could be envisaged, and this made it one of the most serious candidates for reactor development. There was concern about the maximum beta-value consistent with equilibrium and stability, but theoretical models were crude and left ample room for optimism. 'Try-and-see' was the best the experimentalists could do to produce a quiescent plasma, free from turbulent fluctuations. Spitzer suspected that there was an intrinsic anomalous particle loss mechanism, 'pump-out' as it was known in those days, perhaps corresponding with Bohm's empirical rule (section 1.2.3):

$$D_B = T_e/16B \text{ m}^2\text{ s}^{-1}; \qquad \tau_B \approx 3a^2B/T_e \text{ s} \quad [T_e \text{ in eV}]$$

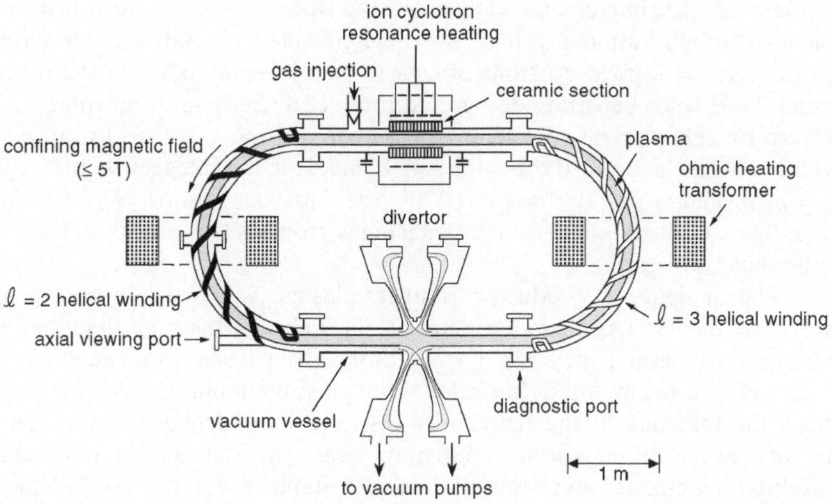

Figure 5.1. A simplified plan view of the C-stellarator showing the location of major components.

where D_B is the Bohm diffusion constant, τ_B is the Bohm diffusion time and the diffusion length is taken to be $a/2.4$.* And if this was the case, the next question was if it really resulted from a diffusion-like process caused by some kind of instability or from a lack of equilibrium. The equilibrium was suspect because there was no mathematical proof that the stellarator did confine individual particles to closed drift surfaces. As to instability, since the plasma was mostly produced by Ohmic heating, current-driven modes of the kind described in Geneva by Lehnert and shortly thereafter explained by Kadomtsev and Nedospasov as well as by Hoh and Lehnert [1] were the first candidates. But theory had also predicted density- or temperature-gradient driven 'universal' modes, which would come into the picture if the pump-out persisted in currentless plasmas.

After the Geneva Conference, Princeton no longer held a monopoly on stellarator experiments. The laboratory had built a series of devices, of which Etude and B3 remained in operation, and was constructing the larger C-stellarator (figure 5.1) which was to come into operation in 1961. Garching and the Lebedev Institute in Moscow were the next to build stellarators, later to be followed by several other laboratories. All adopted $l = 2$ or $l = 3$ helical fields, produced by $2l$ helical windings carrying currents in alternate directions, l being the periodicity of the magnetic surfaces in a poloidal cross-section. A possible advantage of $l = 3$ is that it penetrates not so deep, so that it provides shear in the rotational transform, which

* As for a zero-order Bessel function profile.

tends to stabilize interchange-like modes. But shear entails that the transform passes through rational values, at which magnetic islands can develop. Moreover, $l = 3$ gives no transform near the magnetic axis. On the other hand, $l = 2$ fields penetrate deeper and produce a nearly uniform rotational transform. This can be adjusted so as to avoid the most dangerous rational values of q, which are those with small values of the integers m and n, if $q = m/n$. Garching [2] chose $l = 3$ for their first stellarator, Wendelstein 1-A, but adopted $l = 2$ for later machines from W1-B on, as did most other laboratories.

The challenge to produce currentless plasmas was met in a variety of ways. At the 1965 Culham Conference, the Lebedev Institute described a gun-injected plasma in the L-1 stellarator. The particle loss came out a factor of 3–4 below the Bohm rate, although still anomalous. At the same time, the tokamak at the Kurchatov Institute showed better confinement by one order of magnitude. Garching went even further; in a plasma produced by contact ionization of caesium vapour in Wendelstein W1 they had seen no pump-out at all and the confinement was close to classical stellarator diffusion—one to two orders of magnitude better than Bohm [3]. They had borrowed the technique of exploiting the low ionization potential of alkali atoms from a class of linear devices, the Q-machines.*

Despite all evidence from other experiments, Princeton stuck to Bohm's law for their stellarators. In addition to B/T scaling, they confirmed an a^2-dependence of the particle confinement time by comparing C-stellarator results with those of their smaller devices. While it was possible that the Garching stellarator results could have something to do with the heavy ions used in those experiments, the apparent difference between the Princeton stellarators and the Moscow tokamaks was harder to explain. The plasma current could drive certain instabilities, but in this respect Ohmically heated stellarators should not be worse than tokamaks, as the experimental data suggested. The point could not be settled definitively, but Spitzer and his co-workers in Princeton kept stressing that none of the toroidal systems seemed to deviate from Bohm's formula by a conclusive margin [4].

It was not until the 1968 Novosibirsk Conference that sufficient data piled up, some from stellarators but most from other devices, to finally discredit Bohm's empirical rule. Culham, Kharkov and Novosibirsk had by then joined the stellarator league. To become independent of Ohmic heating, Princeton had now studied gun-injected hydrogen plasmas in Etude, B-3 and the C-stellarator, as well as microwave-produced xenon plasmas in the

* Q-machines played an important role in basic plasma physics. They had an axial field between two hot tungsten plates; the plasma was formed by directing a stream of alkali-metal vapour (sodium, potassium, caesium and lithium were used) on to these plates, from which the atoms returned as ions, while electrons were supplied by thermal emission.

latter. The electron temperature was derived from the electrical conductivity of the plasma, which followed from the current–voltage characteristic—in current-free plasmas a small oscillating voltage was applied—and this indicated that the particle confinement time exceeded Bohm by factors ranging up to ten. But Princeton still remained reluctant to discard Bohm's law and were concerned that the temperature of the bulk electrons might be lower than measured. This would be possible if some of the current was carried by a small fraction of more energetic electrons—an argument they held against the tokamak as well. The Lebedev Institute, Kharkov and Novosibirsk all found anomalous loss, although the latter had $\tau \propto B^2$, rather than $\tau \propto B$. But Culham [5] reported confinement up to 15 Bohm times in the Proto-Cleo stellarator and Garching had now arrived at a factor of 100 with a barium plasma in W2. Similar evidence from tokamaks and theta pinches also disproved Bohm's law, putting this issue aside but at the same time confronting the stellarator with powerful competitors. The Novosibirsk Conference also heard that the issue of single-particle confinement was resolved, in practical terms, by an experiment in Culham in which electrons released by beta-decay of tritium in the small stellarator Clasp were shown to be confined for 10^7 transits around the torus [6].

Finally, Fontenay [7] proposed a new kind of stellarator, which they named 'torsatron' to distinguish it from the 'classical stellarator'. In this new scheme, to which we will return in section 5.4, the helical field was produced, not by pairs of windings with opposite currents, but by unidirectional helical currents. A similar configuration was arrived at from a different direction by a group in Kyoto [8]. In the Heliotron-B and -C devices, Uo and associates had studied toroidal fields produced by circular conductors, arranged in pairs with opposite but unequal currents, similar to a 'picket fence' or 'bumpy torus'. At first the rings were outside the vacuum vessel; later they were submerged in the plasma as in internal-ring devices (figure 5.2). Already in 1961, they had proposed a helical version of this configuration, which they called a 'helical heliotron', and which eventually evolved towards the torsatron.

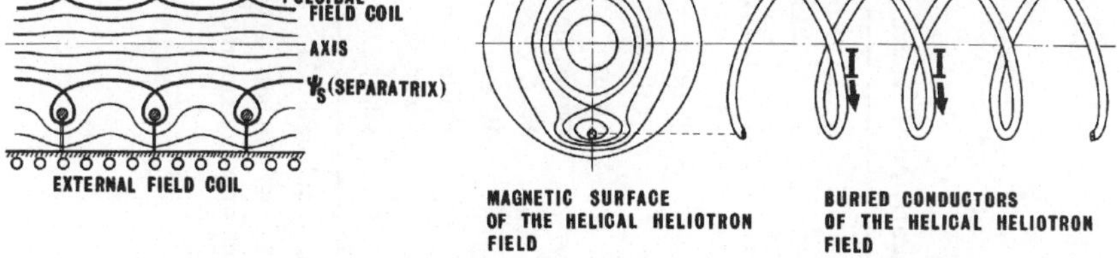

Figure 5.2. Earliest Heliotrons at Kyoto. Left: stabilizing rings submerged in the plasma and corresponding flux surfaces. Centre: flux surfaces of helical heliotron. Right: helical windings carrying unidirectional currents.

Table 5.1 Early post-Geneva stellarators

Location	Name	Year‡	R_0 (cm)	a (cm)	B_t (T)	l/m	Remarks
Princeton	C-stellarator	1961–69	100*	5–7.5	3.5	3/8 + 2/8	Divertor/ICRH
Lebedev Inst.	L1	1963–71	60	5	1	2/14	Magnetic islands
	TOR-1	1967–73	60	5	1	2/14	
	TOR-1	1967–73	63	3.6	0.8–2.5	2/16	
Kharkov	Sirius	1964–75	60†	2.6	1.6	3/–	Turbulence
Novosibirsk	No name	1968	50†	<5	0.2	3/26	Fluctuations
Garching	W-A	1965	35†	2	1	3	Classical loss rate
	W-B	1968	35†	2	1	2	Rational-q resonances
Culham	Proto Cleo	1968	40	5	0.3	3/7, 3/13, 2/6	Neoclassical diffusion
	Twist	1967	32	4.5	0.3	3/4	Turbulent heating
	Clasp	1967	30	5.6	0.1	3/8	Single-part. conf.

* Racetrack; total length 12 m.

† Racetrack; other machines are circular.

‡ Year of first operation (and of closure where known).

After August 1968, reactions to the results that had been presented at Novosibirsk were particularly strong in the US, where the disappointing performance of the C-stellarator led to a reappraisal of toroidal confinement. An experiment with caesium plasma in the C-stellarator failed to reproduce the Garching results, pointing to defective magnetic surfaces as a likely cause of the persistent pump-out. The Americans had misgivings about the far better results claimed for the Russian tokamak T-3, but when these were unequivocally confirmed one year later, they launched an aggressive tokamak programme, the first step in which was to convert the C-stellarator into a tokamak. We return to this in section 5.4.

5.2 Tokamaks: from Geneva to Novosibirsk

The earliest reports on toroidal pinches with strong stabilizing fields [9] from the Kurchatov Institute in Moscow described an 80 cm major-radius porcelain torus, named TMP [10]. Dolgov-Saveliev [11] mentioned five other toroidal chambers, of which the most recent had a continuous metal liner. After Geneva, the first international meeting where fusion physicists met was the Fourth Conference on Ionization Phenomena in Gases, held in Uppsala, Sweden. There, Yavlinski's group [12] introduced the name *tokamak* (from the Russian words ***to**roidalnaya **ka**mera* for 'toroidal chamber' and ***ma**gnitnaya **k**atushka* for 'magnetic coil')* for what had been torus number 5, and would henceforth be named T-1 (figure 5.3).

They had meanwhile installed a diaphragm, similar to the limiters in the Princeton stellarators, to constrict the discharge and to concentrate the plasma–wall interaction mostly on a small surface of refractory metal. They confirmed the Kruskal–Shafranov stability condition that the safety factor q (see boxes 2.4 and 5.1) must exceed unity.

The main characteristics of tokamaks are described in box 5.1 and are illustrated in figure 5.4.

In subsequent years, the Kurchatov team experimented with a succession of tokamaks of about the size of T-1 (see table 5.2) [13]. Gradually, they improved both their diagnostics and their means to control the discharge so that by the time of the Culham Conference in 1965, the main characteristics of standard tokamak plasmas had emerged. The equilibrium had been studied in T-5 through measurements of currents to segmented limiters [14] and had been found to be well described by Shafranov's theory [15]. Minor and major disruptions of the discharge current, associated with negative voltage spikes and bursts of X-rays, had been observed and had appeared to correlate with q reaching the value of one near the centre of the

* Golovin had proposed *tokamag*, but Yavlinsky feared that this would make people think of magic, so they added the word katushka.

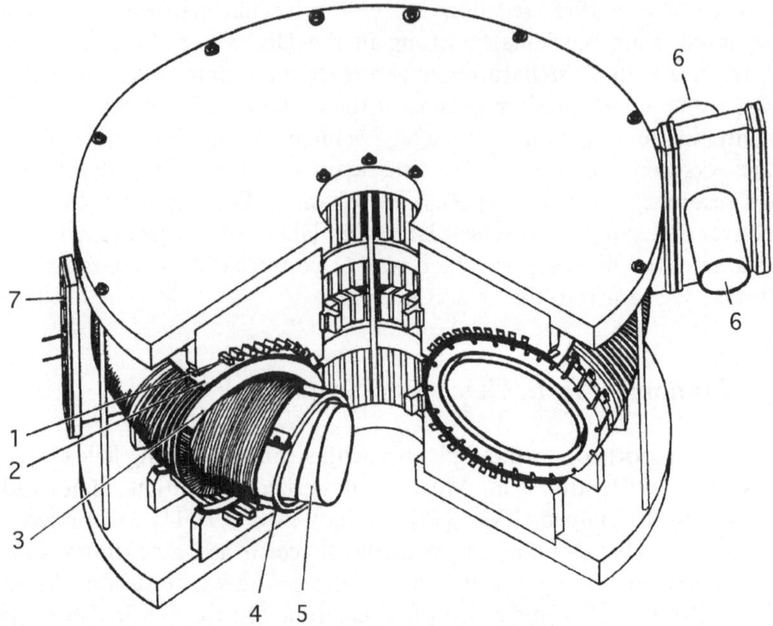

Figure 5.3. The Moscow Torus 5, later renamed Tokamak T-1. 1, coil for primary toroidal current; 2, screening copper shield; 3, longitudinal magnetic field coil; 4, stabilizing copper shield; 5, metal liner, 6, vacuum outlet; 7, port for diagnostics.

discharge (though, as we will see later, further studies showed that disruptions are linked to the value of q at the edge). In 'stable' discharges, they had seen sub-MHz MHD oscillations caused by helical distortions at integral values of q. In the larger T-3 device, the ion and electron temperatures had been raised to 100 eV and the energy confinement time had reached 2–4 ms [16].

With these figures, the tokamak had overtaken competing toroidal devices like ZETA and the C-stellarator. More important, the tokamak (together with the L-1 and Wendelstein stellarators and the Garching octopole) had beaten the dreadful Bohm-loss formula by a significant margin. But the world was not ready to receive the message. Spitzer, in his summary of the experimental work presented at Culham, stressed '*that anomalous particle loss is present and that it is roughly within an order of magnitude of that predicted by the Bohm formula*' and Artsimovich did have reason to complain in his 1968 Novosibirsk paper that his Culham-conference report '*was received with remarkable suspicion*'. Indeed, until 1968 the Kurchatov team was left to pursue the tokamak trail without competition from other institutes both outside and inside the Soviet Union. The only exception was the LT series of pinch experiments at the Australian National University in Canberra which were operated in tokamak-like configurations (though

Box 5.1 The tokamak

The tokamak is an axially symmetric field configuration with closed magnetic surfaces, in which a toroidal field is produced by currents in external coils and a poloidal field by a current in the plasma. The weaker poloidal field determines the plasma confinement and the stronger toroidal field provides the stability. In a tokamak with large aspect ratio $A = R_0/a$ and circular cross-section, the safety factor (box 2.4) $q(r) \approx rB_t/R_0B_p$ so that the value at the plasma boundary $r = a$ is given approximately by $q_a \approx 5a^2B_t/R_0I$ (where the subscripts t and p denote toroidal and poloidal components and the toroidal plasma current I is in MA). The disruption limit, $q_a < 3$, determines the maximum value of the current I in a tokamak of given R_0, a and B_t (box 6.2).

The tokamak plasma current is usually driven by a toroidal electric field, induced by transformer action (section 1.2.1); some tokamaks have iron transformer cores like that shown schematically in figure 5.4. Most of the available flux change is used in setting up the magnetic configuration and driving the current in the early stages when the plasma is cold and resistive. Thereafter, relatively little flux is consumed in maintaining a steady current in a highly conducting hot plasma—allowing tokamaks to be driven inductively for long pulses (up to one thousand seconds for a reactor). The current can be maintained indefinitely using non-inductive drive schemes as discussed in section 7.3.

It is also necessary to have an axial or 'vertical' field to control the tendency of the plasma column to expand in the direction of the major radius. The vertical field strengthens the poloidal field on the outside, and weakens that on the inside of the torus. In early tokamaks a quasi-static vertical field was applied and fine control of position relied on image currents induced by the plasma in a thick electrically conducting copper shell. Modern tokamaks control the plasma position with an active feedback system using external coils to generate the transverse magnetic fields.

There are advantages for confinement and achievable pressure with plasmas that are vertically elongated and slightly D shaped. However, vertical elongation renders the plasma unstable against vertical displacement and additional coils are required to shape the plasma and control its position. Traditionally the toroidal field coils were of circular cross-section but vertically elongated D-shaped coils are better able to resist electromagnetic stresses.

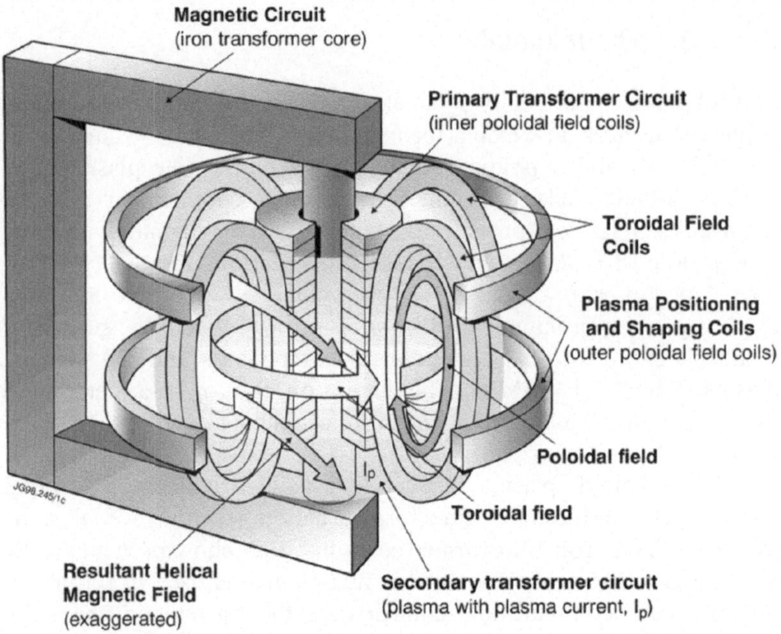

Figure 5.4. Schematic of a tokamak. The transformer induces the toroidal current, whose poloidal field confines the plasma; the outer poloidal-field coils position and shape the plasma; the toroidal-field coils provide the stabilizing field.

not referred to as such at the time) as early as 1965 [17]. With T-3 and TM-3, the Kurchatov team brought the tokamak scheme to a level of sophistication appreciated in full only after Novosibirsk. At that conference, Artsimovich reported further improved performance of T-3, with electron temperatures up to 1000 eV, ion temperatures half as much, and energy confinement times up to 50 Bohm times [18] (figure 5.5).

The Kurchatov team established that the energy loss, although significantly below Bohm's formula, remained anomalous and took place mainly

Table 5.2 Early tokamaks at the Kurchatov Institute in Moscow

Name	Year	R_0 (cm)	a (cm)	B_t (T)	I (kA)	Remarks
TMP	1955	160	13	1.5	260	Porcelain torus; $q < 1$
T-1 (torus 5)		62.5	13	0.5	220	Metal liner; $q > 1$
T-2	<1960	62.5	13	0.6	20	Bakeable liner
T-3	1962	100	20	4	60	Bakeable liner
TM-1	1960	40	10	2	15	Adiabatic compression
TM-2	1961	40	10	2.6	15	Found disruptions
TM-3		40	8	5	40	Modified TM-2

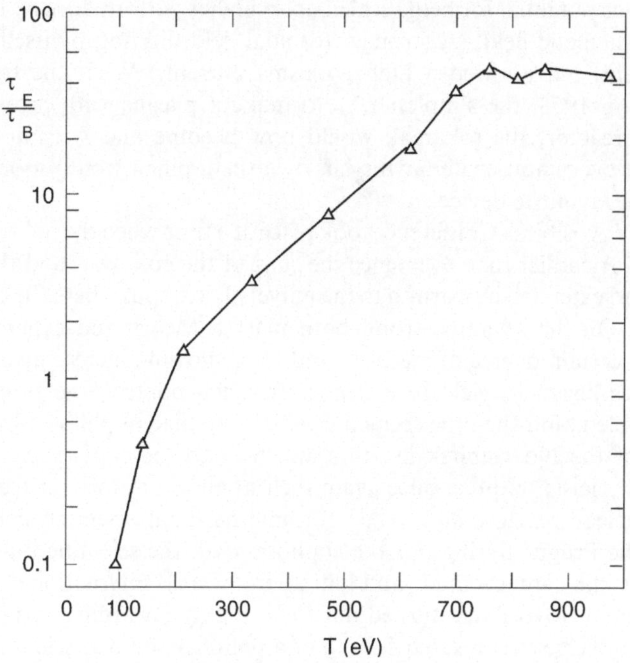

Figure 5.5. T-3 results presented at 1968 Novosibirsk Conference. Here, we reproduce only the data taken with $I_p = 180\,\mathrm{kA}$, $B_{tor} = 3.4\,\mathrm{T}$ and varying filling pressure to produce a range of plasma temperature (note that $T = T_e + T_i$). In the temperature range 700–1000 eV the confinement time τ_E reaches 40–50 Bohm times τ_B.

through the electrons. The 'overall' temperature, $\langle T\rangle = (n_e T_e + n_i T_i)/n_e$, was derived from the diamagnetism of the plasma and, with the ion temperature determined from the spectrum of charge-exchange neutrals, the electron temperature could be deduced by subtraction. The electron temperature determined in this rather indirect way was consistent with values obtained by soft-X-ray measurements or derived from the electrical conductivity. In conditions where the electric field exceeded the runaway limit [19]—at low densities and high electric fields—there was an anomalous resistivity as predicted by the Kadomtsev–Pogutse (1967) theory. For the energy confinement time, Artsimovich tentatively proposed the empirical scaling

$$\tau_E \propto a^2 n_e \langle T\rangle^{3/2}.$$

He noted, however, that his data indicated saturation at high n_e, and that also the dependence on $\langle T\rangle$ was rather uncertain. But τ_E was certainly not decreasing with increasing $\langle T\rangle$ or T_e, as Bohm would have had it. Furthermore, τ_E was proportional to the plasma current, I, as long as the Kruskal–Shafranov stability condition was satisfied. The tokamak was just one among several machines that were reported in Novosibirsk to defy the Bohm law, but it had obvious potential for being scaled up to higher performance by increase in size and magnetic field. (A stronger toroidal field did not, by itself, improve confinement, but allowed a higher plasma current.) What the theta pinch had been in 1958, the simplest way to make a plasma with properties like that in a reactor, the tokamak would now become and Artsimovich's jest [20] about every housewife having her own theta pinch would soon fire back at his own favourite device.

The Novosibirsk Conference took place at a time when the political atmosphere was friendlier than ever since the start of the cold war, and the 'Prague Spring' had extended its warmth to the university campus where the conference was held. On the scientific front, both plasma theory and experiment had reached a certain degree of maturity and were showing increasing agreement; instabilities began to yield to measures for their control, the Bohm barrier had crumbled, and the time seemed ripe for a change in philosophy. Budker, director of the Novosibirsk institute and host to the conference, urged the plasma physicists to direct once again their attention to the construction of a thermonuclear reactor, and called for international scientific cooperation. Neither the Prague Spring nor the euphoria over the scientific outlook were to last, but the contacts made and initiatives towards international collaboration taken at Novosibirsk turned out to be highly rewarding and capable of surviving adverse circumstances, both of a political and of a scientific nature.

5.3 Diagnosing the plasma

Artsimovich's Novosibirsk paper in 1968 described how the ion and electron temperatures were derived from measurements of diamagnetism, soft X-rays, electrical conductivity and charge-exchange neutrals. Neutron yields were still marginal, but charge-exchange neutrals gave fairly convincing ion temperatures [21]. However, the reported electron temperatures had been measured less directly and were therefore more controversial. To derive a coherent picture of the plasma required, as it still does today on a higher level of complexity, a judicious weighing of experimental data and theoretical modelling.

But although no-one was higher regarded for his judgement than Artsimovich, the scepticism with which the tokamak had been received three years earlier at the Culham Conference did not evaporate at once. Particularly in Princeton, it was argued that runaway electrons could confuse the measurements of both the resistivity and the diamagnetism [22]. Although the Kurchatov team countered this objection by referring to their X-ray measurements, which allowed them to distinguish between discharges with and without runaway electrons, the sceptics were not convinced. It would take another year for the remaining doubts to be dispelled in a conclusive manner. Already during the conference at Novosibirsk, Pease and

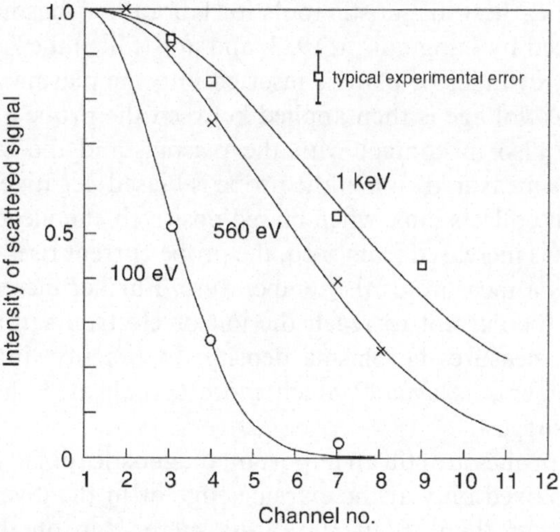

Figure 5.6. Electron temperature in the T-3 tokamak, measured with Thomson scattering of ruby laser light. The short-wavelength wing of the scattered 6943 Å light is collected in channels of 78 Å width. Experimental points are shown for three discharge conditions with matching theoretical curves from which the electron temperature can be determined.

Artsimovich had discussed a plan to send a team from Culham to Moscow with equipment to measure the electron temperature directly with the newly developed method based on Thomson scattering of ruby laser light. This remarkable mission [23] confirmed unequivocally that the Russians had been right (figure 5.6) and, moreover, paved the way for further collaborative projects in fusion research. For those who still had misgivings about the breakthrough of the tokamak at the Novosibirsk Conference, the special conference on tokamak physics organized by Artsimovich in Dubna near Moscow in September 1969, where Robinson and Peacock announced their results, was the decisive moment.

Some years later, a serious debate on fusion research in the House of Lords, the upper chamber of the British Parliament, would be interrupted by the question '*How do they measure a temperature of 300 million degrees?*' which provoked the response, '*I expect that they use a very long thermometer*'. In fact extrapolating the scale on a conventional mercury-in-glass thermometer to fusion temperatures would require a length of 600 km. During the decades after Novosibirsk, the development of reliable diagnostic* methods became an integral part of the work of all experimental groups [24].

* Fusion scientists adopted the term 'diagnostic' from the medical profession to describe the instruments used to measure plasma parameters. The word diagnosis is derived from the Greek words *Δια* which means 'through'—but in this context could be translated as 'thorough' or 'deep'—and *Γνωσις* which means 'knowledge'. Indeed, only the most meticulous attention to all available diagnostics is rewarded by 'a deep knowledge' of the plasma.

One of the earliest diagnostic tools for laboratory plasmas was a metal probe, introduced by Langmuir in 1923, and now called the *Langmuir probe*. A small sphere, cylinder or plate is inserted into the plasma at the point of measurement. A voltage is then applied between the probe and some reference electrode, also in contact with the plasma, and the current flowing between them is measured. When the probe is biased negatively with respect to the plasma, it collects ions; when biased positively it collects electrons. As the bias voltage is increased from zero, the probe current rises linearly at first and the slope is a measure of the temperature. Further increase of the bias voltage causes the current to reach the ion or electron saturation current, both of which measure the plasma density. In practice, interpretation of probe measurements is usually much more complicated than this simple picture would suggest.

Langmuir probes are still an important diagnostic tool in fusion plasmas but can be employed only at the extreme edge or in the divertor where the temperature is less than about 100 eV—a mere 2 km on the mercury-in-glass scale. Elsewhere, plasma diagnostics must rely on remote sensing methods. These fall into two general groups: passive methods that rely on particles or radiation emitted spontaneously by the plasma, and active methods that rely on probing the plasma with beams of particles or radiation.

Typical of the active remote methods is the Thomson laser scattering technique to which we have referred already. When a beam of monochromatic light is directed into the plasma, some photons are scattered by the plasma electrons and their wavelength is changed by the velocity of the moving electrons (box 5.2). Thus the spectrum of the scattered light measures the electron temperature, or more precisely their velocity distribution. The Thomson scattering cross-section is extremely small and requires a powerful laser to ensure enough intensity. The early Thomson scattering systems made a measurement at a single spatial point and at a single time in the discharge. Repetitive measurements over a series of hopefully identical discharges were required to build up a time history or a spatial profile. Important developments in the decades following Novosibirsk included the use of multiple lasers and multi-pulse lasers that follow the temporal development of temperature and sophisticated optical systems that measure a complete profile with each laser pulse (figure 5.8).

Another method to measure electron temperatures uses electron cyclotron emission (ECE). This technique utilizes the fact that the plasma emits as a black body source at the electron cyclotron frequency (usually at the fundamental and at the first harmonic). The development of this diagnostic method came as a result of concern that cyclotron radiation (or synchrotron radiation as it was then called) might be a serious energy loss in fusion plasmas. During the mid-1970s, attempts to measure these losses and to extend the theory of cyclotron emission led to the realization of a powerful new diagnostic technique that is now widely used as a thermometer for hot-electron plasmas.

Box 5.2 A Thomson scattering heuristic

When a beam of monochromatic light is scattered by a moving electron, the change in wavelength contains information about the particle's velocity. If in the laboratory frame the phase of the beam is given by $\exp[i(\mathbf{k}_1 \cdot \mathbf{r} - \omega_1 t)]$, where $\mathbf{k}_1$ is the wave vector, $\mathbf{r}$ is the position and ω_1 is the angular frequency of the wave, and the position of the electron is $\mathbf{r} = \mathbf{r}_0 + \mathbf{v}_e t$, then the Doppler-shifted frequency in the electron frame equals $\omega_1 - \mathbf{k}_1 \cdot \mathbf{v}_e$. If now the electron emits at this frequency, the outgoing signal is received as a wave with frequency $\omega_2 = \omega_1 - (\mathbf{k}_1 - \mathbf{k}_2) \cdot \mathbf{v}_e$. (The velocity change of the electron is relatively small.) In the approximation $v_e \ll c$ (a hot plasma really requires a relativistic treatment), the frequency and wavelength changes are also relatively small and, with $\omega = kc$ and $\lambda = 2\pi/k$, one obtains:

$$\Delta\lambda/\lambda \approx -\Delta\omega/\omega \approx -\Delta k/k \approx (\mathbf{e}_1 - \mathbf{e}_2) \cdot \mathbf{v}_e/c$$

where $\mathbf{e}_1$ and $\mathbf{e}_2$ are the unit vectors in the directions of the in- and outgoing beams. This shows that the method detects the electron velocity component in a direction determined by the lines of sight of the emitted and the detected signals and these can be arranged so as to measure either the parallel or the perpendicular (relative to the magnetic field) velocity distributions.

Looking at this process from a particle–dynamics point of view and invoking conservation of momentum and energy in the collision of a photon and an electron, one has:

$$\Delta\mathbf{p}_e + \Delta\mathbf{p}_r = 0 \qquad \text{and} \qquad \Delta E_e + \Delta E_r = 0$$

where $\mathbf{p}$ and E stand for momentum and energy, and e and r for electron and photon. For a non-relativistic electron, $E_e = p_e^2/2m_e$, whereas for a photon $E_r = p_r c$. Then, the energy equation yields $\mathbf{p}_e \cdot \Delta\mathbf{p}_e + m_e c \Delta p_r \approx 0$, and with $\Delta\mathbf{p}_r \approx p_r(\mathbf{e}_2 - \mathbf{e}_1)$ one finds

$$\Delta p_r/p_r \approx \mathbf{v}_e \cdot (\mathbf{e}_1 - \mathbf{e}_2)/c$$

which is equivalent with the result from the wave model.

The particle model also shows that $\Delta v_e/v_e \approx E_r/(E_e E_0)^{1/2}$, where E_0 is the electron's rest-mass energy. This confirms that with visible or infrared light, that is with $E_r \approx 2\,\text{eV}$, E_e in the keV range and $E_0 = m_e c^2 = 511\,\text{keV}$, the relative velocity change of the electron is indeed small.

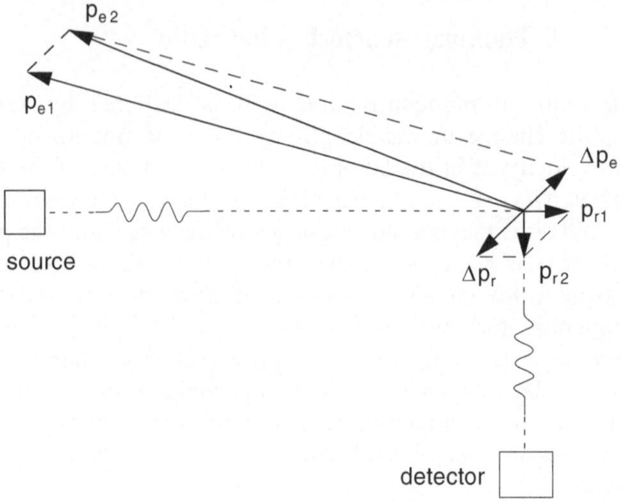

Figure 5.7. Exchange of momentum between a photon and an electron in Thomson scattering. From conservation of momentum and energy one can derive (box 5.2) how the frequency of scattered light relates to the electron velocity.

A particularly important application of ECE has been to the study of the propagation through the plasma of heat pulses resulting from internal disruptions ('sawteeth', section 6.4), which allows a direct measurement of the local thermal transport coefficients.

The ion temperature in T-3 as reported at Novosibirsk was derived from the energy spectrum of neutral atoms emitted passively from the plasma by charge-exchange between plasma ions and atoms coming from the wall. This technique was developed into a very sophisticated measurement, particularly by Afrosimov's group at the Ioffe Institute in Leningrad. In small tokamaks, the temperature so obtained is an average over the whole plasma volume, but in bigger tokamaks thermal neutrals are ionized in a thin outer layer, so that the flux of charge-exchange neutrals from the core is very small and the temperature is weighted to the edge. However the active variant of this technique, which uses the enhanced charge-exchange emission induced by an injected neutral beam, is suitable for larger tokamaks.

Determined efforts were made to develop collective Thomson scattering as an ion temperature diagnostic—the equivalent to the very successful electron diagnostic. Likewise collective scattering was considered one of the prime candidates for measurements of the distribution function of confined alpha particles. The method relies on the ions perturbing the electron density within their Debye sphere. However, although the principle has been demonstrated experimentally, the application and the interpretation of the

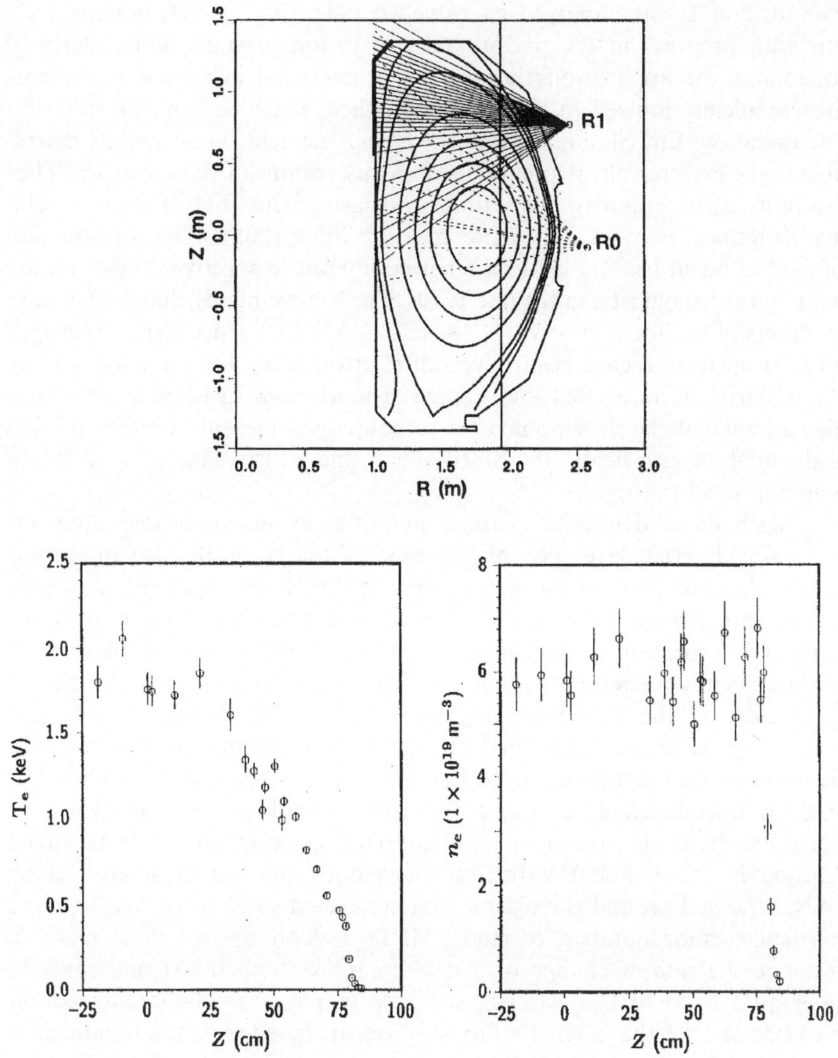

Figure 5.8. Multiple Thomson scattering diagnostic for DIII-D. Top: the laser beam passes through the plasma along a vertical line at major radius $R = 1.9$ m. As many as 40 detector channels simultaneously receive light scattered towards the points R_0 and R_1. Bottom: temperature and density profiles along the vertical chord illuminated by the laser at one instant of time. Repeated laser pulses allow the evolution of these profiles to be studied.

measurements have proved so difficult that the method has not been applied as a routine diagnostic.

Optical spectroscopy had long been used to measure characteristics of laboratory and astrophysical plasmas such as electron and ion temperatures, bulk motion of the plasma, charged particle concentrations and atomic composition. It was employed extensively in the early days of fusion research, but with progress in the performance of fusion plasmas, particularly of tokamaks, the applications became more restricted. The low-Z elements are completely ionized in the interior of these machines and do not emit line radiation. Emission lines from the high-Z elements are generally detectable in the extreme ultraviolet and soft-X-ray regions of the spectrum. They are utilized for impurity concentration measurements but at these shorter wavelengths it is more difficult to measure line profiles. The introduction of neutral beam heating led to a renewed interest in spectroscopy as an ion temperature diagnostic in the late 1970s. The beams induce charge-exchange reactions of the form $H^0 + A^{+q} \longrightarrow H^+ + [A^{+(q-1)}]^*$, in which a hydrogen atom from the injected beam gives an electron to an impurity ion leaving the impurity in an excited lower stage of ionization. In the late 1980s this also stimulated the development of techniques to measure current profiles and radial electric fields (the Stark effect) and density fluctuations (beam emission spectroscopy).

Methods to determine plasma quantities by magnetic measurements were already well developed by the time of the Novosibirsk Conference. Rogowski coils (wound on belts encircling the plasma column) were used to measure plasma currents. Similar coils wound with alternating pitch (sine and cosine coils) could register horizontal and vertical displacements of the current channel. Flux loops and diamagnetic loops (poloidal windings lying close to the flux-conserving shell and to the plasma column, and measuring, respectively, the small changes in poloidal and toroidal magnetic fields) were used to measure beta. For short-pulse, high-density pinches, Los Alamos introduced a technique to measure field distributions within a plasma column by means of miniaturized coils in thin quartz tubes. Mirnov in the Kurchatov Institute developed the use of small pick-up coils, arranged around the plasma and connected so as to be sensitive to a particular mode number, to study MHD oscillations. All these methods measure the rate of change of a field component, so the signals must be integrated to yield field strengths. These magnetic methods allowed the measurement of the magnetic flux surfaces at the edge of the tokamak, in particular the edge safety factor q_a. It was to prove much more difficult to measure these important quantities in the interior of tokamak plasmas until the techniques based on far-infrared polarimetry and the motional Stark effect were developed in the late 1980s.

Millimetre wave interferometry had been applied on some of the earlier fusion devices to measure electron densities. Since high-frequency waves propagate as if the plasma had a dielectric constant $\varepsilon_0(1 - \omega_p^2/\omega^2)$, where ω_p is the plasma frequency and ω is the frequency of the wave, the wave cannot propagate where $\omega_p > \omega$, which means that the 8 mm waves that were used in the first experiments worked only below $n_e = 2 \times 10^{19}\,m^{-3}$. Clearly, higher frequencies were needed to measure higher plasma densities and the development went via 4 and 2 mm microwave oscillators to far-infrared and infrared lasers. Measurements along multiple chords were employed to determine density profiles by inversion techniques. Microwave techniques for density measurements returned in the 1980s as reflectometers working at $\omega < \omega_p$, that could probe a plasma along a single sight-line and these diagnostics were developed also for fluctuation studies.

In the nuclear fission laboratories where much of fusion research originated, neutron measurements were the order of the day. So, when the first hot plasmas were produced, there would be neutron detectors around, both for health protection and to see if the plasma had produced any fusion reactions. Of course neutrons by themselves would not prove that thermonuclear reactions had occurred, but their absence would definitely prove the opposite. To make sure that neutrons are produced in truly *thermo*-nuclear fusion reactions, one has to prove that they originate in the plasma, not at the wall, and that their time dependence is consistent with the lifetime of the hot plasma, short bursts of neutrons generally having their origin in some form of instability. Moreover, their energy spectrum has to be symmetric around the reaction energy, not shifted as a result of a beam–plasma interaction, and to have a width corresponding to the ion temperature. If all these requirements are satisfied, the absolute yield of neutrons is a sensitive measure for the ion temperature, certainly at the threshold of the thermonuclear regime where the yield goes up very steeply with the temperature (around 1 keV, the D–D yield scales as T_i^8). However, since at these temperatures most of the neutrons are produced by ions in the tail of the Maxwellian distribution, the neutron yield is extremely sensitive to distortions of the thermal distribution or to non-thermal ions.

As neutrons are electrically uncharged, their detection must be based either on nuclear transformations or on the detection of nuclei—specifically protons—recoiling when struck by fast neutrons. Early experiments were monitored by Geiger counters surrounded by thin foils of silver or indium, which after neutron capture emit beta rays with a half-life in the order of minutes. This activation technique therefore lacks both time and energy resolution. Boron fluoride proportional counters and uranium fission chambers respond instantaneously to low-energy (a few eV) neutrons. But the MeV neutrons from the fusion source must first be slowed down with a suitable proton-rich moderator like paraffin or polyethylene and all record of the primary neutron energy spectrum is lost. Photographic emulsions or cloud chambers, which register recoil protons, have sufficient energy resolution to measure the thermonuclear spectrum but the emulsions lack time resolution and both methods have narrow windows of sensitivity—too few neutrons and the measurement is lost in noise, too many and the detectors saturate.

Many of these techniques require thick shielding against hard X-rays and gamma rays, emitted by the plasma or by auxiliary equipment like spark gaps. Moreover, to obtain sufficient spatial resolution to determine from which part of the discharge the neutrons originate requires strong collimation and this severely reduces the signal strength.

We are now convinced that the first really strong evidence of thermonuclear neutrons came from a theta pinch but, around the time of the Geneva Conference, there had been many tentative or outspoken claims of thermonuclear neutrons. Most were to be followed by retractions based on better measurements, the best known being the ZETA episode in 1958. Ironically, the ZETA neutrons were unmasked with equipment that had been available all the time in another corner of the same building at Harwell. Sadly, the perpetrators of the cold fusion myth thirty years later failed to learn from these salutary lessons and, misled in part by their neutron measurements, rushed into publication with results that any expert would have dismissed.

Artsimovich in his summary talk at the Salzburg Conference had felt obliged to censure his audience: '*I do not wish to deal here with the question of the nature of the neutron radiation observed in some cases of fast compression of a deuterium plasma because this problem is of interest only as a sport*', Yet, neutron measurements could be useful to corroborate other diagnostics. In the summer of 1969, the neutron flux from discharges in the T-3 tokamak was measured with boron fluoride proportional counters. This yielded a temperature of 300 eV, in agreement with the spectrum of charge-exchange neutrals [25]. But although this became a standard diagnostic for tokamaks [26], neutron yields of sufficient intensity to permit reliable time, space, and energy-resolved measurements with neutron spectrometers had to await the big tokamaks in the 1990s.

Particularly important factors in the development of plasma diagnostics were the parallel developments in data recording, storage and computing techniques that had taken place during this period. As an example, figure 5.9 shows one frame of a tomographic sequence displaying the course of events in a sawtooth crash. In the early days of fusion research, data were displayed on an oscilloscope, recorded on photographic film, measured with a pair of dividers and calculated on a slide rule. Tales abound of the critical results that were lost to the world by errors overnight in the photographic department. The development of the polaroid camera and the storage oscilloscope in the early 1960s greatly improved the productivity of the experiments. But measurements off the polaroids generally continued to rely on the judicious use of a ruler for another decade or so, limiting not only the accuracy but also the amount of data that could be recorded and processed. After electronic recording methods became available in the late 1970s, there was a rapid explosion in the amount of data that could be taken and analysed per shot. Fortunately the developments in computing technology have kept pace, though only just, with the voracious data appetites of diagnosticians. On-line data analysis with results displayed in real-time in the control room and digital feedback control of plasma

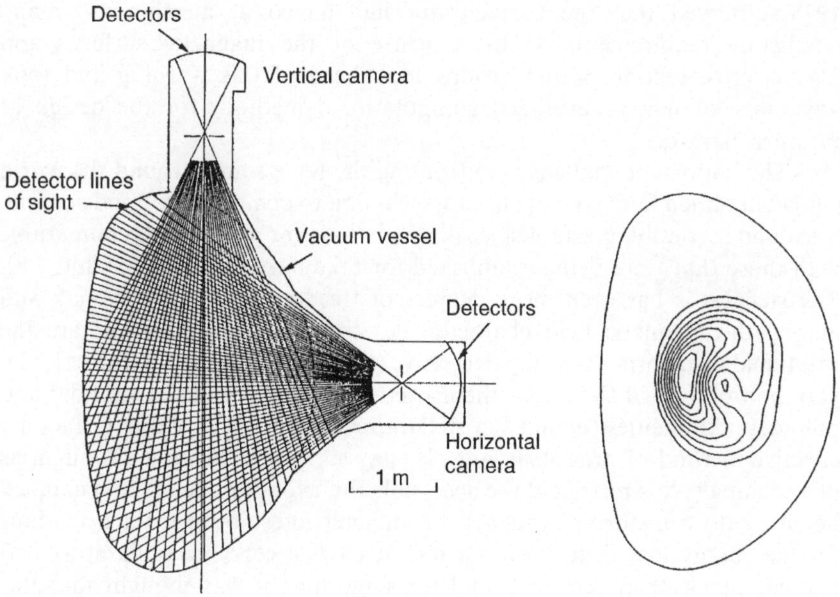

Figure 5.9. X-ray tomography on JET. Left: arrangement of the detectors in a vertical and a horizontal array. Right: contour plot of reconstructed X-ray image of a limiter discharge, taken at a time roughly halfway through a sawtooth crash. The crescent shape is a hot, high-emission region and the bubble (which in a series of such images is seen to move inwards) is a cool, low-emission region. (See figure 6.4.)

parameters are the tools of today's generation of fusion scientists who will look in amazement at what was achieved by the crude techniques of their predecessors.

5.4 Stellarators trailing tokamaks

Although the US suspended stellarator research after 1969, stellarators maintained a niche elsewhere as the main alternative line to the increasingly ubiquitous tokamak and new ideas continued to emerge. Full analysis of the C-stellarator results, which unfortunately did not take place until after the machine had been converted into a tokamak [27], suggested that the cause of the poor confinement might be due to the destruction of the magnetic surfaces by the 'circularizer coils' that were supposed to match the half tori with helical windings into the linking straight sections without windings. The C-stellarator (figure 5.1) had consisted of eight sections: a bend with an $l = 3$ helical winding, a straight section for ion cyclotron heating, a bend with an $l = 2$ (or $l = 3$) helical winding and a straight divertor section, all separated by circularizer sections. Miyamoto, who reviewed stellarators in

1978, observed that the C-stellarator had aimed at meeting too many conflicting requirements at the expense of the magnetic surfaces and plasma cross-section. Other groups took heed of this warning and took advantage of newly developed computational methods for the design of future stellarators.

The important challenge confronting the few groups around the world who maintained faith with stellarators was first to demonstrate good confinement and secondly to establish scaling relations for stellarators comparable with those that were being established for tokamaks (sections 6.9 and 7.8). The stellarator has even more degrees of freedom than the tokamak and among the magnetic field characteristics that affect confinement are the rotational transform, its radial derivative (shear), the well depth, the poloidal variation of $\int dl/B$ (box 4.2), the aspect ratio (R_0/a) and the toroidal and poloidal periodicities (m and l in stellarator terminology). At first, the only effective method of producing hot plasma was Ohmic heating although, as in a tokamak, this restricted the accessible range of parameters and coupled heating with transform, so that it was difficult to establish scaling relations for the energy loss. But it was clear that current-carrying stellarators had anomalous electron heat loss, and for some time it was thought that this corresponded with Artsimovich's pseudo-classical scaling [28].

Various ways were explored to produce hot, current-free plasmas but it took until about 1980 before this deadlock was finally broken by the application of powerful heating methods* to the largest stellarators then in existence. At the 8th IAEA Fusion Conference in Brussels in 1980, Garching reported an experiment in W7-A in which 180 kW of neutral beam injection (NBI) had sustained a plasma with several hundred eV ion and electron temperatures after the Ohmic heating current was turned off and, simultaneously, the helical currents were brought up to maintain the rotational transform. Culham also had a currentless plasma in CLEO using 12 kW electron cyclotron resonance heating (ECRH), but their ions remained cold and electron energy losses were high. ECRH at 150 kW was used in JIPPT-II, a hybrid machine at Nagoya that could operate as either tokamak or stellarator. Before long, Kyoto [29] followed suit with 200 kW ECRH in Heliotron-E and from that time, W7-A and Heliotron-E took stellarator research into the parameter range where the tokamak had operated since 1968. Eventually, both were equipped with several MW of heating power, divided between ECRH, ion cyclotron resonance heating (ICRH), and neutral beam injection (NBI), and with pellet injectors for density profile control. How this point was reached is discussed below. The parameters of major stellarators of this period are summarized in table 5.3.

In the years after Novosibirsk, much progress was made in understanding the wide range of coil and field configurations that formed the stellarator

* Tokamaks began to apply additional heating from 1973; the techniques will be described in chapters 6 and 7.

Table 5.3 Stellarators of the 1970s and 1980s

Location	Name	Operation	R_0 (cm)	a (cm)	B_t (T)	l/m	Type
Moscow (Lebedev)	L2	1975	100	11.5	2	2/14	Classical stellarator
Kharkov	Saturn-1	1970–80	36	6.3	0.7	3/8	Classical stellarator/torsatron
	Vint-20	1972–80	31	7.5	1.0	1/13	
	Uragan	1968–76	110	6.7	1	3/10	Racetrack
	Uragan 2	1976–81	110	6.7	2	3/18	Racetrack
	Uragan 3	1981–90	100	12	1.5	3/9	Helical divertor torsatron
Nagoya [30]	JIPP Ia	1970	50	5	0.4	3/8	
	JIPP Ib	1973	50			2 + 3	
	JIPP T-II	1979	91	17	3	2/4	Classical stellarator*
Kyoto [31]	Heliotron D	1971	108	7	0.3	2/25	Torsatron
	Heliotron DM	1976	45	<6	1	2/21	Torsatron
	Heliotron E	1980	220	20	2	2/19	Torsatron
Tokyo [32]	SHATLET-M	1988	42	~5	0.15	2/12	Torsatron
Garching	W2-A†		50	5	0.6	2/5	Classical stellarator
	W2-B†	1971	50	5	1.25	2/5	Classical stellarator
	W7	1975	200	10	3.5	2/5	Classical stellarator
Grenoble	WEGA***	1978–83	72	15	1.4	2/5	Classical stellarator‡
Culham	CLEO	1974	90	12.5	2	3/7	Classical stellarator‡
	TORSO	1972	40	5	2	3/12	Torsatron
Madison	Proto Cleo§	1974	40	5	0.3	3	Classical stellarator/torsatron
	IMS	~1984	40	4	0.7	3/7	Modular stellarator
Canberra	Shiela	1984–93	20	3	0.2	–/3	Heliac

* JIPP T-2 operated as a tokamak (see table 6.1) as well as a stellarator. In 1983 it was converted to the JIPP T-2U tokamak.

† W-2-A operated with Ba plasma, W-2-B with hydrogen plasma. ‡ WEGA and Cleo also operated as tokamaks (see table 6.1).

§ Proto Cleo was previously at Culham (see table 5.1). *** WEGA was moved to the University of Stuttgart and later to Greifswald.

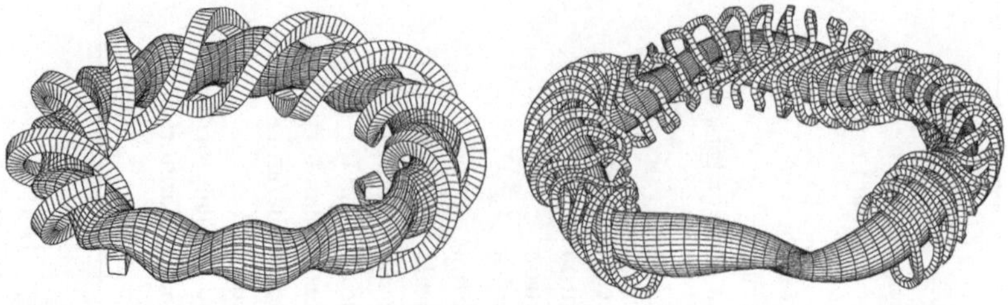

Figure 5.10. Coils and magnetic surfaces for: left, $l = 2$ Torsatron; right, five-period advanced stellarator heliacs with modular coils.

family.* The classical stellarator had a toroidal magnetic field produced by toroidal field coils, as in a tokamak, and a poloidal field with l-fold symmetry produced by a set of $2l$ toroidally continuous helical windings with currents flowing in opposite directions in adjacent windings. Adding poloidal-field coils to produce a vertical field allowed the magnetic axis to be shifted towards a minimum-B configuration. This arrangement has the advantage of flexibility and the possibility to vary the toroidal and poloidal fields independently but the disadvantage that the helical and toroidal coils sets are inter-linked, making assembly and maintenance difficult.

The currents in these coils are to some extent interchangeable. Thus the 'torsatron' scheme, proposed at the Novosibirsk Conference by Gourdon from Fontenay-aux-Roses,† dispensed with the toroidal-field coils and has the helical currents flowing in a unidirectional set of l helical coils to generate both rotational transform and toroidal field. The basic torsatron retains the poloidal-field coils to cancel the vertical field produced by the helical coils. To construct an 'ultimate torsatron' with only a helical coil (figure 5.10), one must specify the 'winding law'—the way in which the pitch angle varies in the poloidal cross-section (helical coils with a smaller pitch angle on the inside or the outside have the effect of poloidal-field coils located on the inside or on the outside, respectively). The torsatron loses some of the flexibility of the classical stellarator but it gains considerably in engineering simplicity.

New stellarators were designed to explore these new configurations. The torsatron scheme was utilized in different forms in Kharkov, Culham and

* There is no uniformly accepted usage, within the fusion community, of generic and specific names. Spitzer invented the two basic schemes to generate rotational transform in toroidal vacuum fields—helical axes and helical coils—and called both of them stellarators, so we use this as the generic name for all devices based on these principles—but we note also that 'helical systems' has come into use as an all-embracing term. Stellarators with unidirectional helical currents were called torsatrons by the French inventors and are known as heliotrons in Japan. For consistency of style, we will refer to torsatrons/heliotrons simply as 'torsatrons' and to devices with currents in alternate directions in the helical coils as 'classical stellarators'.

† The torsatron appears to have been suggested independently in 1961 by Aleksin (unpublished) from Kharkov.

Nagoya. It was also adopted in Kyoto as a natural development of the heliotron configuration (section 5.1). Heliotron-E, which came into operation in 1980, provided a solid experimental basis for the torsatron line and would lead to the construction of the Compact Helical System (CHS) in Nagoya and later of the Large Helical Device (LHD) in Toki. In the USSR, the Lebedev Institute had brought into operation in 1975 the L-2 machine, an $l = 2$ stellarator with unusually steep, short-pitch ($m = 14$) helical windings giving fairly strong shear compared with other $l = 2$ stellarators. Kharkov, after having worked with the small torsatrons Saturn and Vint and the racetrack machines Uragan 1 and 2, commissioned the $l = 3$ torsatron, Uragan 3, in 1982.

As stellarators grew in size and field strength, helical coils interlinked with each other and with the plasma column became ever more difficult to construct. The rotational transform could also be produced, however, either by planar coils with their centres offset with respect to the toroidal axis, or by non-planar 'warped' coils—Wobig coils (figure 5.9).

Letting the magnetic axis twist around the toroidal axis (box 2.3) yields a further degree of freedom. An extreme example of such 'heliacs' is TJ-II at CIEMAT in Madrid, which has a ring-shaped conductor along the toroidal axis for additional flexibility in adjusting the field configuration.

After studying the conventional stellarator W7-A, Garching turned to the modular coil concept for future machines and developed sophisticated computational techniques that enabled them first to specify a field shape optimized for its plasma properties and then to design the coils that would produce the desired field. This approach, which had been initiated by Arnulf Schlüter, revolutionized stellarator design [33]. By making the field approach Palumbo's isodynamic property [34] (the magnetic field strength being constant on a magnetic surface, or along a field line), they could greatly reduce the secondary (Pfirsch–Schlüter) currents* resulting from poloidal variation of $\int \mathrm{d}l/B$, while at the same time narrowing the width of the banana orbits that are responsible for neoclassical transport (section 6.8). Also minimizing the bootstrap current, they succeeded in largely eliminating the longitudinal currents that plagued earlier stellarator configurations. In line with their successful earlier Wendelstein experiments, they opted for low-shear fields to avoid major resonances, relying on magnetic well depth for stabilization. As a first step, they started the design of an 'Advanced Stellarator' W7-AS to succeed W7-A.

Stellarators even made a cautious return to the USA where fusion research expanded rapidly during the 1970s and several universities either entered the field or initiated new programmes. The University of Wisconsin at Madison launched an experimental stellarator programme in the mid-1970s by taking over Proto-Cleo from Culham and also included a modular-coil stellarator

* The longitudinal current equalizing the transverse charge separation shown in figure 2.9.

in its series of fusion reactor design studies. A joint Los Alamos–Princeton study likewise assumed modular coils, while MIT and Oak Ridge, along with Kharkov, conducted reactor studies based on the torsatron. Madison and Oak Ridge both proposed large experiments and eventually Oak Ridge received support for construction of the Advanced Toroidal Facility (ATF). This machine was designed to optimize beta in the so-called second stability regime, in which the plasma was predicted to modify the magnetic field so as to dig its own magnetic well. A further point to which, particularly in the US, reactor designers were sensitive was the size of a reactor, and in this respect the torsatron offered the prospect of a more compact torus with lower aspect ratio than current, more slender stellarators. Thus, while constructing ATF with aspect ratio $A = 8$, Oak Ridge studied the design of an ATF II with $A = 4$, though this was never built.

In terms of their absolute performance, stellarators were completely overshadowed after 1983 by the results from the big tokamaks, but they continued to make steady progress in terms of detailed physics. Indeed there was an upsurge of interest because of their potential advantages as steady-state reactors and because of the potential contributions that they could make to the understanding of toroidal confinement.

Three major new experiments came into operation in 1988: ATF in Oak Ridge and CHS at Nagoya, both torsatrons, and W7-AS at Garching, an advanced stellarator with modular coils. ATF, although at the time hailed as the largest stellarator in the world, had a somewhat troubled and brief life. Its start-up having been delayed by engineering problems,* it became a victim of tightening budgets for fusion in the USA. First mothballed and then restarted for a brief campaign, it was closed down permanently after only a few years of actual operation. CHS was successful in obtaining high beta plasmas (2.1% average beta) and also clarified the important role in stellarator confinement played by the electric field. CHS operated successfully in Nagoya until 1999 when it was moved to the new laboratory of the National Institute for Fusion Studies (NIFS) in Toki.

These experiments fall into two broad classes: the torsatrons with helical windings and significant shear (Heliotron-E, ATF, CHS, Uragan 2M and 3M) and the modular-coil machines with low shear (W7-AS, H-1 and TJ-II). W7-AS could vary the shear, plasma shape and location of the magnetic axis with supplementary toroidal and vertical-field coils. Similar coils allowed Uragan 2M and Heliotron-E to change the toroidal component of the field and thus the transform.

A main issue for stellarator research, as for tokamaks, is energy confinement. To clarify this, an international database of energy confinement in stellarators was compiled [35] in 1995 from ATF, CHS, Heliotron-E, W7-A

* One of the disadvantages of stellarators is the more complex engineering compared with tokamaks.

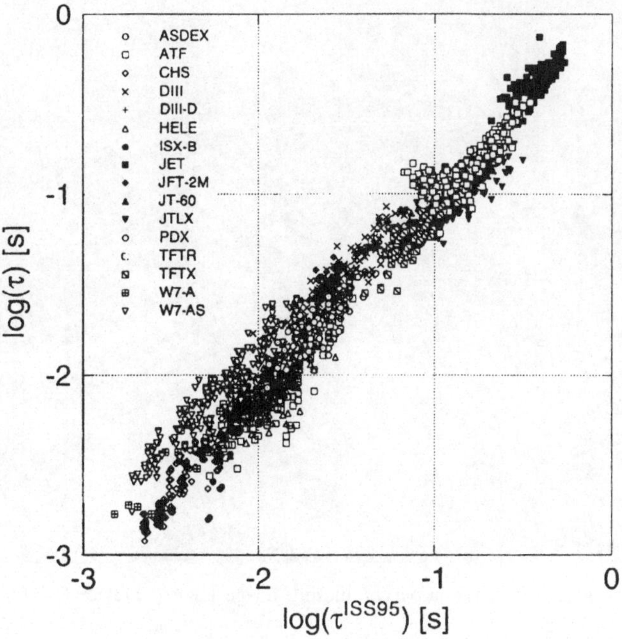

Figure 5.11. Experimental confinement times from the stellarator and L-mode tokamak data bases plotted against the predictions of the ISS95 stellarator scaling expression given in the text.

and W7-AS. It was found necessary to introduce an additional arbitrary parameter in order to combine the data from the heliotron/torsatrons and the shearless stellarators into a single scaling (figure 5.11) of the form

$$\tau_E = 0.079a^{2.21} R_0^{0.65} P_{tot}^{-0.59} n_e^{0.51} B_t^{0.83} \iota^{0.4}.$$

In this expression, ι denotes the rotational transform at 2/3 of the minor radius. One notes, for comparison, that the L-mode scaling for tokamaks (section 7.6) if expressed in terms of ι, would yield

$$\tau_E = 0.04a^2 R_0^{0.35} P_{tot}^{-0.5} n_e^{0.1} B_t^{1.05} \iota^{0.85},$$

which is in reasonably good agreement. Improved design should, however, yield more favourable scaling and the results from W7-AS and LHD suggest that this may indeed be the case.

The experience and results from H-E, ATF and CHS were used to optimize the design of the Large Helical Device (LHD) (figure 5.12). This new machine—the world's largest stellarator—which started operation at the Toki laboratory in 1999, uses superconducting coils for all windings. The first results [36] obtained within the first year of operation were impressive with electron and ion temperatures above 3 keV, density in the range 10^{19} to 10^{20} m^{-3}. The energy confinement time, in the range of 0.1 to 0.3 s, is 50% higher than the empirical stellarator scaling.

Figure 5.12. Overview of the superconducting Large Helical Device (LHD) at the Toki Laboratory in Japan.

The torsatron Uragan 2M in Kharkov came into operation in 1995 and the heliacs, H-1 in Canberra and TJ-II in Madrid, in 1993 and 1999 respectively. All had external heating in various combinations of NBI, ECRH and ICRH. The parameters of these machines, as well as those of Uragan 3 and Heliotron-E, both of which have remained in operation, and W7-X (figure 5.13), the big new machine under construction in a new laboratory at Greifswald in Germany, are summarized in table 5.4 [37].

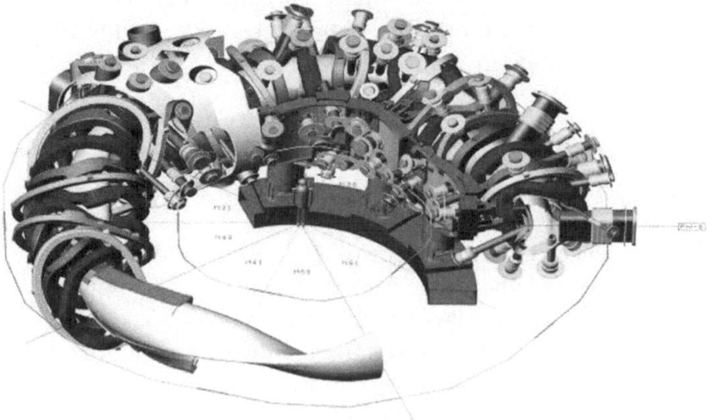

Figure 5.13. Schematic of the W7-X modular stellarator due to come into operation at Greifswald, Germany, around 2006.

Table 5.4 Major stellarators after 1988*

Location	Device	Year†	Type	R_0 (m)	a (m)	B (T)	l/m	Remarks
Nagoya	CHS	1988	Torsatron	1.0	0.2	1.5	2/8	Moved to Toki in 1999
Oak Ridge	ATF	1988	Torsatron	2.1	0.27	2	2/12	Closed early 90s
Garching	W7-AS	1988	Modular	2.0	0.17	2.5	–/5	Advanced stellarator
Kharkov	Uragan 2M	1995	Torsatron	1.7	0.22	2.4	2/4	
	Uragan 3M	1990	Torsatron	1.0	0.125	1.5	3/9	Heliotron divertor
Canberra	H-1	1993	Heliac	1.0	0.2	1	–/3	
Madrid	TJ-II	1999	Heliac	1.5	0.22	1	–/4	Central ring
Toki	LHD	1998	Torsatron	3.6–3.9	0.65	3	2/10	Heliotron divertor
Greifswald	W7-X	~2006	Modular	6.5	0.65	3	–/5	In construction‡

* Heliotron-E, IMS, L-2 and Shiela (see table 5.3) also remained in operation after 1988.

† With the exception of ATF (now closed) and W7-X (under construction) all are operating at the time of writing.

‡ W7-X is expected to reach deuterium plasma parameters comparable to those obtained around 1990 in the large tokamaks.

Chapter 6

The dash to tokamaks

The 1968 IAEA Conference in Novosibirsk had been a great success for the Russian organizers, not least because it focused everyone's attention on the spectacular performance of their tokamaks. Surely, there remained room for questions about the temperature measurements, as was stressed in particular by the colleagues and competitors at Princeton. But there was no question that, at least for the moment, the Moscow team had shown the clearest path towards thermonuclear conditions and the tokamak became what the theta pinch had been since the 1958 Geneva Conference: the best way to produce a hot plasma for the study of magnetic confinement. This chapter covers the performance of tokamaks from 1968 to 1978, along with some general aspects of tokamak development that span a longer time interval, but have been collected here.

6.1 The tokamak goes abroad

After Novosibirsk, two events would finally convince even the strongest critics that the performance claimed for the tokamak was genuine. As we saw, the Culham expedition to Moscow removed any doubt about the electron temperature, but in visits to Western Europe and the US, which culminated in a lecture series at MIT, Artsimovich had already presented a very strong case.

At MIT, Allis, Brown and Rose had established a strong plasma physics tradition, and had been looking into fusion research from their backgrounds in gas discharges, microwaves and nuclear engineering. Moreover, the National Magnet Laboratory had expertise in producing strong magnetic fields. Now, a new generation was preparing to launch a coherent fusion-research programme and to construct their own experimental device, so Artsimovich's lectures fell into fertile soil. His visit became a triumphant success [1]; not only did MIT start construction of the first in the Alcator series of high-field tokamaks, which would push Ohmic heating to the limit, but other laboratories in the US as well [2] entered the tokamak arena.

Princeton converted their existing C-stellarator into a tokamak, renaming it ST, and built a smaller new tokamak, ATC, to study heating by both adiabatic toroidal compression and neutral beams. Over the next few years, Princeton embarked on an aggressive programme of tokamak building with two larger machines: PLT, a 1 MA machine due in 1975, and PDX, to study poloidal divertors, due in 1978. Oak Ridge National Laboratory, hitherto devoted mainly to mirror machines, also moved into tokamaks, initially building ORMAK, equipped with beam heating, and later ISX, specifically constructed to study impurity problems. At San Diego on the west coast of the USA, the General Atomics laboratory initiated an original line of extremely elongated tokamaks with strongly indented cross-section—so strongly indented in fact that the so-called 'doublet' looked almost like two tokamaks one on top of the other. This configuration had grown out of the toroidal multipoles when Tihiro Ohkawa proposed adding a toroidal plasma current. The Doublet was expected to have higher stable beta values than tokamaks with circular cross-section because of its higher magnetic shear. The first small experiment Doublet I [3] was followed by the larger Doublet II, in which the magnetic boundary was shaped by a copper shell, and Doublet IIA, which used field shaping coils to provide more flexibility [4].

The same thing happened in Europe, Japan and other countries, as well as in the Soviet Union. The boldest step forward was taken by the French who, without preliminaries, built TFR, the Tokamak at Fontenay-aux-Roses. TFR had a thick copper shell for stability and was similar in physical size to T-3 but, with a higher toroidal magnetic field, could reach record-breaking plasma currents of 400 kA. Other laboratories practised on a smaller scale before committing themselves to such major investments. The German fusion laboratory at Garching built a smaller device, Pulsator, capable of 100 kA, and the Italians at Frascati built the small TTF, devoted to turbulent heating, and the high-field Frascati Tokamak FT. The British at Culham interrupted the construction of their long-awaited new stellarator CLEO, leaving out the helical winding, and operated it in 1972–73 as a tokamak with neutral beam heating while they constructed their first purpose-built tokamak DITE.

The Russians continued to set the early pace with an expanding and diverse programme of machines. T-3 was rebuilt with enhanced current capacity as T-4, and a number of new tokamaks were constructed to study specific issues. The main theoretical predictions concerning plasma equilibrium were confirmed experimentally in T-5, which was equipped with special coils to set up a time-varying vertical magnetic field. Feedback control of the plasma position using a vertical magnetic field and no copper shell was demonstrated for the first time in the TO-1 tokamak and this technique was adopted for the majority of subsequent tokamaks. The series of smaller TM tokamaks led to much fruitful research: TM-1 was

Table 6.1 Tokamaks built in the first decade after Novosibirsk

Location	Name	Operation	R_0 (m)	a (m)	B_t (T)	I (kA)	Remarks
USA							
MIT [6]	Alcator A	1972–78	0.54	0.10	9.0	300	High B_t, high j
	Alcator C	1978–86	0.64	0.16	13	800	
Princeton	ST	1970–74	1.09	0.14	4.4	130	Formerly C-stellarator
	ATC [7]	1972–76	0.88	0.11	2.0	50	Initial plasma
			0.38	0.17	4.7	118	Compressed plasma
	PLT [8]	1975–86	1.32	0.42	3.4	700	First '1 MA' tokamak
	PDX [9]	1978–83	1.4	0.4	2.4	500	4 and 2 null divertors. Became PBX (table 7.1)
Oak Ridge [10]	ORMAK	1971–76	0.8	0.23	2.6	230	Neutral beam heating
	ISX-B	1977–84	0.93	0.27	1.6	250	$\kappa = 1.2$–1.6
GA [11], San Diego	Doublet II	1972–74	0.63	0.08	0.8–0.95	90–210	Cu field shaper, $\kappa \approx 1.5$
	Doublet IIA	1974–79	0.66	0.15	0.76	<350	Field shaping coils
	Doublet-III	1978–85	1.45	0.45	2.6	0.61	Operated jointly with JAERI. Became D-IIID (table 7.1)
Europe							
Fontenay [12]	TFR	1973–78	0.98	0.20	6.0	400	Became TFR-600
	TFR-600	1978–86	0.98	0.22	6.0	600	With Cu shell removed
Grenoble [13]	WEGA	1975–78	0.72	0.15	2.2	80	Also stellarator (table 5.3)
	Petula-B	1974–86	0.72	0.18	2.7	230	
Garching	Pulsator [14]	1973	0.7	0.12	2.7	125	

Frascati	TTF	1973	0.3	0.04	1.0	5	
	FT [15]	1978	0.83	0.20	10.0	800	High field
Culham	CLEO	1972–73	0.90	0.18	2.0	120	Rebuilt as stellarator
	Tosca [16]	1974	0.3	0.1	1.0	20	
	DITE [17]	1975–89	1.17	0.27	2.7	260	Bundle divertor
Japan							
Naka [18]	JFT-2	1972–82	0.9	0.25	1.8	170	
	JFT-2a/DIVA	1974–79	0.60	0.10	2.0	70	Divertor, $\kappa = 1.4$
Nagoya	JIPP-T2	1976	0.91	0.17	3.0	160	
Tokyo	TNT-A	1976	0.4	0.09	0.42	20	$\kappa = 1.4$
USSR							
Moscow [19]	T-4	1974–78	0.90	0.16	5	300	Was T-3, replaced by T-7
	T-7	1979–82	1.22	0.35	2.4	390	Superconducting coils. Moved to Hefei
	T-5	1962–70	0.625	0.15	1.2	60	Moved to St Petersburg as FT-1
	T-6	1970–74	0.7	0.25	1.5	220	Ultra-low q—replaced by T-11
	T-11	1975–84	0.7	0.25	1.0	170	Moved to Troitsk
	T-9	1972–77	0.36	0.07	1.0		Elongated cross-section
	T-12	1978–85	0.36	0.08	1.0	30	
	T-10	1975–date	1.5	0.39	5.0		
	TO-1	1972–78	0.6	0.13	1.5	70	Feedback control
St Petersburg [20]	TUMAN-2	1971–75	0.4	0.08	0.4–1.2	8	Minor radius compression
	TUMAN-2A	1977–85	0.4	0.08	0.7–1.5	12	Compression
	FT-1	1972–2002	0.62	0.15	0.7–1.2	30–50	Was T-5 at Kurchatov

built to study heating by adiabatic compression, TM-2 (later modernized to TM-3) discovered the main tokamak instability—the disruption.

In Japan, two tokamaks—JFT-2 and DIVA—were built at the JAERI site near Mito and others at Nagoya and Tokyo; some characteristics of the main tokamaks of the period 1968–78 are listed in table 6.1 [5].

Not only did most institutes for fusion research turn at least part of their efforts to tokamak physics, but several university laboratories also entered this field and by 1986 the IAEA [21] counted more than 70 tokamaks world-wide, some of which were small devices devoted primarily to teaching. During the next three decades, the tokamak would become the mainstream of fusion research, resulting in continuously increasing sizes and levels of performance (figure 6.1) and in a rapidly expanding data base, which in

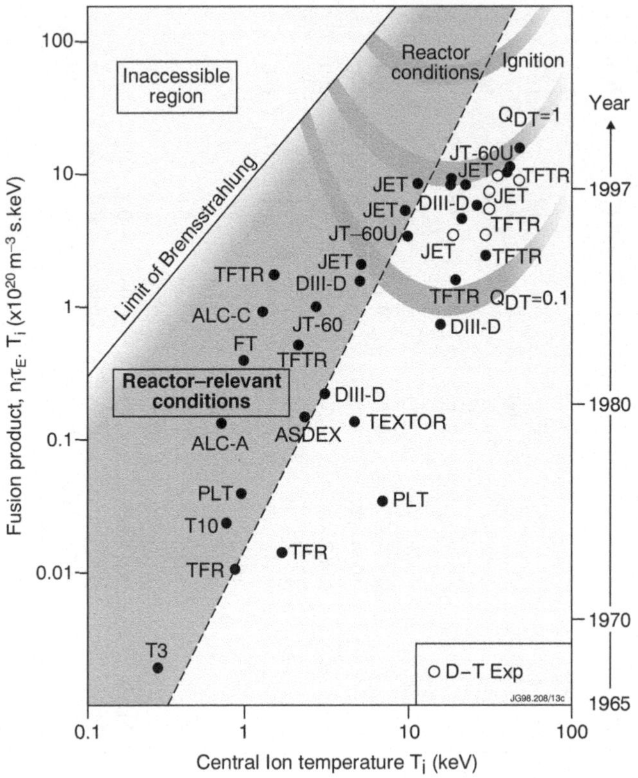

Figure 6.1. Tokamak performance in the period 1965 to 1997 is illustrated by plotting the fusion triple product $n_i\tau_E T_i$ (see box 8.2) against the central ion temperature $T_i(0)$ for representative experiments. The U-shaped regions at the top right hand corner indicate the values for ignition, for $Q_{DT} = 1$ and for $Q_{DT} = 0.1$ respectively. Note that most of the data (those represented by solid points) are in D–D plasmas and have been scaled to D-T using known values of the fusion cross-sections—only the open points were measured directly in D–T plasmas.

turn supported the design of still larger devices. The construction of ever bigger machines soon exceeded the capacities of all but the largest institutes and would eventually exceed the capacity of even the national programmes. So, the design studies for proof-of-principle or break-even experiments that were initiated in the early 1970s first required and indeed were accompanied by the strengthening of national—and within Europe international—coordination and in the next stage brought about world-wide cooperation.

6.2 Neutral beam heating

In contrast to rapidly pulsed discharges, in which shock heating or adiabatic compression may yield high transient temperatures, steady-state devices like stellarators need special provisions for producing a plasma of any temperature. Tokamaks belong to an intermediate class of quasi-steady-state devices in which the currents needed to sustain the field configuration deliver at least some power to the plasma. Early tokamak experiments relied entirely on Ohmic heating and achieved temperatures in the 100 to 1000 eV range.

Box 6.1 Neutral beam injection (NBI) heating

A beam of energetic neutral atoms injected into a plasma is not deflected by a magnetic field. When the particles are ionized by collisions with the plasma particles, the resulting energetic ions become trapped and the plasma is heated as these 'fast ions' slow down and transfer their excess energy by collisions with the thermal ions and electrons. The most critical requirements for efficient heating are that the beam must penetrate to the plasma core and that the fast ions must be confined for long enough to transfer a high fraction of their energy to the thermal plasma. The depth of penetration is determined by the beam energy and by the temperature and density. Generally penetration is not a problem in small and medium sized experiments (indeed a serious concern in the early experiments was the risk of over-penetration and damage to the far wall) but increasingly higher beam energies are required for larger devices. In most cases the beams are oriented to inject along a direction more or less tangent to the toroidal direction since this reduces the problem of losses of fast ions trapped in the ripple of the toroidal field. In smaller tokamaks the confinement of the fast ions was a concern because their orbits in the poloidal magnetic field were comparable with the plasma dimensions, but this was amply overcome in the larger tokamaks designed to confine the alpha particles.

Although the high-field tokamaks at MIT and Frascati could get to higher temperatures with Ohmic heating only, Coppi's claim [22] that this path would lead to ignition assumed a favourable scaling of τ_E which had not been established. Moreover, such a reactor would require magnetic fields as high as 20 T, which would preclude the use of superconducting coils. Even copper coils would be extremely difficult to engineer in the large sizes needed for a high-field reactor. The mainstream of tokamaks stayed with toroidal fields below about 5 T and clearly would require additional heating to get to burning temperatures. Two main routes were developed: first the injection of energetic neutral atom beams and, slightly later, the resonant absorption of radio-frequency electromagnetic waves (section 7.2).

Powerful neutral beam sources were being developed already at Berkeley and Oak Ridge [23] for mirror research, their main purpose being to fill up the magnetic bottles. Conveniently for the fledgling tokamak programme, their particle energy of some 15 to 20 keV was in the range required for heating the plasmas in current tokamaks. By 1973 the groups working on ORMAK at Oak Ridge, ATC at Princeton (using sources from Berkeley) and CLEO at Culham (using sources descending from the Oak Ridge design) [24], were all installing beam heating. In these preliminary experiments, short pulses of neutral beams with typical powers of 80 kW were injected into target plasmas with typically 200 kW of Ohmic heating and the increase in the ion temperature was small (typically in the range 10–15%). An important observation was that the measured energy spectrum of the injected ions showed that beam trapping, fast-ion confinement and slowing-down were all in good agreement with theory. Progress was rapid and in a year ATC had increased its beam power to 100 kW and had improved the efficiency so that the ion temperature rose from the Ohmic-heating value of 200 eV to over 300 eV. This was followed up at Fontenay [25], the Kurchatov [26] and many other institutes, and all found neutral beams to be effective in raising the ion temperature. But the bigger tokamaks of the future would require far higher particle energy and beam power, and great efforts were made in several laboratories to develop ever more powerful sources.

The most spectacular application came a few years later in the Princeton Large Torus (PLT), when the ion temperature was pushed above 5 keV with 2 MW of beam heating [27]. The Princeton results were particularly interesting because the trapped-ion instabilities predicted by theory failed to show up as the plasma entered into the regime of low collisionality where they were expected to arise. True, the confinement time was somewhat disappointing, but this was not immediately recognized as a new trend. At about the same time, MIT established a favourable scaling of the energy confinement time with density in Alcator [28] and reached $n\tau \approx 3 \times 10^{19}\,\mathrm{m}^{-3}\,\mathrm{s}$ with Ohmic heating only. So, within ten years after Novosibirsk the American tokamak programme had come to fruition.

6.3 Disruptions and density limits

From the earliest experiments it was clear that tokamaks were subject to a variety of macroscopic instabilities, arising from the profiles of plasma current and plasma pressure. The MHD modes fell into two categories: ideal modes that were unstable in a perfectly conducting plasma and resistive modes that were unstable only as a result of the finite conductivity of the plasma.

A major disruption of the plasma column showed up as a sharp reduction of the current, accompanied by a negative voltage spike, and usually resulted in an abrupt termination of the discharge (box 6.2). Minor

Box 6.2 Disruptions

A major disruption in a tokamak is a dramatic event in which the toroidal current abruptly terminates and confinement is lost. It is preceded by a well defined sequence of events with four main phases. First there are changes in the underlying tokamak conditions—usually an increase in the current or density, but not always so clearly identified. When these changes reach some critical point, the second phase (the precursor) starts with the onset (or increase) of MHD activity. Usually this is most obvious as a growing $m = 2$ mode (detectable with external magnetic pick-up loops), whose growth time is variable but typically of the order of 10 ms. The sequence then passes a second critical point and enters the third phase, where events move on a much faster time scale—typically of the order of 1 ms. The confinement deteriorates and the central temperature collapses. The toroidal current profile flattens and the change in inductance gives rise to the characteristic negative voltage spike—typically 10 to 100 times larger than the normal resistive loop voltage. Finally comes the current quench—the current decays to zero at rates that can exceed 100 MA s^{-1}. Whether the current and the plasma recover from a disruption depends in part on the volt-seconds remaining available in the transformer. Disruptions cause large forces on the vacuum vessel that increase with machine size. In the big tokamaks forces of several hundred tonnes have been measured and a reactor would have to withstand forces at least an order of magnitude higher.

Attempts to control disruptions have met with only limited success. Applying feedback control to the growing $m = 2$ mode has delayed the onset of the third phase but only marginally. Reactor designers have proposed that disruptions should be avoided at all costs, but this seems unrealistic. A possible remedy is the injection of a so-called 'killer pellet' to cool the plasma by radiation before the onset of the current quench.

disruptions were characterized by small, often repetitive dips in the current. As we have seen already, early work in the Kurchatov Institute had established a link with the q-value. First suggestions were that a steepening of the current profile, by which q on the axis (q_0) decreased to one (thereby violating the Kruskal–Shafranov limit) was responsible for the major disruptions. However, Mirnov [29] noted that perturbations with poloidal mode number m equal to q at the limiter (q_a) developed progressively when q_a passed through integral values, 6, 5, 4, 3, 2 during the current rise and that those with $m = 3$ or 2 would grow and move inwards during the current plateau, eventually leading to the current termination. The Princeton ST team confirmed Mirnov's observations that q at the limiter rather than the plasma centre was linked to disruptions and, along with other groups, refined the picture and found it to be in broad agreement with resistive MHD theory [30]. However, the detailed physics of disruptions is complex and some aspects still remain incompletely understood, even though the basic sequence of events is now fairly well described [31].

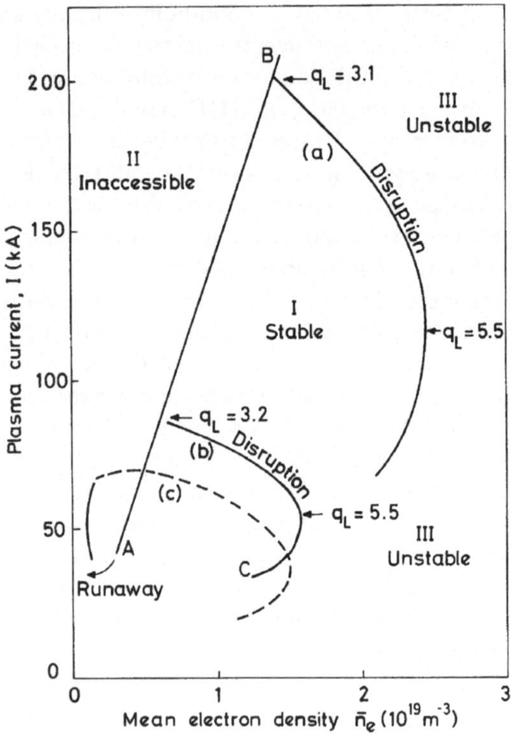

Figure 6.2. Hugill diagram showing a typical operating regime limited by disruptions at low edge q_a (q_L in the figure) or at high average electron density density. The diagram can also be plotted in a normalized form if both axes are multiplied by R_0/B_t—so that the vertical axis becomes $1/q$ and the horizontal axis becomes $n_e R_0 B_t$.

In clean discharges with moderate densities, disruptions usually occurred when the current was increased so that q_a fell under 3 and certainly by the time it reached 2. But in less clean discharges, disruptions frequently occurred even when $q_a > 3$. Murakami at Oak Ridge [32] recognized that there was a density limit that scaled proportional to B_t/R_0 and decreased with increasing impurity content. This suggested that these disruptions were caused by Ohmic heating failing to balance radiation loss (which increases as n_e^2) so that the effective plasma edge shrank inside the limiter radius. Hugill at Culham saw that the critical density was proportional to the current and proposed a universal diagram (figure 6.2) of normalized current, $1/q_a$, versus normalized density, $n_e R_0/B_t$, which clearly showed the boundary between stable and disruptive discharges at both the high-density and high-current limits [33]. The radiation-cooling model predicted that the density limit would go up with additional heating. This was observed to some extent already in ORMAK but would become more apparent in later years when more heating power was available. The model also predicted that the density limit would increase as impurities were reduced, and this was confirmed by titanium gettering experiments (section 6.5) and later by experiments with pellet refuelling. A new empirical scaling, the Greenwald limit, was established in the 1980s.

6.4 Sawteeth

The first time a significant new tokamak phenomenon was discovered outside Moscow was when a soft X-ray detector with good spatial and time resolution was installed on ST. The X-ray emission from the plasma core (figure 6.3) showed periodic 'sawtooth-like' cycles [34] which were identified as being due to a slow ramp-up of the central electron temperature followed by a rapid flattening of the temperature profile in the plasma core. These 'sawtooth oscillations' were reported also in T-4 and ATC and soon were seen in every tokamak. Indeed, sawteeth became the accepted sign that a new tokamak had reached respectable operating conditions where the core was hot enough to become unstable. Instead of being associated with major disruptions as first thought, the value of q in the plasma core, $q_0 < 1$ became the condition for the occurrence of sawteeth. It was realized that the observed collapse time ($\sim 100\,\mu s$) was much shorter than the characteristic resistive time associated with the radius at which the sawteeth were observed—this was assumed to be the $q = 1$ magnetic surface. Kadomtsev [35] developed a theory in which magnetic field lines in a narrow annular layer would reconnect rapidly—within the geometric mean of the resistive diffusion and Alfvén times. Many authors did further theoretical and computational work along this line and generally found their results to be in good agreement with current experimental results [36].

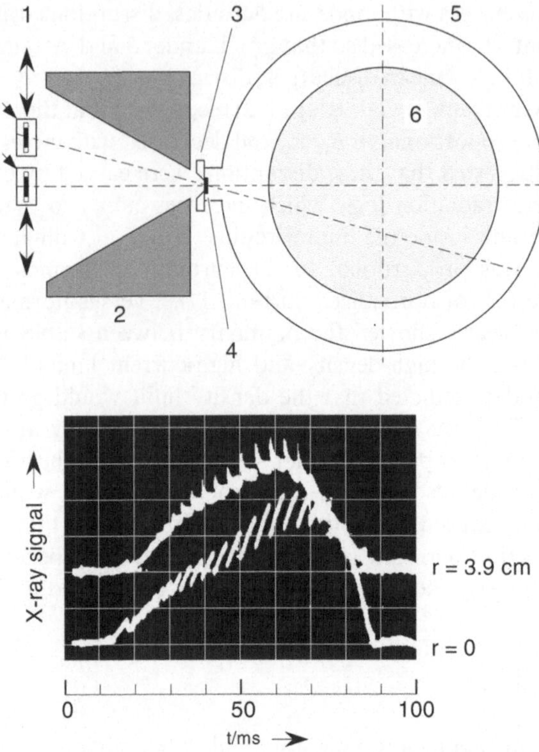

Figure 6.3. Sawtooth oscillations. Top: experimental arrangement used in the first observations in the ST experiment with 1, movable soft X-ray detector; 2, lead shield; 3, slot aperture; 4, Be window; 5, vacuum vessel; 6, plasma. Bottom: oscillograms taken at different positions of the detector.

The first spatial scans on ST were effected on a shot-by-shot basis by moving the detector behind a collimating slot. Later, tomographic methods using arrays of detectors and multiple viewing directions made it possible to invert the X-ray emission—that is to reconstruct the spatial distribution of the radiation source. The oscillations could also be observed with other diagnostic methods, in particular electron cyclotron emission (figure 6.4). The sawtooth crash was preceded by a growing temperature oscillation localized at the $q = 1$ surface. Tomography of the core displayed hot and cold plasma 'bubbles' interchanging position, more or less as described by Kadomtsev's model. After the crash, the temperature outside the $q = 1$ surface rose and a 'heat pulse' propagated towards the plasma edge. With increasing sensitivity of ECE and X-ray diagnostics, the propagation of this heat pulse became an important indicator for the local transport coefficients.

During the 1970s, sawteeth appeared to be one of the success stories in tokamak physics; they were extensively studied experimentally and seemed

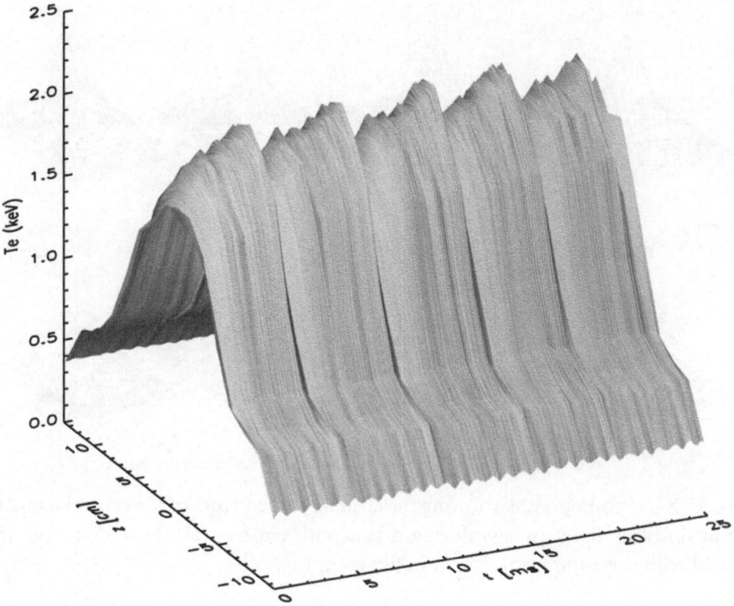

Figure 6.4. Electron temperature profiles in the Textor-94 tokamak, measured with an electron cyclotron imaging (ECEI) technique, operating at the second harmonic of the electron cyclotron frequency. The measurements show sawtooth oscillations with a periodicity of about 5 ms as well as a 1 ms radial oscillation of the plasma outside the temperature pedestal.

in good agreement with Kadomtsev's theory. However, additional heating in some of the larger tokamaks and improved diagnostics threw new light on the phenomenon. The first discrepancy with Kadomtsev's model showed up in TFR [37]. The central temperature after the crash fell below the initial island temperature, prompting a new turbulent-transport model [38]. Further discrepancies appeared in JET (figure 6.5). Kadomtsev's model predicted that the crash time would increase with machine dimensions and should be ~10 ms in a machine as large as JET but the crash was observed to take place on a hundred times faster time scale. Moreover JET sawteeth were not accompanied by the usual precursor oscillations that signalled the growth of the kink instability at the $q = 1$ surface but showed oscillations after the crash [39]. JET sometimes observed sawteeth with very long stable flat-top periods extending for several seconds, the so-called 'monster sawteeth'. Some of these observations were consistent with a new theory, proposed by Wesson, who assumed that a region of nearly uniform q between two $q = 1$ surfaces allowed rapid growth of a kinklike interchange mode.

However, this quasi-interchange model was itself put into doubt when new direct measurements of the poloidal field in TEXTOR [40], using

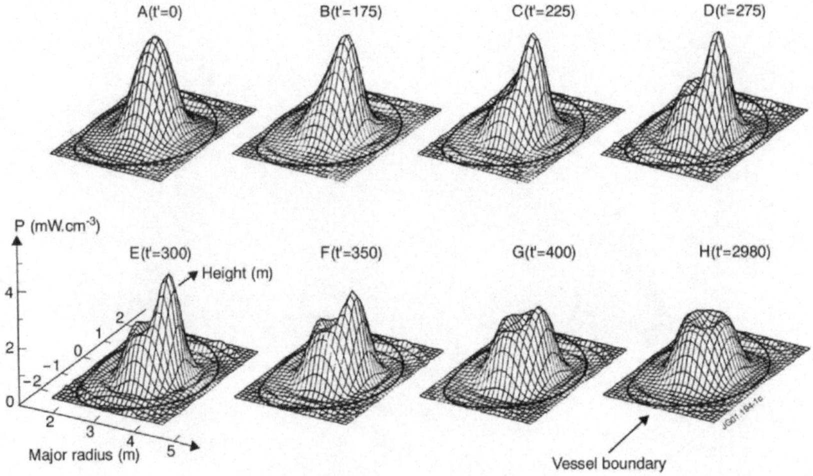

Figure 6.5. X-ray tomography showing the eroding and collapse of the central temperature profile at different times (in μs) during a sawtooth crash in a limiter discharge in JET (measured with the equipment shown in figure 5.9).

Faraday rotation of an infra-red beam, showed that q reached unexpectedly low values ($0.6 < q < 1$) in the core and changed very little during a sawtooth cycle. The observation that $q = 1$ was not a hard stability limit shook the very core of plasma theory. It is remarkable, but characteristic of fusion research, that a phenomenon which appeared initially to have a simple and elegant explanation should be found on closer study and in different experiments to be so complex and so difficult to understand.

6.5 Passing through purgatory

> *It is now clear to all that our original beliefs that the doors into the desired region of ultra-high temperatures would open smoothly at the first powerful pressure exerted by the creative energy of physicists, have proved as unfounded as the sinner's hope of entering Paradise without passing through Purgatory. And yet there can be scarcely any doubt that the problem of controlled fusion will eventually be solved. Only we do not know how long we shall have to remain in Purgatory. We shall have to leave it with an ideal vacuum technology ... bearing in our hands the high temperature plasma, stable and in repose, pure as a concept in theoretical physics when this is still unsullied by contact with experimental fact.*

So Artsimovich concluded his summary speech at the Salzburg Conference in 1961.

T-3 had established that in order to attain good plasma performance it was necessary to control the position of the plasma so that it remained centred in the torus and did not lean excessively on the vessel walls. The toroidal geometry leads to a hoop force in the outward direction—tending to expand the major radius of the plasma ring. This is counteracted by a 'vertical' magnetic field, which in T-3 and other early tokamaks was produced by a combination of currents in external coils and a thick copper shell surrounding the plasma. Image currents induced by the plasma in the copper shell interacted with the plasma current itself to provide the forces needed to keep the plasma centred in the vessel. This was a transient effect and the discharge duration was limited by the skin penetration time of the copper shell. It became impractical to increase the copper shell thickness to keep pace with the requirements of longer discharge duration. The first tokamaks outside the USSR, for example TFR, used a copper shell but later machines dispensed with this and controlled the plasma position relative to the vacuum vessel with an active feedback system using external field coils to generate the transverse magnetic fields and position sensors to determine the plasma position.

The influx of impurities (box 6.3) into T-3 had been reduced by initial baking of the vessel to remove surface contaminants, followed by conditioning

Box 6.3 Sources of impurities

In a reactor, the fusion process is an internal source of helium 'ash'. Moreover, impurities are released from the material surfaces surrounding the plasma by a variety of processes. Under normal conditions, surfaces in vacuum are covered by a layer of molecules adsorbed during exposure to the atmosphere or to previous plasma discharges. There are also impurities, particularly carbon, that were entrained in the bulk of the metal during manufacture and migrated to the surface. These surface contaminants are released by radiation from the plasma, or as a result of sputtering, arcing and evaporation. Sputtering is a process in which energetic ions or neutrals knock atoms from a surface. Arcing is driven by the potential difference between the plasma and the surface. Evaporation occurs when the power load is sufficient to heat surfaces to temperatures near to their melting point—this is often localized in hot spots on prominent edges exposed to plasma. All three mechanisms are important at the limiter and the divertor which are subject to direct plasma contact, but generally the walls are shielded from charged particles and are subject only to sputtering by charge-exchange neutrals. (In a reactor, though, a large fraction of the fusion power will arrive at the wall as radiation.)

using repetitive 'training discharges' (a process now referred to as 'discharge cleaning') and by the use of a metal limiter to localize the contact between the plasma and the wall to a defined point that could be designed to withstand high power loads and could be kept well conditioned.

Until about 1975 most tokamaks had stainless steel walls with molybdenum or tungsten limiters. The plasmas in many of the main experiments of the day, ST, TFR-400, JFT-2, T-4 and Ormak, were characterized by Z_{eff} values (box 6.4) in the range 3 to 8, mainly determined by the low-Z impurities, oxygen and carbon, although high-Z impurities as well contributed to the radiated power loss [41]. Moreover, the average electron densities could not be raised above about $5 \times 10^{19}\,m^{-3}$. However, a few experiments, notably ATC and Alcator, had demonstrated already that it was possible to attain values of Z_{eff} close to unity. Alcator had also reached much higher densities pointing to the causal link between impurity radiation and the limit on density imposed by disruption (although the detailed sequence of events that leads to a disruption at the density limit (box 6.2) was not fully established until later). By the late 1970s the study of plasma–surface interactions had become an important branch of fusion research, to which most tokamak groups made their contributions [42]. The effect of a small concentration of impurity on thermonuclear ignition of a deuterium–tritium plasma was calculated in 1974 by Meade [43] who showed that there is a critical concentration for each impurity species at which the radiation cooling becomes so large that ignition would be impossible.

Just as ceramic tori had begun to give way to metallic ones already before 1958, other parts of the vacuum systems of tokamaks were constructed to new standards of ultra-high vacuum technology to reduce surface contamination. Vacuum technology had made great strides during the war, stimulated by the development of electromagnetic mass separators, and in the years thereafter when vacuum tubes and particle accelerators set the requirements. The issue in fusion experiments was to eliminate all sources of organic molecules that would be broken up in the hot plasma. So, rubber gaskets and vacuum grease gave way to systems with metal seals or welded joints and oil diffusion pumps were replaced by turbomolecular pumps. These vacuum systems could moreover be baked to much higher temperatures and in some tokamaks the walls were even kept at elevated temperatures during operation in order to reduce the problem of re-contamination between discharges.

After the initial bake-out, the more tightly bound molecules had to be removed from the vacuum surfaces by some form of discharge cleaning. Early tokamaks had tended to rely on relatively high-current, unstable tokamak-like discharges, but it turned out that such energetic discharges left impurities in ionized states which could not be pumped out of the torus. Also metal atoms were sputtered from the walls creating fresh active sites to which impurities could immediately reattach. Improved procedures

Box 6.4 Impurity radiation

Impurities in tokamak plasmas introduce a variety of problems. The most immediate effect is the radiated power loss. A convenient parameter to characterize the impurity content is the effective ion charge, $Z_{\rm eff} = \sum_{\rm i} n_{\rm i} Z_{\rm i}^2 / n_{\rm e}$ where the summation is taken over all the ionization states of all the ion species. The bremsstrahlung per ion scales as $Z_{\rm i}^2$, but for given $n_{\rm e}$ the total bremsstrahlung loss scales as $Z_{\rm eff}$. To calculate this quantity, one needs to know all the ion concentrations, so usually it is measured directly from the enhancement of the bremsstrahlung radiation in a spectral region free from line radiation. If $T_{\rm e} = T_{\rm i}$, the margin between bremsstrahlung loss and the fusion energy deposited into the plasma by charged reaction products is not more than a factor of 30 for pure D–T (figure 1.9). Because there are also other loss processes, ignition becomes impossible well before $Z_{\rm eff}$ reaches this value.

Line radiation from partially stripped impurity ions greatly exceeds bremsstrahlung, but it ceases at the temperature where the atoms are fully stripped. Hence, radiative cooling is particularly significant in the start-up phase and at the plasma edge and, as we saw in section 6.3, edge cooling may lead to disruptions. 'Low-Z' impurities such as carbon or oxygen are completely stripped in the core of a reactor, but to reach ignition one must overcome the radiation peak, which for carbon lies around 10 eV. Next, the plasma must burn through the radiation barriers presented by 'medium-Z' impurities such as iron and nickel and 'high-Z' impurities such as molybdenum and tungsten—around 100 eV and 1 keV, respectively. A hydrogen plasma with one part of tungsten in 500 would emit some 100 MW m^{-3} in line radiation at $n_{\rm e} \approx 10^{20}$ m^{-3} and $T_{\rm e} \approx 1$ keV, which is far beyond the heating power—less than 1 MW m^{-3}—that will be installed to reach ignition. In fact, the hydrogenic and helium-like states of refractory metals, in which the nucleus has retained only one or two electrons (tungsten for example, with $Z = 74$, has a K-shell binding energy of 60 keV), continue to radiate very strongly at electron temperatures in the thermonuclear range.

From the point of view of radiation cooling, much larger concentrations of carbon and oxygen could be tolerated in a fusion reactor but then the problem of fuel dilution would arise. The impurity ions produce many electrons and in view of the operating limits on density and pressure (section 6.3), this has the effect of replacing fuel ions. For example, at given $n_{\rm e}$ each fully ionized carbon ion replaces six fuel ions, so that a 7% concentration of fully ionized carbon in the plasma core, corresponding to $Z_{\rm eff} \approx 3$, would reduce the fusion power to one half of the value in a pure plasma.

using pulsed discharges at much lower energies, glow discharges and discharges excited by radio frequencies were tried—the objective in each case being to optimize the desorption rate of the contaminant molecules and their pump-out from the torus. Taylor [44] made a systematic study of the cleaning process in the small tokamaks Microtor and Macrotor and prescribed a universal formula, but generally discharge cleaning tended to remain something more akin to cooking than physics and each tokamak group maintained its own favourite recipes. Many different gases were tried for discharge cleaning, but hydrogen and deuterium became most generally used because of their reactivity and ability to form molecular compounds with carbon and oxygen that could be pumped away.

A break-though in the struggle to achieve low-Z discharges was made in 1974 in the ATC tokamak when a thin layer of titanium was evaporated on to the plasma-facing wall [45]. The active metallic layer reduced the impurity influx by chemically trapping some, such as oxygen, and physically burying the others. This process, known as gettering, was subsequently applied in other tokamaks [46] using various active metallic layers. Later gettering experiments in DITE showed that it was possible to raise the plasma density to much higher values when the oxygen concentration was reduced to a low level.

6.6 Hydrogen recycling and refuelling

Early tokamaks generally had recycling coefficients (box 6.5) significantly greater than unity; the majority of the plasma originated from molecules previously adsorbed on the walls rather than from the filling gas, so that it was impossible to exercise any real control over the plasma density. A clear demonstration was provided by isotope exchange experiments carried out on several tokamaks [47]. After a period of operation with hydrogen, long enough to saturate the walls, the filling gas would be changed to deuterium (or vice versa) and the plasma isotopic composition monitored spectroscopically. Generally it was found that the change-over to the new isotope took many discharges as the gas already loaded in the walls remained the dominant source of plasma particles. The quantity of hydrogen attached to the walls was typically one hundred times the amount of hydrogen required to fuel the plasma so it was common for the density to rise uncontrollably until the discharge terminated in disruption. When titanium gettering was introduced in ATC to reduce the impurities, there was a dramatic change in the density behaviour as well. The uncontrolled rise in density previously observed throughout the discharge was replaced by a falling density—indicative of a recycling coefficient below unity. Gettering began to be used widely to control recycling as well as to reduce impurities.

With the recycling coefficient reduced to less than unity, it became a matter of concern to maintain the plasma at a constant density. Experiments

Box 6.5 Hydrogen recycling

An issue closely related to impurity influx is recycling of the hydrogen ions. In most tokamaks the pulse length is much longer than the average time that a particle spends in the plasma. Thus on average each particle leaves the plasma and returns or is replaced by another particle several times during the course of a single discharge. This process is called recycling and the ratio of the flux of particles returning to the plasma to that leaving is called the recycling coefficient. When an ion or neutral atom escapes from the confined plasma and arrives at a solid surface, it may either be scattered back into the plasma immediately or be trapped in the surface. Trapped atoms can return to the plasma after diffusing back to the surface through the solid lattice; also the bombardment of the wall may release adsorbed gas. These processes can lead to a recycling coefficient in excess of unity.

For a limiter tokamak, the recycled particles enter the plasma predominantly as neutrals in the vicinity of the limiter but are ionized and trapped on closed magnetic surfaces after traversing only a small fraction of the plasma radius. The situation is more complex in the case of a divertor tokamak where most of the ionization may take place near to the target so that under some circumstances there may be no particle source on the closed surfaces. Although the recycling flux density is largest at the limiter or divertor target, recycling also occurs all around the walls due to the charge exchange flux and, because of the much larger areas, the integrated flux from the walls can be of the same magnitude as the integrated flux from the limiter or divertor.

in Alcator showed that first setting up a low-density discharge and then admitting neutral hydrogen gas from fast-acting valves placed around the walls was effective not only in raising the density but also in keeping impurities at bay by cooling the plasma edge. This technique of 'strong gas puffing' became widely applied in other tokamaks. But the effects of gas admission at the plasma edge were not always found to be beneficial. In some cases the interaction between the edge plasma and the wall enhanced the influx of impurities and, if the edge were cooled too strongly, there was a disruption or a 'MARFE'—a radiation instability localized on the inside of the torus. The ideal would be to bypass the plasma edge and refuel the core plasma directly. To some extent this was achieved with beam heating, which as we saw came into widespread use; when beam powers increased and confinement times lengthened, the beam became a significant particle source in the core as well. Later, deuterium–tritium plasmas would be produced in

JET and TFTR in this way by injecting tritium neutral beams into a deuterium target plasma, but for the time being the only way to inject fuel directly into the core was pellet injection.

To refuel the plasma core, many tokamaks employed techniques for producing millimetre-sized pellets of solid hydrogen and accelerating them to velocities in the range of 1 to 2 $km\,s^{-1}$ [48]. Both gas-driven guns and centrifuge accelerators [49] were developed for this purpose. To deposit the pellets deep into reactor-sized plasmas requires much higher velocities, but it may not be necessary to make them penetrate all the way to the centre. Core refuelling with pellets allowed tokamaks to reach much higher densities than edge refuelling with gas and it was found in many cases that energy confinement improved. A possible heuristic explanation is that the temperature profile is largely invariant, so adding density in the hot core is more efficient in increasing the energy content than adding density at the colder edge. In most applications, pellets were injected from the more accessible outside of the plasma—the low-magnetic-field side—but experiments in ASDEX–U in the late 1990s would show improvements when the pellets were injected from the high-magnetic-field side [50].

6.7 Divertors

A tokamak plasma is confined on a set of nested, closed magnetic surfaces, the outermost of which is called the 'last closed flux surface'. This is determined either by a limiter which intercepts all the surfaces outside it, or by the magnetic geometry itself when this has a separatrix outside which the surfaces are open, allowing the plasma to flow towards a divertor target. The plasma that is in contact with the material limiter or divertor target, is called the 'scrape-off layer'. The radial width of this layer is determined by the balance between transport across the magnetic surfaces and flow along them. This width is of critical importance for the power flux per unit area of the limiter and divertor target.

The divertor was first proposed by Spitzer [51] in 1951 as a means of impurity control in the early stellarators (figure 6.6). The original purpose was to separate the source of impurity production from the main plasma by moving the plasma contact with a material surface to a separate chamber. Also impurities released from the walls of the main chamber would be ionized in the scrape-off layer and swept into the divertor before they could penetrate the main plasma. The early stellarators diverted the toroidal field all around the poloidal circumference at one toroidal location but this caused a severe perturbation of the toroidal magnetic field and was later recognised as bad for confinement. Modern stellarators make use of the 'natural' helical divertor inherent in the field configuration—magnetic islands can guide the diverted plasma towards divertor plates. Likewise FRCs have natural divertors.

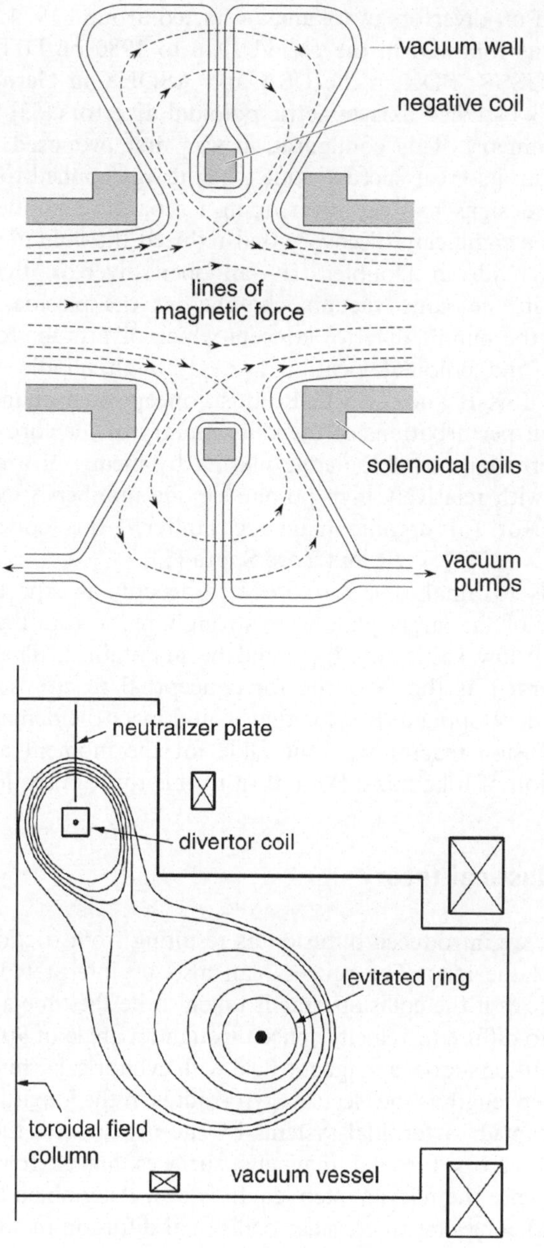

Figure 6.6. Toroidal and poloidal divertors. Top: toroidal divertor for the Princeton C-stellarator; bottom: poloidal divertor for the Livermore levitron.

Research on divertors in tokamaks started around 1974 with DIVA in Japan and was followed in the period 1976 to 1980 on DITE in the UK, T-12 in the USSR, PDX in the USA and ASDEX in Germany. Most of these tokamaks studied axisymmetric poloidal divertors [52] that preserved the axial symmetry. This configuration was first proposed for a levitron (figure 6.5), but its great success came when it was applied to the tokamak. The original designs had the divertor in a separate chamber—the closed divertor. But a significant discovery that reduced the volume taken up by a divertor was made in Doublet III—an open divertor, that is with the target plate in the same vacuum chamber as the plasma, also worked. DITE tested the bundle divertor concept which diverts the toroidal field at one toroidal and poloidal location but, although bundle divertors were designed for ISX-B and TEXTOR, this concept was abandoned due to concern about perturbations of the axisymmetry of the core plasma. But a deliberate perturbation of the edge plasma by means of a set of external helical coils with relatively high poloidal mode number is exploited in the ergodic divertor (or ergodic magnetic limiter)—this concept has been tested on TEXT, TEXTOR and Tore Supra [53].

The early tokamak divertors operated broadly as expected. Impurities characteristic of the target plate were strongly reduced in the main plasma and relatively low values of Z_{eff} could be maintained. The axisymmetric divertor emerged as the favourite for conceptual reactor studies, but the challenge of developing a divertor that would meet the demanding requirements of a fusion reactor was put aside for the moment and left to the next generation of tokamaks. We return to this topic in section 8.3.1.

6.8 Neoclassical theory

In chapter 1, we introduced diffusion as resulting from friction between the electrons and the ions. The process can also be interpreted as a random walk of particles if the collision time is taken to be the time a particle takes, on average, to diffuse in velocity space through an angle of 90°. The classical theory of diffusion across a magnetic field with cylindrical symmetry, in which the typical step length is the electron gyro-radius in the longitudinal magnetic field, was adapted to toroidal systems by the neoclassical theory, in which particle drift across toroidal magnetic surfaces makes for a larger step length. In essence the relevant step length is now determined by the poloidal magnetic field. The first to consider collisional diffusion in toroidal systems were Budker and Tamm in 1951 in Moscow, who found that guiding-centre drift, rather than cyclotron gyration, determined the excursions of particles with respect to magnetic surfaces. These excursions exceed the gyro-radius by approximately a factor $q \approx rB_{\mathrm{t}}/RB_{\mathrm{p}}$, which by itself would cause the diffusion constant, $D \approx \delta^2/\tau_{\mathrm{c}}$, where δ is the excursion length and τ_c is the collision time, to increase by a factor of order q^2 in a $q \gg 1$ tokamak or stellarator. From a different point of view Pfirsch and Schlüter described in 1962 a double vortex induced by the charge separation caused by the same grad B and curvature drift. When the equalizing electron flow along the field lines is impeded by finite resistivity, the resulting vertical electric field likewise causes an increase of the net outflow of plasma by roughly a factor q^2 as compared with classical diffusion in a straight cylinder.

More subtle effects were reported by Galeev and Sagdeev [51] in 1968 to be caused by the trapping of particles between the magnetic mirrors they encountered when travelling along a helical field line from the outer, weak field, to the inner, strong field region of the torus. These mirrors, whose relative strength is determined by the inverse aspect ratio $\varepsilon = r/R$, divide the particles into a trapped population that is reflected by the mirror regions and a passing one that is not reflected. The orbits of the trapped particles when projected on to a poloidal plane have a characteristic shape that coined the term 'banana orbits' (figure 6.7). Collisions between particles of these two classes give rise to an enhanced outward diffusion of the trapped

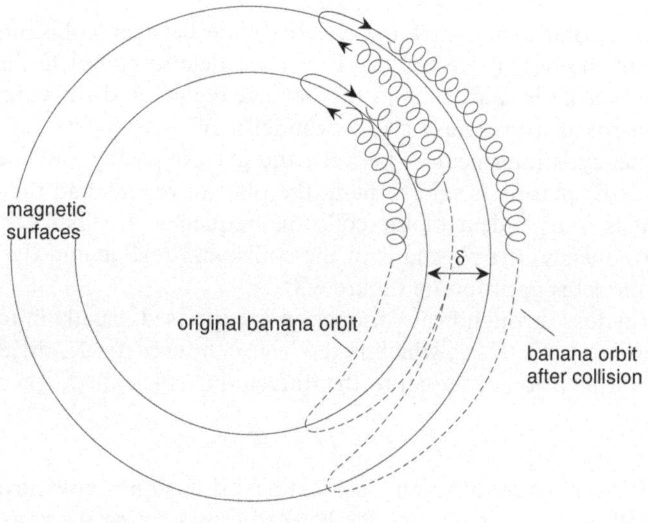

Figure 6.7. Banana diffusion. In a first approximation the particles follow helical field lines. The orbits are projected toroidally on to a poloidal cross section, so that up and down movement goes along with back and forth movement in the toroidal direction. A banana orbit results when a particle continually drifts in the same vertical direction (upwards in this figure) and thereby moves outwards with respect to the magnetic surfaces when in the top half and inwards when in the lower half of the torus. A collision makes the orbit jump over a distance δ, which can exceed the gyro-radius by a factor q (see text).

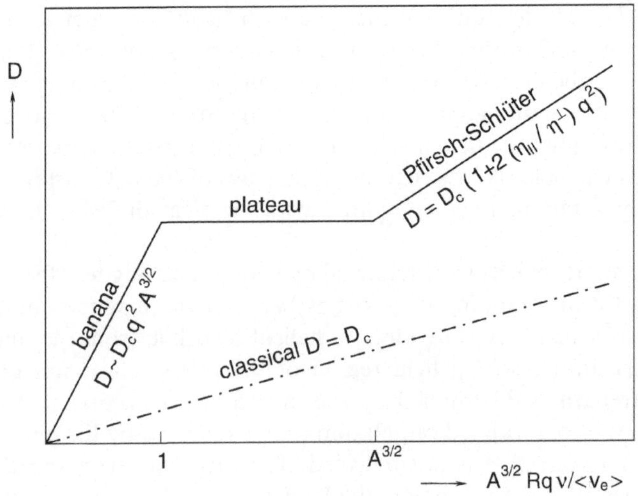

Figure 6.8. Schematic representation of the diffusion coefficient in a torus as a function of the collisionality. Note that in the notation used in this figure, $A = \varepsilon^{-1}$ where $\varepsilon = r/R$ is the inverse aspect ratio (but elsewhere in the text $A = R_0/a$).

particles. In conditions where the effective time between collisions (ε/v) is longer than the time ($\sim qR/\varepsilon^{1/2}\langle v_e \rangle$) for a particle to complete the banana orbit, the plasma is said to be in the banana regime and the enhancement factor compared with diffusion in a cylinder is of order $q^2/\varepsilon^{3/2}$. If the collision frequency is intermediate between the gyrofrequency and the banana frequency, the plasma is said to be in the plateau regime and the diffusion coefficient is independent of the collision frequency. If, finally, it exceeds the gyrofrequency, the plasma is in the collisional regime and the Pfirsch–Schlüter model is appropriate (figure 6.8).

Energy loss through heat conduction is expressed usually in terms of a heat diffusivity, $\chi = K/n_e$, which is the heat conductivity, K, divided by n_e to make χ dimensionally equal to the diffusion coefficient, D. Then,

$$\tau_E = nL^2/K \propto a^2/\chi.$$

In contrast to particle diffusion, classical heat diffusion is governed by like-particle collisions, for which the Rutherford cross-section scales as $m^{-2}v^{-4}$. Except for numerical factors of order unity, the neoclassical cross-field electron heat diffusivity is equal to the electron–ion diffusion coefficient, but the ion–ion heat diffusivity is greater by a factor $(m_i T_e/m_e T_i)^{1/2}$.

Attempts to compare these theoretical predictions of particle and energy transport with experimental results met with mixed success. The particle loss rates measured experimentally, although by now much smaller than according to Bohm, were much larger than the rates calculated from the neoclassical theory. The ion and electron heat diffusivities could be compared separately against the measured heat losses through the ion and electron channels. Energy from Ohmic heating enters the system via the electrons and is transferred to the ions by collisions. Either species can receive energy directly from additional heating. The electrons lose energy by conduction, particle diffusion and radiation; the ions by conduction, diffusion and charge-exchange. Although at this stage full profiles of the ion temperature were not generally available, analysis of ion temperature data in ATC, ORMAK and TFR in the period 1974 to 1976 indicated that the ion thermal conductivity was close to the level predicted by neoclassical theory and this conclusion was subsequently confirmed on many other tokamaks. On the other hand, the electron heat loss was found to be much higher than the theory predicted.

Further development of neoclassical transport theory soon led to the prediction of two other important effects [52] that were observed in experiments only much later. The Ware pinch, an inward plasma flow associated with a toroidal electric field, was proposed by Ware in 1970 [53]. A toroidal electric current maintained by the radial pressure gradient was predicted [54] by Bickerton, Connor and Taylor and independently by Galeev and Sagdeev in 1971. It was called the 'bootstrap current' because it offered a way for the tokamak plasma to generate its own steady-state current. In the full neoclassical theory, reviewed in detail by Hinton and Hazeltine in 1976, the radial particle and heat fluxes for each species and the toroidal electric current are related by a transport matrix to the radial gradients of the density and the temperature and to the toroidal electric field. The diagonal elements of the matrix are the familiar particle and heat diffusion coefficients and the parallel conductivity, but there are also non-zero off-diagonal terms, including those responsible for the Ware pinch and the bootstrap current.

Generally the neoclassical theory [55] has given values for the ion diffusion and thermal conduction corresponding to the lowest that have been achieved in experiments. The parallel resistance and the more subtle effects like the bootstrap current and the inward pinch* also have been confirmed in many experiments. But neoclassical theory underestimates electron particle diffusion and heat conduction by between one and two orders of magnitude. Attempts to find refinements to the theory that remove this discrepancy have not met with success. One is left with the conclusion that the theory is only correct in its description of transport carried by the ions in a quiescent plasma. The remaining anomalous electron transport must then be due to fluctuations which either do not affect the ions

* It would be found more than two decades later that α particles also appeared to be confined neoclassically (see chapter 8).

because of the larger ion orbit size or whose effect on the ions is simply too small to be seen against the much larger neoclassical ion conduction.

6.9 Empirical scalings

For a while, anomalous (i.e. faster than neoclassical) transport was discussed in terms of a so-called pseudoclassical theory [56] but this could not explain the experimentally observed dependence on, in particular, the density. In the absence of an adequate theoretical understanding of confinement, it became necessary to rely on empirical methods. The simplest was to accumulate data from one or several tokamaks over a range of different conditions and to use statistical methods to determine the dependence of the confinement time on the plasma parameters. These 'scalings' connect empirical confinement times* with machine and plasma parameters like R_0, a, B_t, I, n_e and either P or T, along with other geometrical parameters and profile functions, the ion mass and charge numbers M_i and Z_i. This approach is of course already well established in other areas of science and engineering—the performance of aeroplanes and ships can be reliably predicted using similar scalings without a detailed understanding of turbulent hydrodynamic flow. Many authors have proposed such empirical procedures in relation to plasma devices as illustrated by Artsimovich's early proposal $\tau_E \propto a^2 n_e T$.

The attainment of very pure plasmas in Alcator in 1976–77 allowed the global energy confinement time τ_E to be measured over a wider range in density than had been explored previously. It was found that τ_E increased with density (figure 6.9), in conflict with the decrease of confinement with density predicted by neoclassical theory, but enabling Alcator to reach record values of $n_e(0)\tau_E \approx 2 \times 10^{19}\,\mathrm{m^{-3}\,s}$. These results stimulated the study of confinement scaling in other Ohmically heated tokamaks including TFR, Pulsator and, at lower magnetic fields, in ISX-B. Results compiled by Jassby, Cohn and Parker [57] from several experiments produced an overall scaling for Ohmically heated tokamaks of the form† $\tau_E \propto n_e a^2 q^{1/2}$.

The bigger Alcator-C tokamak was built to take advantage of the high density limit permitted by its high magnetic field and to exploit the a^2 dependence of the τ_E scaling to go to record values of $n_e\tau_E$—this was expected to get into the $10^{20}\,\mathrm{m^{-3}\,s}$ range. However, although τ_E increased linearly with density up to about $2 \times 10^{20}\,\mathrm{m^{-3}}$, it began to saturate at higher densities and Alcator-C failed to extend the $n_e\tau_E$ performance of its smaller predecessor (now known as Alcator-A) by quite the margin that

* After pump-out ceased to be the main problem, it became the general practice to use the global energy confinement time τ_E rather than the particle confinement τ_p as a measure of confinement.

† This scaling in a slightly modified form was adopted for INTOR reactor design studies (see chapter 9) as a benchmark for confinement scaling.

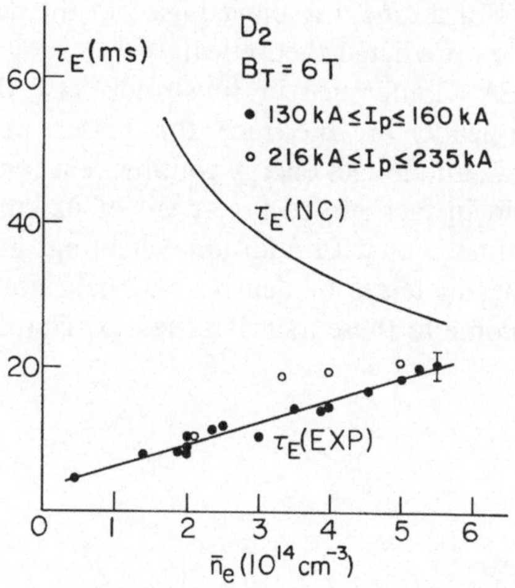

Figure 6.9. Energy confinement time versus density for Ohmic discharges in Alcator. The experimental data points τ_E(EXP) show confinement increasing approximately linearly with density in contrast to neoclassical theory τ_E(NC) which predicts a deterioration of confinement with density. As discussed in the text, later experiments in Alcator found that this favourable scaling of confinement stopped at higher density.

had been extrapolated. Two regimes of confinement in Ohmically heated tokamaks became well identified [58]. The global energy confinement time increases linearly with density up to a critical density where it saturates. The critical density decreases with increasing machine size. The conventional understanding of the increase in τ_E in the linear regime is that this is due to the decrease in temperature and the resulting reduction of anomalous transport from either collisionless or dissipative trapped electron modes. The saturation of τ_E is explained as an increase in enhanced ion transport, possibly due to turbulence driven by the ion temperature gradient.

It had been recognized for some time that the coupling between heating and confinement inherent in Ohmic tokamak discharges gave rise to an ambiguity in the scalings [59] that could be resolved only when additional heating became available. Then in principle the temperature would be uncoupled from the plasma current, and it would become possible to tackle the question of how τ_E depended on T_e or T_i. However, understanding confinement did not turn out to be quite as simple as might have been expected. Indeed, in spite of substantial efforts, many aspects of this important issue remain unresolved. The first neutral beam heating experiments had insufficient power to see any departure from confinement with Ohmic heating. At the next stage, high-power heating experiments concentrated on achieving record temperatures in low-collisionality regimes at high plasma current and relatively low density. The low-density Ohmic target plasmas already had relatively poor confinement, so τ_E did not appear to change drastically with beam heating. For a time this allayed fears of the negative scaling of τ_E with T_e that had been predicted theoretically.

The 1978 IAEA Conference in Innsbruck saw the tokamak firmly established as the leader in the race for better performance—higher plasma pressure, nkT, and longer energy confinement time, τ_E. In the following decades it would in fact run so far ahead of its competitors that even degradation of confinement with additional heating and tightening of the operating constraints in terms of density and beta could not threaten its position. We shall come to these issues in the next chapters.

Chapter 7

The next generation

Already in the early 1970s, when the results from the first generation of western tokamaks started to come in, plans were being formulated for the construction of three very large tokamaks aiming at significant α-particle heating in D–T plasmas or at simulating this condition in normal hydrogen or deuterium. Meanwhile, a new generation of small and intermediate-sized tokamaks was planned in parallel to the construction of the big machines. New tokamaks, usually with more specific objectives than their predecessors, continued to be built, though at a much slower rate, into the 1990s. The most important of these machines are summarized in table 7.1 [1]. We devote this chapter to these intermediate experiments and come to the big tokamaks in chapter 8. There were many interactions between the two categories that were important for the development of tokamak physics, so there is some arbitrariness in where we treat specific points.

7.1 New machines

In the USA, Princeton replaced ST and ATC with PLT. This was the first tokamak designed to reach 1 MA; it came into operation in 1975 and would continue until TFTR took over. In 1979, Princeton also launched the PDX tokamak to study poloidal divertors as in the European ASDEX. At MIT, the first Alcator high field tokamak, now known as Alcator-A, was replaced by Alcator-C. Much later, in 1993 came Alcator C-mod—a completely new machine with a divertor. At Oak Ridge, ORMAK was superseded by ISX, in which the emphasis was on impurity studies, and the University of Texas entered the field with TEXT, devoted to turbulent heating. At the General Atomics laboratory in San Diego, the doublet programme of studying elongated and indented cross-sections continued with the construction of a larger device Doublet III which came into operation in 1978. After being operated in the doublet configuration as a

Table 7.1 Tokamaks built after 1980

Location	Acronym	Operation	R_0 (m)	a, b (m)	B_t (T)	I (MA)	Special features
USA							
Princeton	PBX	1984–85	1.45	0.3	2.4	0.6	Was PDX—indented plasma
	PBX-M	1987–93	1.65	0.3	2.0	0.6	
	TFTR	1982–97	2.62	0.97	5.9	3.0	See table 8.1
	NSTX	1999–date	0.85	0.68	0.3–0.6	1.0	Spherical tokamak
MIT	Alcator-C-mod	1993–date	0.67	0.22	9.0	0.64	Divertor
San Diego	DIII-D	1986–date	1.67	0.67, 1.36	2.1		Divertor, was Doublet III
Austin	TEXT	1981–96	1.0	0.27	2.8	0.34	
Russia							
Moscow	T-15	1988–date	2.43	0.7	3.5	1.0	Superconducting coils
Troitsk	TSP	1991–date	1.0	0.32	2.0		*R* and *a* compression
St Petersburg	TUMAN-3	1980–date	0.55	0.24	0.5–1.2	0.16	Confinement and compression
	FT-2	1982–date	0.55	0.08	2.2–3.5	0.06	LHCD, heating, diagnostics
	GLOBUS-M	1999–date	0.36	0.24	0.62	0.5	Spherical tokamak
Europe							
Cadarache	Tore Supra	1988–date	2.37	0.80	4.5	2.0	Superconducting coils
Garching	Asdex [4]	1980–90	1.65	0.40	2.8	0.5	Divertor (closed geometry)
	Asdex-U	1991–date	1.65	0.50, 0.8	3.9	1.0	Divertor (open geometry)
Jülich	Textor†	1983–date	1.75	0.5	2.6	0.6	Plasma–wall interaction
Culham	Compass	1989*	0.56	0.21, 0.38	2.1	0.28	
	JET	1983–1992	3.0	1.25	3.5	7.0	With limiter (see table 8.1)
	JET	1994–date	~2.9	~1.0	4.0	5.0	With divertor (see table 8.1)
	START	1991*	0.2	0.15	0.6	0.12	Spherical tokamak
	MAST	1999–date	0.85	0.65	0.5	2.0	Spherical tokamak

Frascati	FT-U	1990–date	0.93	0.30	8.0	1.3	High B_t; additional heating
Nieuwegein	RTP	1989*	0.72	0.16	2.5	0.16	Formerly Petula
Lausanne	TCA [5]	1980*	0.61	0.18	1.5	0.17	Alfvén wave heating
	TCV	1992–date	0.875	0.24	1.43	1.2	Highly elongated cross-section
Japan							
Naka	JT-60	1985–89	3.0	0.95	4.5	2.7	Divertor on low-field mid-plane
	JT60-U	1991–date	3.4	1.1	4.2	5	See table 8.1
	JFT-2M	1983–date	1.3	0.35, 0.53	2.2	0.5	
Fukuoka	TRIAM-1M	1986*	0.8	0.12, 0.18	8	0.1	Superconducting coils
China							
Hefei	HT-6B	1982*	0.45	0.125	1.2	0.04	
	HT-6M	1984*	0.65	0.2	1.5	0.12	
Leshan	HL-1	1985*	1.02	0.2	5.0	0.135	

* First operation

† Since 1996, Textor has been operated by the Trilateral Euregio Cluster, TEC, formed by the Research Centre at Jülich (D), the Royal Military School in Brussels (B) and the FOM-Institute Rijnhuizen at Nieuwegein (NL).

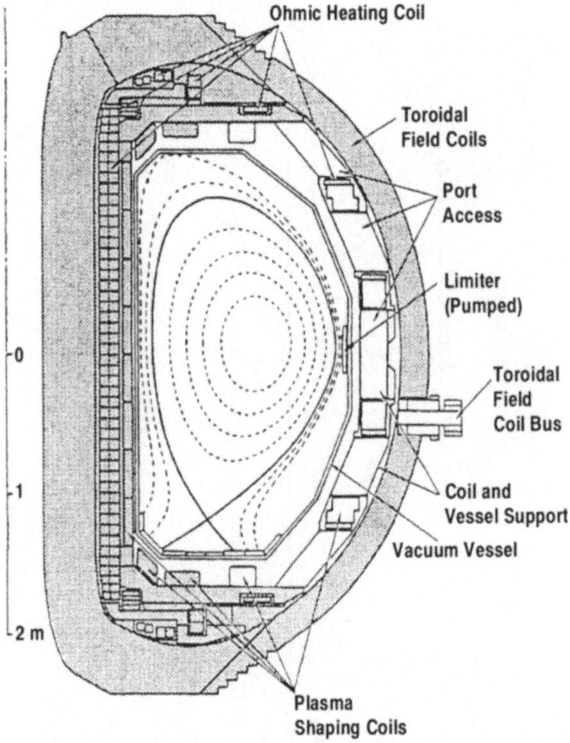

Figure 7.1. Cross section of the D III-D tokamak after its rebuild in 1986 showing the open poloidal divertor at the bottom of the torus.

joint US–Japan venture, it was converted in 1986 into a large D-shaped tokamak with a poloidal divertor and renamed D III-D (figure 7.1). This became a very productive machine, testing many new ideas and standing its ground even in competition with the still larger tokamaks to which we come in chapter 8.

In Europe, the French TFR was upgraded in minor radius and current capacity to emerge in 1980 as TFR-600. The copper stabilizing shell in the original machine—referred to thereafter as TFR-400—was removed and the plasma position was controlled with a feedback system, though this turned out to be less effective than expected due to problems with disruptions and impurities. The superconducting tokamak, Tore-Supra, set up to address problems of steady-state operation, came on line in 1988 after the French fusion effort was relocated from Fontenay-aux-Roses and Grenoble* to Cadarache. In the UK, DITE continued to operate until 1989 when it was

* The French fusion programme had started on the outskirts of Paris at Fontenay-aux-Roses and at Saclay. The Saclay group moved first to Grenoble and later both groups moved to Cadarache.

closed to make way for Compass-D and START—the first of Culham's spherical tokamaks (section 8.3.3). In Italy, the Frascati laboratory near Rome had started with a high-field tokamak FT, which followed the Alcator line of work on Ohmic heated plasmas at high density. This was succeeded by FT-U, which had larger cross-section and plasma current and was used to study radio frequency heating in high density plasma. At Jülich in Germany, TEXTOR was built specifically to study plasma–wall interactions. Garching's poloidal divertor tokamak ASDEX was replaced by a new and bigger machine ASDEX-U.* The original ASDEX was dismantled and shipped to China where it was rebuilt as HL-2A at the South Western Institute of Physics; a similar rebirth awaited Petula from Grenoble which was rebuilt as RTP at the FOM institute Rijnhuizen in the Netherlands. The Swiss at Lausanne entered the tokamak league with TCA, designed to study Alfvén wave heating, and later built TCV with a highly elongated plasma and strong ECRH.

In Japan, in parallel with the construction of JT-60, the medium-sized tokamak JFT-2 was rebuilt as JFT-2M, with an elongated plasma cross-section and an open poloidal divertor in the upper part of the vacuum vessel. JFT-2M came into operation in 1985 producing non-circular plasmas with ellipticity of 1.7 and currents up to 500 kA. The smaller tokamaks TRIAM-1 and WT-2 were also upgraded (to TRIAM-1M [2] and WT-3 [3]) with programmes devoted to the study of steady-state tokamak operation with RF current drive. A unique feature of the superconducting TRIAM was the 8 T magnetic field. Aware of gaps in other parts of their programme, the Japanese also entered into an agreement with the USA to participate in the Doublet III experiment (before its upgrade to DIII-D) and this led to a long-standing close relationship.

Several new tokamaks came into operation at the Kurchatov Institute around 1975. The largest, with major radius 1.5 m, minor radius 0.4 m and 5 T toroidal field, was T-10. Initially optimized to obtain maximum ion temperatures in Ohmic heating, this machine was later used for experiments with electron cyclotron heating and current drive. T-10 is one of the few tokamaks of this era remaining in operation. The smaller T-11 was equipped with neutral-beam injection heating. In 1986 this tokamak was moved to the Troitsk branch where *Tokamak Silnoye Pole* 'with a strong field' (TSP) (section 9.7) was under construction. In 1972 Artsimovitch and Shafranov had proposed that a tokamak with a vertically elongated plasma could operate at higher currents for the same toroidal field compared with a circular cross-section. The stability of plasmas with elongation up to 2 was demonstrated in the T-9 tokamak and in a new machine, T-12, which was equipped with a poloidal divertor.

* This was a completely new machine—the confusing practice of calling new projects 'an upgrade' emerged during the 1980s.

The first step to steady-state operation of a tokamak was taken at the Kurchatov Institute in 1979 with the world's first superconducting tokamak T-7, which was equipped with lower hybrid current drive. A much large superconducting tokamak T-15 was under construction but did not enter operation until 1988 and its subsequent operation would be limited by technical and budgetary problems.

The second major centre for tokamak research in the Soviet Union was the Ioffe Institute in Leningrad with the TUMAN series of machines. Adiabatic compression in combination with RF heating was studied in TUMAN-3, which first came into operation in 1979, but was rebuilt in 1990.

7.2 Radio-frequency heating

The successful development and application of neutral beam heating that had started in the early 1970s was matched by equally determined efforts to heat tokamaks with radio-frequency methods (box 7.1). Radio-frequency heating transfers energy from an external source of electromagnetic waves to the plasma. When a wave travels through a plasma there is always some absorption of energy due to particle collisions but this is generally small in a hot plasma as the collisional absorption scales as $T_e^{-3/2}$ in the same way as Ohmic heating. However, there are certain frequencies at which electromagnetic waves in a plasma are absorbed resonantly. As this is a collisionless process, its efficiency does not decrease with increasing temperature. A magnetized, multi-species plasma has a number of resonances, of which three are particularly suitable for heating; in descending order of frequency: electron cyclotron resonance heating (ECRH) at $\omega \approx \omega_{ce}$, lower hybrid resonance heating (LHRH) at $\omega \approx \omega_{LH}$ with $\omega_{ci} < \omega_{LH} < \sqrt{\omega_{ci}\omega_{ce}}$, and ion cyclotron resonance heating (ICRH) at $\omega \approx \omega_{ci}$.

Some of these ideas had been tested already on stellarators and now were applied to tokamaks. The Russians took an early lead in ECRH because the Applied Physics Institute in Nizhniy Novgorod (known as Gorki in the Communist era) had developed reliable high-power gyrotron sources in this frequency range, well before these were available elsewhere. But after they had demonstrated the method on the TM-3 tokamak in 1976 and had carried out a comparison of X- and O-mode heating in FT-1, ECRH was added on to many tokamaks while the heating powers were pushed up progressively to the MW level. Central electron temperatures up to 10 keV were measured when 2.5 MW at 81 GHz and 1.1 MW at 75 GHz were applied in T-10. In DIII-D a central temperature of 5 keV was reported with 1 MW at 60 GHz. One of the factors that would prevent the application of ECRH on the three big tokamaks was the lack of gyrotrons at frequencies matching the magnetic fields on these machines. Also the modular size of gyrotrons remained relatively low until the late 1990s, so that many parallel units

Box 7.1 RF heating

A magnetized plasma can support a wide variety of oscillations and waves. At high frequencies, only the electrons move under the influence of oscillating electrostatic or electromagnetic fields, pressure gradients and constant magnetic fields. At low frequencies, the ions experience the same kinds of force and, in addition, carry the electrons along so that in compressional waves the electron pressure comes in addition to that of the ions. At intermediate frequencies, both species play their respective roles. When waves travel across gradients in the density or the magnetic field strength, they may encounter cut-offs where they are reflected, or resonances where the wave energy is dissipated into kinetic energy of either species. In a tokamak, the toroidal field falls off as $1/R$ so that resonant waves may be used for localized as well as for global heating. This may require the wave to be launched from the high-field side, that is from the inside of the torus.

The electron cyclotron resonance, $\omega = \omega_{ce}$, occurs at 28 GHz T^{-1} (Gigahertz per Tesla), so it requires frequencies up to 200 GHz for a reactor. Electron cyclotron waves heat the electrons, which in turn transfer energy to the ions by collisions. Thus, ECRH has been applied to control sawteeth and disruptions.

The ion cyclotron resonance, $\omega = \omega_{ci}$, depends on the charge to mass ratio of the ion (Z/M) and on the magnetic field. The fundamental resonance lies at $15.2(Z/M)$ MHz T^{-1}, which places the range of ion cyclotron frequencies within that used for commercial broadcasting. The wave propagates to the resonance region as the fast magnetosonic (the compressional Alfvén) wave. In a plasma with a single ion species, heating is possible at the second harmonic resonance, $\omega = 2\omega_{ci}$. Alternatively, with two ion species, the wave can be absorbed at the fundamental resonance of the minority species.

The lower hybrid resonance is intermediate between the ion and electron cyclotron frequencies. It falls in the range of 1–8 GHz, corresponding to free space wavelengths in the microwave region. A problem with LHRH is that the position of the resonance depends on the plasma density as well as on the magnetic field so that tuning the heating to a predetermined radius is not straightforward.

would have been required to deliver power in the multi-MW range required for these big machines.

Results with lower hybrid waves as a heating system have been less spectacular. The main interest in lower hybrid waves has become to drive currents in the plasma, this will be discussed in the next section.

Box 7.2 Alfvén waves

Unlike high-frequency waves, in which electrons oscillate and ions stay at rest, low-frequency waves involve motions of the plasma as a whole. Without a magnetic field, a plasma can carry only longitudinal (acoustic) waves. These are not affected by a magnetic field parallel to the material displacement, but a perpendicular field must move with the plasma so that the magnetic pressure adds to the kinetic pressure, making for a higher phase velocity (magneto-acoustic or compressional Alfvén waves).

Alfvén discovered that a magnetic field also makes it possible for transverse waves to propagate.

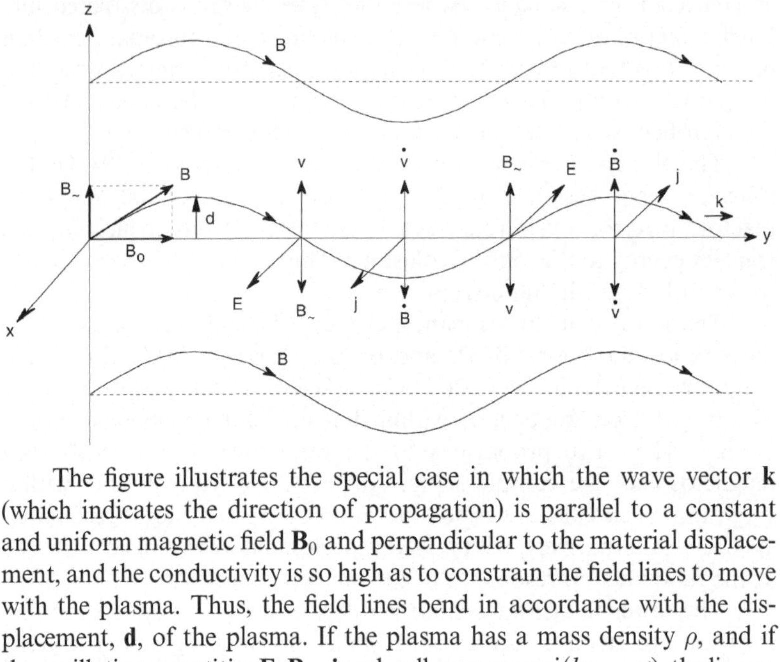

The figure illustrates the special case in which the wave vector **k** (which indicates the direction of propagation) is parallel to a constant and uniform magnetic field $\mathbf{B}_0$ and perpendicular to the material displacement, and the conductivity is so high as to constrain the field lines to move with the plasma. Thus, the field lines bend in accordance with the displacement, **d**, of the plasma. If the plasma has a mass density ρ, and if the oscillating quantities **E**, $\mathbf{B}_\sim$, **j** and **v** all vary as $\exp \mathrm{i}(kz - \omega t)$, the linearized equations $\nabla \times \mathbf{E} = -\partial \mathbf{B}_\sim/\partial t$, $\mathbf{E} + \mathbf{v} \times \mathbf{B}_0 = 0$, $\nabla \times \mathbf{B}_\sim = \mu_0 \mathbf{j}$ and $\mathbf{j} \times \mathbf{B}_0 = \rho\, \partial \mathbf{v}/\partial t$ together yield the phase velocity $\omega/k = B_0/\sqrt{\mu_0 \rho}$. Note that Maxwell's displacement current is usually neglected in ideal MHD.

Ion cyclotron heating had been investigated by Thomas Stix on the Princeton C-stellarator already in the 1950s and the method was applied to the TFR-400 tokamak in 1979. Pulses of 200 kW lasting 20 ms were launched from the high field side into a plasma of mean density $\sim 3 \times 10^{19}\,\mathrm{m}^{-3}$, resulting in an increase of the ion temperature by about 150 eV. The following year,

500 kW were launched into the upgraded TFR-600 and by 1982 the ICRH power had been increased to between 1.5 and 2 MW. The TFR team studied and successfully applied minority-species healing, which was used extensively thereafter on JET and other machines. On the other side of the Atlantic, after low power tests on ST and 145 kW on ATC, PLT launched 2.6 MW of ICRH producing a central ion temperature of 3.6 keV at $5 \times 10^{19}\,\mathrm{m}^{-3}$.

Alfvén waves (box 7.2) have been proposed to heat plasma at frequencies below the ion cyclotron frequency but thus far has not yielded notable plasma heating. The method involves launching a compressional Alfvén wave which converts to a torsional Alfvén wave at the resonant surface. Electrons are heated by Landau damping. Ion Bernstein wave heating, which requires frequencies above the ion cyclotron resonance, led to significant heating in the HT-7 superconducting tokamak in Hefei as well as in FTU in Frascati.

7.3 Non-inductive current drive

From a reactor designer's point of view, steady-state operation would offer great advantages in avoiding cyclic thermal and mechanical stresses. The tokamak's inductive current drive, which requires a steadily increasing primary current, must inevitably reach saturation at some stage, because of either electrical or mechanical limitations, though the transformer's flux swing (its volt-seconds) may last for as long as several thousand seconds. And JET showed that a discharge could be re-started immediately with the current running in the reverse direction. Nonetheless, the finite pulse length has long been held to be a serious disadvantage for the tokamak as a reactor system. But there are prospects for maintaining a tokamak current indefinitely by a combination of the inherent bootstrap current and various mechanisms for non-inductive external current drive.

The first experimental evidence for the theoretically predicted bootstrap current was obtained not in a tokamak but in the Levitated Toroidal Octupole in Madison [6]. Four years later, Princeton obtained discharges in TFTR which showed strong evidence for one third to one half of the total current being driven by this mechanism. This was to be confirmed in other tokamaks, for example DIII-D, JET and JT-60 [7], and to become an important element in tokamak reactor design (as well as something to be avoided in stellarators). However the bootstrap current will not provide all the non-inductive current and its radial distribution may not match the profiles required for MHD stability or confinement enhancement (see section 8.3.2). This leads to a requirement for an additional external non-inductive drive to supplement or tailor the inherent bootstrap current.

Various heating mechanisms give rise to asymmetric velocity distributions in the ion or electron populations. Ohkawa had pointed out in 1970 that injection of neutral beams with a velocity component parallel or anti-parallel to the toroidal current (referred to as co- or counter-injection) would impart toroidal momentum in the direction of the beam preferentially to either the ions or the electrons in the plasma. Below a critical electron temperature—which depends on the injection energy—mainly the electrons are accelerated; above the critical temperature, the ions are driven. The first experimental evidence for this effect was found in the Culham levitron and the DITE tokamak [8].

Current-drive effects may also be produced by RF heating methods. Before 1953 when pinches became transformer driven, Thonemann [9] had used travelling magnetic fields, produced by phased currents in external coils, to drive electron currents. In the same vein, Wort [10] proposed to drive a tokamak current by pushing ions and electrons in opposite directions by means of Alfvén waves with suitably chosen phase velocities. The first systematic study of RF current drive was carried out by S Yoshikawa and his collaborators in the C-stellarator in the mid-1960s [11]. It was thought [12] that direct transfer of the momentum of electromagnetic waves to electrons would cost too much energy, but Fisch [13] pointed out that the efficiency of fast waves would be higher than it appeared to be in a fluid theory, because they would drive streams of high-velocity (runaway) electrons, which would undergo less friction with the ions than would a slowly drifting bulk electron plasma. Fisch suggested the use of lower-hybrid waves and this was taken up by several experimental groups. The theory was carried an important step further when Fisch and Boozer showed that instead of transferring axial momentum from the wave to the electrons, one could also drive a current by producing an asymmetry in the electron velocity distribution [14]. If a wave imparts perpendicular energy to electrons moving in one toroidal direction, their interaction with the ions is reduced because of the v^{-4} dependence of the Rutherford cross-section. Thus, the ions are driven in the opposite direction by the electrons whose perpendicular energy is not affected. This opened the way for electron cyclotron waves, tuned to a Doppler-shifted resonance, to be used for highly localized current drive, a significant point because other schemes to drive tokamak currents did not provide the current–density profiles required to satisfy MHD stability theory.

There followed an explosion of experimental activity, first of all on LH current drive. Japanese groups detected the effect in the tokamaks JFT-2 at JAERI and WT-2 in Kyoto [15] and more than one full session was devoted to the subject at the 1982 Baltimore Conference. Up to 420 kA were driven in PLT [16]. The effect was observed at densities up to $5 \times 10^{20}\,\mathrm{m^{-3}}$ in Alcator, which also had the highest efficiency, $\eta_{\mathrm{CD}} = 0.12 \times 10^{20}\,\mathrm{A\,m^{-2}\,W^{-1}}$, in terms of the figure of merit $\eta_{\mathrm{CD}} = \Delta I n_e R / P_{\mathrm{RF}}$, where ΔI is the driven current. (In some early work, the figure of merit is defined as $\Delta I n_e / P_{\mathrm{RF}}$.) In addition to those mentioned already, groups in Garching, Grenoble, Moscow and Nagoya contributed to the development of LH heating and current drive. By 1986, JT-60 [17] achieved 1.7 MA with 1.2 MW of RF power at 2 GHz, with η_{CD} up to $0.28 \times 10^{20}\,A\,m^{-2}\,W^{-1}$ and went on to demonstrate a discharge with all the current driven by LHCD [18]. The JAERI group also found a linear scaling of η_{CD} with T_e. The small superconducting tokamak, TRIAM-1M [19], has been kept in a steady state for two hours and the larger Tore Supra for 70 s TCV has recently achieved full current drive with EC waves with LH current drive.

These tantalizing results, nevertheless, fall short of what is needed in a commercial reactor, which will require an efficiency of a few times $10^{20}\,A\,m^{-2}\,W^{-1}$, if the current is to be maintained by an external non-inductive driving mechanism. The ITER concept definition in 1989 noted that steady state current drive represented a more demanding requirement than heating to ignition. It was estimated that in the technology phase about one third of the nominal current of 20 MA would be provided by the bootstrap current and that about 100 MW of auxiliary power (assumed to be a mixture of NBI, ERCH or ICRH in the plasma core and LHCD at the edge) would be required to drive the other two thirds [20].

7.4 The switch to carbon

The Kurchatov tokamak T-3 and its early western progeny had refractory metal limiters of molybdenum or tungsten. High-Z limiters worked well in Ohmic-heated tokamaks, particularly at high plasma densities. Problems began to emerge with additional heating, especially when powerful neutral beams were directed into low-density plasmas. In PLT it had been found that the electron temperature was very sensitive to the choice of limiter material. The first heating experiments in late 1977 had used tungsten limiters but the power lost from the plasma core by tungsten line radiation had prevented any significant rise in temperature. Under some conditions, the cooling effect of the tungsten radiation was even large enough to produce hollow electron temperature profiles—the core was cooler than the outside. The tungsten limiters in PLT were therefore replaced, first with stainless steel but soon thereafter with carbon [21]. Further tests in TFR [22] and T-4 [23] eventually led to the installation of carbon limiters in most tokamaks. Only the high-field devices Alcator and FT, which relied predominantly on Ohmic heating and which by operating at high densities maintained cold edge plasmas that minimized impurity release, retained high-Z limiters.

Carbon limiters undoubtedly produced a marked reduction in the concentrations of metallic impurities, but these did not disappear completely. Detailed examination of samples taken from carbon limiters removed from tokamaks such as ASDEX [24], JET [25], T-10 [26] and TFR [27] showed that the surface became contaminated with metals that migrated to the limiter from the walls during discharge cleaning or following disruptions.

Contamination of the limiters could be alleviated to some extent by carbonization of the walls but a more permanent cure would be to cover the exposed metal walls with carbon tiles. The first tokamak to operate with walls fully covered with carbon was the Russian TM-G, a modification of TM-3, which reported good performance at the Baltimore Conference in 1982 [28]. Other groups proceeded more cautiously and installed carbon progressively. By 1988 about 50% of the metal wall in JET [29], 50% in DIII-D and 40% in JT-60 would be covered with carbon tiles and a few years later the coverage became almost 100% but, as we shall see in section 8.2.2, solid carbon brought its own problems.

Carbonization—the deposition of carbon layers—was experimented with first in DIVA [30], developed further in TEXTOR and applied extensively on JET. Boronization and siliconization were pioneered in TEXTOR and subsequently used to coat the walls of many other tokamaks [31]. A glow discharge in a gaseous carbon compound such as methane deposits a thin layer of amorphous carbon on the wall, which at the optimum wall temperature of about 300°C forms a hard adherent film. The technique had the advantage of reducing the metal influx into the plasma but the disadvantage of loading the walls with hydrogen—again making recycling more difficult to control. Boronization and siliconization were applied

Box 7.3 Limiters

A limiter has to play a number of roles in a tokamak: it must protect the wall from damage in conditions when there are runaway electrons, disruptions or other instabilities; it must localize the plasma–material interaction so that the high power and particle fluxes to the limiter rapidly remove adsorbed gases and reduce the impurity influx; and finally it must by the same mechanism localize the particle recycling. Materials for limiters must first of all withstand thermal shock since very high power rates may occur during disruptions, secondly they must have good thermal conductivity to allow effective cooling, thirdly they must have low sputtering and arcing rates to minimize the influx of impurities and fourthly they should preferably have low atomic number so that any impurities that are released have minimum impact on the plasma. All these criteria cannot be satisfied simultaneously. Of the low-Z materials, only carbon and beryllium can withstand the high heat loads, but they have relatively high sputtering rates and beryllium has a relatively low melting point. In contrast the high-Z refractory metals have good thermal properties and low sputtering rates, but only very low concentrations can be tolerated in the plasma.

using boranes (B_2H_4 and B_2H_6) and silane (SiH_4)—all requiring careful handling as they are both explosive and toxic. Boron and silicon are effective getters and thin boron films were found to pump both oxygen and hydrogen, thus controlling both the impurity influx and the hydrogen recycling.

7.5 Beta limits

Until high-power auxiliary heating methods came available, tokamaks had volume-averaged beta values in the order of 1% or less. If the power balance is determined by Ohmic heating and bremsstrahlung radiation only, the electron pressure, expressed in terms of beta, reaches [32]

$$\beta_e = 0.16(q_a R_0 B_t)^{-1}.$$

With line radiation and conduction or convection loss, β_e stays even lower. Hence, the need for auxiliary heating to exploit the potential for confining hot and dense plasmas and to explore the beta limit as larger (in terms of $R_0 B_t$) machines were installed. As to stability, theory generally predicted that ballooning modes would grow unstable at beta values of a few per cent. Economic considerations [33] indicated that, in a reactor, β in the range 5 to 10% would be the optimum dictated by the cost of initial instalment (which decreases with rising β) and the need for periodic replacement of radiation-damaged components (which increases with rising β).

From the theoretical side, numerical codes were developed, like PEST in Princeton and ERATO in Lausanne, to calculate the permissible beta for various pressure and current density profiles as well as for shaped plasma cross-sections. A combination of ellipticity and triangularity resulting in a D-shaped plasma cross-section turned out to be particularly favourable for stability. This shape had engineering advantages also as it helped to minimize electromechanical stress in the toroidal field coils and was compatible with a poloidal divertor. Even more extreme deviations from circularity were studied at General Atomics in the Doublet series of experiments, in which the plasma had a kidney-shaped cross-section with an indentation on the outside; at Princeton in PBX, which had the indentation on the inside; and at the Kurchatov Institute where an extremely elongated D-shape was produced in the 'finger-ring' tokamak T-9 [34].

At the time of the Innsbruck Conference in 1978, some experiments had achieved volume-averaged beta values around 1% [35]. This remained well below the theoretical ballooning-mode limit, but with more powerful heating Doublet III reached $\langle\beta\rangle = 4.5\%$ in 1982 [36]. The urgency of systematic tests of these limits was emphasized by Troyon [37], who had gone over a wide range of plasma shapes and profiles in a numerical analysis of ideal-MHD stability. Specifically for INTOR (see chapter 9) and JET, this work indicated stability limits for $n = 1$ free-boundary modes well below those assumed in the design studies. The Troyon limit*

$$\beta_{\max}(\%) = \beta_N I_N, \qquad \beta_N = 2.8, \qquad I_N = I/aB_t \qquad [I \text{ in MA}]$$

has since been confirmed in similar studies, albeit that the numerical factor can be slightly increased by appropriate choice of density, temperature, and current profiles and by surrounding the plasma with a conducting wall. Experimentally, these profiles can only be varied within limits, so that the 'Troyon factor', β_N (sometimes written as g or C_t), to be assumed for reactor-design studies remains somewhat uncertain.

The 1982 Doublet III result corresponded to $\beta_N \approx 3.5$. Later, the cross-section of the plasma was given a D-shape in the modified DIII-D. Moreover, the normalized current, I_N, was increased by reduction of B_t and increase of the elongation, which gives a higher value of I for the same value of the safety-factor q_a. Finally, the current and pressure profiles were further optimized. All this brought β_N in the range 3.5–5 and made for a record beta value of 11% [38].

7.6 Confinement degradation

A less favourable picture of confinement began to emerge when increasingly powerful beams were combined with higher density plasmas in order to explore the β limit. The good news was that in some cases β could be pushed close to the limits predicted by theory, but the bad news was that in other cases β was found to saturate below the theoretical limit with increasing beam power. Various possible causes were considered and it was concluded that the energy confinement time decreased when the beam power was raised. For example, in ISX-B, τ_E decreased by as much as a factor of 2 with 2.5 MW of neutral beam heating [39]. These concerns prompted experiments with beam heating in ISX-B, PDX, D-III, ASDEX and other machines in the period 1981–83 to investigate scaling of confinement against different parameters. In essence, it was observed that τ_E with beam heating did not depend significantly on the plasma density (in contrast to the scaling in Ohmic plasmas) or toroidal field, but it did increase strongly with plasma current. Most seriously, there was a clear deterioration of τ_E with beam power that could be interpreted in two ways.

Plots of τ_E against total heating power $P_{tot} = P_{OH} + P_{add}$ (figure 7.2) indicate a progressive deterioration of τ_E with increasing power that can be fitted with a smooth curve of the form $\tau_E \propto P_{tot}^{-\alpha}$. Reviewing confinement scaling in tokamaks at the EPS fusion conference in Aachen in 1983, Goldston [40] suggested that, if one were to choose an average scaling from the

* The original paper has $(\beta A)_{\max} = 2.2\mu_0 IA^2/R_0B_t$, where A is the aspect ratio. This corresponds to $\beta_N = 2.8$ when I is expressed in MA.

range of values of the exponent α that had been obtained on different tokamaks, '*we might settle on* $P_{\text{tot}}^{-1/2}$' and this convenient value of the exponent has proved to be remarkably resilient over the years. The L-mode scaling [41], derived from the world-wide confinement data base developed for ITER is

$$\tau_{\text{E}}^{\text{ITER-89P}} = 0.048 I^{0.85} R_0^{1.2} a^{0.3} n_{\text{e}}^{0.1} B_{\text{t}}^{0.2} (M\kappa/P_{\text{tot}})^{0.5}$$

where I is the plasma current (MA), R_0 is the major radius (m), a is the minor radius (m), n_{e} is the line-average electron density ($10^{20}\,\text{m}^{-3}$), B_{t} is the toroidal field (T), M is the mass of the hydrogenic species, κ is the elongation of the plasma cross-section and P_{tot} is the total heating power (MW).

The implications of the $\tau_{\text{E}} \propto P_{\text{tot}}^{-1/2}$ scaling are quite serious. In a tokamak of given size, nT scales as $P_{\text{tot}}\tau_{\text{E}}$ and thus the fusion triple product* $nT\tau_{\text{E}}$ as $P_{\text{tot}}\tau_{\text{E}}^2$. So one cannot just add more and more powerful heating in order to push the plasma to ignition: as more power is applied to raise the temperature, so the confinement time decreases and $nT\tau_{\text{E}}$ stays approximately constant. Heating allows control of the plasma temperature, but with this scaling the triple product is an inherent feature of the size of the tokamak.

The effect of additional heating may also be described in a different way. When the normalized stored energy E_{tot} is plotted against the total input power (figure 7.2), the beam-heated points appear to lie on a straight line of different slope to the line connecting the origin to the Ohmic value. This suggests the concept of an incremental confinement time τ_{inc} for the additional heating, lower than the Ohmic confinement time τ_{OH} but not further deteriorating with increasing input power. Then the global confinement time follows from $E_{\text{tot}} = \tau_{\text{E}} P_{\text{tot}} = \tau_{\text{OH}} P_{\text{OH}} + \tau_{\text{inc}} P_{\text{add}}$.

The alternative interpretations of confinement degradation gathered their own protagonists and provoked many heated debates at conferences. The difference between the two models when extrapolated to higher power levels is substantial—but the reality was, and remarkably still is, that the experimental scatter of the data points is too large to make a definitive choice between the two models.

The first results suggested that confinement depended on the additional heating method, but this turned out not to be the case. This in itself was quite surprising given the substantial differences between the heating methods and their ability to heat either the electrons or the ions preferentially. But in fact even plasmas with $T_{\text{e}} \gg T_{\text{i}}$ or $T_{\text{i}} \gg T_{\text{e}}$ were found to have similar confinement properties. Another surprising feature of tokamak confinement is the isotopic dependence—confinement is better in deuterium than in hydrogen. This had already been established [42] for Ohmic plasmas by Hugill and Sheffield in 1978 and now was observed also with additional heating. Results

* Strictly speaking the triple product contains the ion density and temperature.

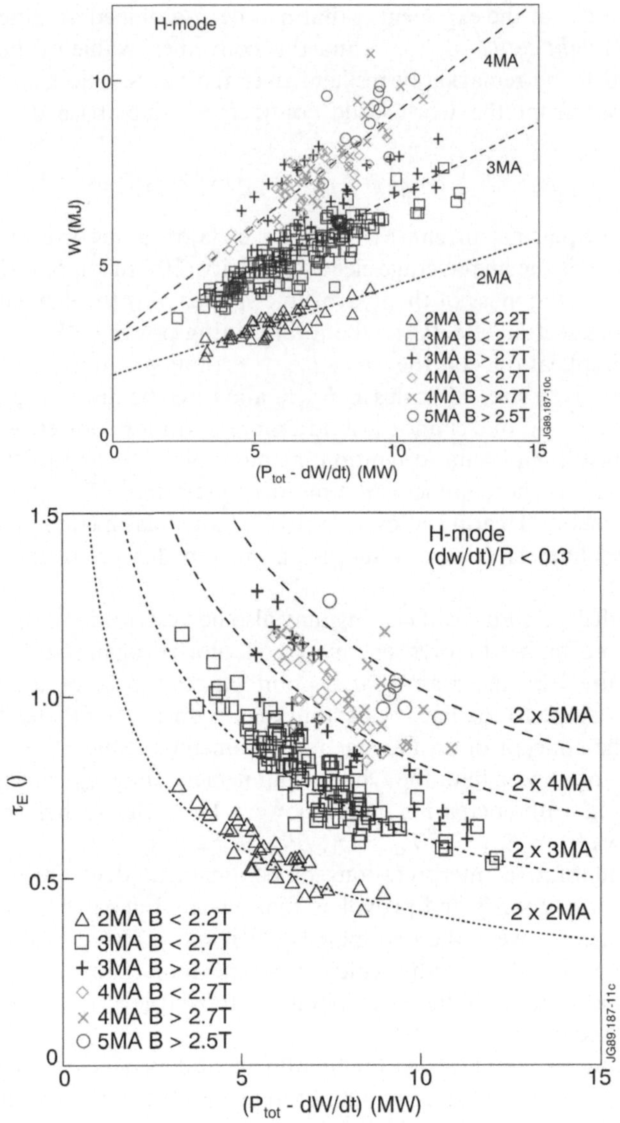

Figure 7.2. Confinement degradation with increasing heating power. This is illustrated with data from JET in H-mode plasmas with NBI at various values of I and B_t. In the lower panel, the data are plotted as τ_E against the effective total heating power P_{tot} (the term dW/dt corrects for the change in plasma energy in non-steady state conditions). The broken curves show the L-mode scaling $\tau_E \approx P^{-0.5}$ (multiplied by a numerical factor 2 for the H-mode). The upper panel shows the same data plotted as stored energy W against the effective total heating power. The broken straight lines illustrate the concept of an incremental confinement time. As discussed in the text, either interpretation appears consistent within the scatter of the experimental data.

from TFTR and JET would later show a further improvement in confinement with tritium. This is quite the opposite of what would be expected from most theories—increasing the ion gyro-radius would be expected to increase the diffusive step length and thus to decrease the confinement.

7.7 The H-mode

Against the gathering gloom of confinement degradation, there were some rays of hope. The year before the Aachen meeting, Wagner found two different operational regimes in beam-heated divertor plasmas in ASDEX (figure 7.3), which he named L- and H-type discharges, for low and high confinement [43]. The L-mode was the normal, degraded confinement obtained with additional heating, but under certain conditions the discharge switched itself into a mode of better confinement characterized by τ_E typically twice the L-mode values. The H-mode was subsequently reproduced in many other tokamaks and it was established that $\tau_E \propto P_{tot}^{-1/2}$, similar to L-mode scaling, though with a higher starting point.

The transition to the H-mode requires that the heating power be above a certain threshold, which was found empirically to increase with plasma size, density and toroidal magnetic field—though the reasons for this are still not

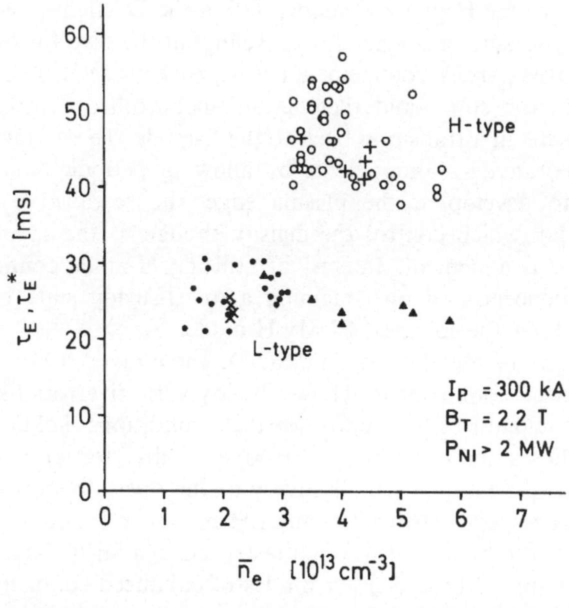

Figure 7.3. Data showing the discovery of the H-mode in Asdex. Confinement time τ_E is plotted against average density showing a substantial improvement for the H-mode compared with the L-mode.

fully understood. There are several other qualitative conditions that affect the ease with which the H-mode can be entered. For example, it is more easily obtained in a divertor configuration, and in particular when the toroidal magnetic field is in the direction so that the ion drift (box 1.3) is towards the divertor. But these are not rigorous conditions; H-modes have been obtained, albeit with more difficulty, with limiters [44] and even with Ohmic-heated plasmas. The change in confinement appears as an increase in the pressure gradient at the plasma edge, mainly through an increase in the edge density. This has the effect of raising the profiles over the whole plasma on a 'pedestal' and thereby increasing the energy content. Although there is evidence for further improvement in confinement across the whole plasma (opinions on this point vary to some extent between the different groups), the major improvement is undoubtedly at the edge. The H-mode caused a renewed interest in edge physics [45], which hitherto had been regarded by many as less important than core physics. But, in spite of intensive efforts, the H-mode remains incompletely understood and attempts to model the phenomenon theoretically have never been entirely successful. While there is considerable support for the paradigm of the suppression of turbulence by a sheared rotation induced by a radial electric field as an element of the L–H transition, there are also reservations and no single theory accounts for all the observations [46].

The improved energy confinement of the H-mode is accompanied by an improvement in the particle confinement. This is responsible for the characteristic signature of the H-mode, a sudden fall in the D_α signal (recombination radiation—a measure of the particle recycling) at the divertor or the limiter. But the improved particle confinement has negative consequences—impurities accumulate in the core—and there is an uncontrolled density rise which ultimately results in a transition back to the L-mode. These unwanted effects can be ameliorated to some extent by allowing periodic small relaxation instabilities to develop at the plasma edge, the so-called edge localized modes (ELMs), which control the density though at the expense of some loss of energy confinement. Interest in utilizing H-mode confinement in a reactor has concentrated on obtaining a quasi-steady state by means of repetitive ELMs—the so-called ELMy H-mode.

The success of the H-mode in ASDEX encouraged other tokamaks to pursue enhanced confinement. Those already with divertors like PDX and Doublet III could most easily reproduce these conditions. Tokamaks without divertors either sought means to introduce a divertor or a divertor-like configuration (the Doublet III discovery of the open divertor (section 6.7) helped in this respect), strove to obtain H-modes in limiter configurations (this was shown to be feasible though restricted to a fairly narrow operating range), or attempted to find other modes of enhanced confinement. At the London Conference in 1984, the Alcator group reported improved confinement with pellet fuelling [47]. They noted a rapid readjustment of the temperature and density profiles within a few hundred μs after injection of the pellet. Later JET would successfully combine the improved confinement in the core following pellet injection with the improved edge confinement of the H-mode—the 'pellet enhanced performance', or PEP mode. The tokamak plasma is a highly non-linear system that evidently can exist in several semi-stable states; subtle differences in the way it is set up apparently determine which path it takes at a bifurcation and at what point in parameter space it chooses to settle. The search for enhanced modes of confinement became an important issue in tokamak research and we will return to this topic in sections 8.3.2 and 9.7.1.

7.8 Attempts to understand confinement

Density profiles in tokamaks are generally observed to be strongly peaked even in the source-free core regions where simple models predict that profiles should be flat. This indicates that there must be an inward convective particle flux competing with the outward diffusive flux [48]. This component could include some or all of the non-diagonal terms in the transport matrix (section 6.8) that contribute to particle fluxes, including the terms driven by temperature gradient and by toroidal E-field (the Ware pinch). The necessity to include a convective inward term to model particle transport is not in dispute, but there has been less unanimity over the existence of a corresponding term in energy transport. The offset linear scaling of the energy confinement time formally translates into an inward convective energy flux [49], which seems to have been observed in some experiments [50].

Several groups had noted a tendency for electron temperature profiles in tokamaks to have a strikingly uniform shape, independent of the heating power deposition profile. This 'profile resilience' (or 'profile consistency') suggested that there may be a canonical equilibrium profile for the electron temperature which is hard to change [51]. Although this line of thinking was challenged to some extent by the discovery of the H-mode, theories abounded to explain the phenomenon. Generally, some kind of micro-instability was supposed to lead to enhanced transport, returning the plasma to a state of marginal stability. Foremost among the candidates were the ion temperature gradient, or η_i modes, which develop when $\eta_i = \mathrm{d}\ln T_i/\mathrm{d}\ln n_i \gg 1$. Alternatively, Rebut proposed and subsequently developed a model involving regions of stochastic magnetic fields between regular magnetic islands into a critical temperature gradient model [52]. Yet another view [53] was that actually it is the current density profile $j(r)$ rather than the electron temperature profile that is constrained by instabilities, so that the normal sequence whereby transport determines $T_e(r)$, which in turn determines the electrical conductivity and thus the current density profile, is reversed. All of these theories fit the empirical data in certain respects and disagree in others.

It is fairly well accepted that the anomalous transport results from some form of turbulent fluctuation, but it has proved to be very difficult experimentally to determine unambiguously if these are electrostatic or magnetic in nature. In the former case the magnetic field would be unaffected, but the fluctuations would lead to enhanced cross-field transport by $\mathbf{E} \times \mathbf{B}$ drift; in the latter case, parallel transport along stochastic field lines would be responsible for the losses. Direct measurements were possible only at the plasma edge and the results were overwhelmingly in favour of electrostatic fluctuations. But also in the plasma core, the evidence—though less direct—seemed to be consistent with electrostatic fluctuations.

Theories of heat conductivity often invoke turbulent drift-wave spectra, in which instabilities are excited whose phase velocity corresponds with the electron drift velocity:

$$\frac{\omega_{\text{wave}}}{k_{\text{wave}}} \approx v_{\text{drift}} = \frac{j}{en_{\text{e}}} \propto \frac{\nabla p}{n_{\text{e}}B} \propto \frac{T}{aB}.$$

If electron heat conduction were determined by a diffusion-like process involving scattering of charged particles by waves, with the step-size and effective collision frequency related to the k and ω of the waves, one would have for the heat diffusivity

$$\chi = \frac{\lambda^2}{\tau} \propto \frac{\omega}{k^2} \propto \frac{T}{kaB}.$$

The value of k at which the drift-wave spectrum peaks should be somewhere between $k \approx a^{-1}$, corresponding to the largest disturbance that will fit in the plasma, and $k \approx \rho_{\text{i}}^{-1}$, determined by finite gyro-radius stabilization of the waves. These upper and lower extremes of the scaling range for the heat diffusivity,

$$\frac{C_2 T}{B} > \chi > \frac{C_1(\rho_{\text{i}}/a)T}{B}$$

are named Bohm scaling and gyro-reduced Bohm (gyro-Bohm) scaling, respectively. In terms of experimental parameters,

$$\tau_{\text{E}} \propto \frac{a^2 B}{T} \qquad \text{(Bohm scaling)}$$

or

$$\tau_{\text{E}} \propto \frac{(a/\rho_{\text{i}})a^2 B}{T} \propto \frac{a^3 B^2}{T^{3/2}} \qquad \text{(gyro-Bohm scaling).}$$

In box 7.4, we discuss the constraints imposed by theoretical models on empirical scaling relations. Similarity rules and dimensionless numbers

Box 7.4 Similarity rules and dimensionless quantities

Similarity rules compare plasma behaviour in geometrically similar devices. If all the linear dimensions are scaled like the minor radius, a, and the plasma parameters as well as the field strengths like appropriate powers of a, one expects similar plasma behaviour if in each governing equation all the terms scale in the same way. But the complete set of equations needed to describe a fully ionized plasma, that is to say Maxwell's equations combined with either Boltzmann's kinetic equation for ions and electrons, or the Navier–Stokes fluid-dynamical equation, does not allow such scaling. A simplified system consisting of the collisionless Boltzmann equation and Faraday's law—which neglects the displacement current—does permit similar solutions with a and B independent. When collisional effects are included, the magnetic field must scale as $B \propto a^{-5/4}$, whereas the complete Maxwell equations require $B \propto a^{-1}$.

Similarity rules put constraints on how confinement can vary as a function of a, B_t, n and T. Solutions of the simplified set of equations may be expressed in terms of reduced velocity v/aB, temperature T/a^2B^2, electric field E/aB^2, density na^2, current density ja/B, etc. as functions of reduced position $\mathbf{r}/a$ and time tB. Then, with two independent variables, a characteristic time like a confinement time τ will be related to other characteristic quantities by equations such as:

$$\tau B = F\left(\frac{T}{a^2B^2}, na^2\right) = F\left(\frac{T}{a^2B^2}, \frac{nT}{B^2}\right).$$

In the low-β approximation, where nT/B^2 vanishes, τB should therefore depend on T/a^2B^2 only. Then, the question whether tokamak and stellarator plasmas display Bohm or gyro-Bohm scaling comes down to the choice between the functional dependences:

$$\tau_B B \propto \frac{a^2B^2}{T} \qquad \text{while} \qquad \tau_{g\text{-}B} B \propto \left(\frac{a^2B^2}{T}\right)^{3/2}.$$

A similar approach is to consider dimensionless quantities like the ratios between characteristic lengths, times, velocities, energy densities, etc. (In principle, one may add the temperature ratio, T_e/T_i and the mass ratio m_i/m_e.) The dimensionless pressure, β and the dimensionless ion gyroradius, $\rho^* = \rho_i/a$, stay constant if $n \approx a^{-2}$ and $T \approx a^2B^2$. To keep the dimensionless electron mean free path, λ_e/a, constant one must prescribe $B \propto a^{-5/4}$, whereas $B \propto a^{-1}$ would be required to keep the reduced Debye length, λ_d/a, constant.

for plasmas were known in the early 1960s [54], and were applied in cost–performance optimization of experiments, but they were not connected with scaling relations for energy confinement until they were rediscovered by Kadomtsev in 1975 and by Connor and Taylor in 1977. Dimensionless quantities and similarity rules and their bearing on scaling relations were reviewed more recently by Kadomtsev and by Connor [55].

The study of scaling relations promoted the so-called dimensionless scaling or wind tunnel experiments. The diffusivity was written in a form with either a macroscopic* Bohm-like ($\rho^* v_{\rm th}$) pre-factor or a microscopic gyro-Bohm-like ($\rho^{*2} v_{\rm th}$) pre-factor (where $\rho^* = \rho_{\rm i}/a$ is the dimensionless ion gyro-radius) multiplied by a function $F(\beta, v, q, {\rm etc})$ containing all the other dimensionless parameters on which transport could depend. Discharges were then carefully selected with identical parameters in F and differing only in the dimensionless gyro-radius ρ^*. Studies in tokamaks as well as in stellarators mostly seem to indicate that gyro-Bohm scaling gives the best fit to the experiments [56].

7.9 Transport codes

In order to compare experimental data with theoretical predictions or to predict the performance of future experiments, it is necessary to solve a set of coupled differential equations describing the cross-field particle and energy transport. It is also necessary to solve the equations determining the magnetic equilibrium, which in turn depend on the plasma profiles. In the core, it is usually safe to assume that the plasma parameters such as density and temperature are constant on the magnetic flux surfaces, so that the equations reduce to one dimension† and can be solved numerically by finite difference methods. Of course, a serious limitation is that the transport coefficients are not known reliably.

Two complementary approaches emerged during the 1970s. The first, exemplified by the transport code developed by Düchs, Post and Rutherford [57], either adopted a transport theory or determined its own empirical model by adjusting the transport coefficients to a set of experimental data. The model could then be used to extrapolate to new experimental conditions and to predict the performance of new machines. Although this approach met with considerable success and was widely employed, it

* Although modern jargon uses the term 'Bohm-like' in referring to transport processes that are determined by macroscopic effects with length scales comparable with the overall dimensions of the plasma, the quantitative values of confinement times are much higher than the Bohm times that had plagued the early years of fusion research.

† Modelling the plasma edge and interactions with limiters and divertors requires two or three dimensions.

was often difficult to isolate a single facet of a problem; there were too many parameters to fit, too many knobs to turn, and the end result was not always unique.

The alternative approach, exemplified by the TRANSP data analysis codes [58], used experimental data whenever available to calculate local transport coefficients. This approach had the advantage that the interpretation of the data and comparison with the model was more direct and the effects of assumptions more explicit. These codes have been refined exhaustively over the years with the inclusion of ever more sophisticated algorithms to model the plasma. A particularly important application of the TRANSP code has been to extrapolate data from experiments in D–D to predict performance in D–T.

Chapter 8

The era of the big tokamaks

Scientific feasibility, technical feasibility, commercial feasibility—these were the milestones on the planning charts that research administrations in the 1970s had to produce to justify their growing budgets for new energy sources. For fusion, 'proof of scientific feasibility' was arbitrarily defined as $Q = 1$, where Q is the ratio between the thermonuclear power produced and the external heating power applied to the plasma.* Discussions on the Joint European Torus (JET) started in Europe as early as 1971, but a final decision to construct this machine did not come until 1978. The Americans, though slightly slower off the mark in starting the design of their big D–T machine, the Tokamak Fusion Test Reactor (TFTR), were quicker to reach a decision and in 1974 Princeton won the assignment to construct the device [1]. The Soviet Union planned to build a test reactor, T-20, but later abandoned this and settled for a more modest superconducting physics experiment, T-15. Japan also opted for physics-oriented experiments in hydrogen and deuterium in JT-60, to be built at the JAERI laboratory at Naka.

8.1 Building the big tokamaks

The single parameter that comes closest to determining the size of a tokamak reactor is the plasma current. This provides the poloidal magnetic field that determines the size of the orbits of energetic α-particles whose containment is vital for the fusion energy balance. Also, although there was no reliable theoretical understanding of the energy and thermal particle loss processes from the plasma, the experimental data available in the early 1970s

* In the case of the D–T reaction, 80% of the fusion power is in the neutrons and 20% in the α-particles. Because only the α-particles are confined, the plasma does not 'ignite' or become energetically self-sustaining at $Q = 1$. Ignition of a D–T plasma requires a value of $n\tau T$ six times greater than the value to reach $Q = 1$. At ignition Q is infinite as the external heating power can be reduced to zero.

showed that both the plasma temperature and the energy confinement time increased with plasma current. So, after the dark 1960s there was light at the horizon and the scientists once more felt encouraged to develop forward-looking plans. And when the time came to ask for their approval, in the wake of the oil crisis of 1973, new energy sources enjoyed great popular and political support, so that the prospects for obtaining the necessary funding were better than ever. However, extrapolating to the plasma conditions for a fusion reactor from the conditions observed in the relatively small tokamaks in operation at the time was a very uncertain procedure. There was much scatter in the data and both the plasma pressure and the confinement time had to be extrapolated over two orders of magnitude, not to speak of all the technical problems that remained to be solved.

The tokamaks then in operation were heated Ohmically and temperatures would extrapolate to the range required for fusion only with very high toroidal magnetic field. But this was difficult to engineer to reactor size and would preclude the use of superconducting magnets. So, although very high fields were considered for experiments in which the physics of ignited plasmas might be studied (see section 8.4), most plans for scaling up the tokamak concept to a reactor assumed more modest fields and would require powerful additional heating. A few preliminary experiments with low-power additional heating had started by 1973 as we have discussed in chapter 6, but it was still too early to say how confinement would scale when the additional heating power exceeded the Ohmic power. The Alcator scaling for Ohmically heated tokamaks was not proposed until 1976 and the Goldston scaling for beam-heated tokamaks not until 1983. The projections available in the early 1970s suggested that a tokamak with a current of about 3 MA would reach the 'scientific feasibility' regime $Q = 1$ and, but only under the most optimistic assumptions, might just reach ignition [2]. Even if the operation fell short of these projections, a 3 MA experiment would be a crucial step to extend the data base much closer to the reactor regime and to pave the way for the design of the next generation—the test reactor that would demonstrate the technical feasibility of fusion energy.

8.1.1 JET—the Joint European Torus

Fusion research in Europe had started in five out of the six original member countries of the European Communities* already before the Geneva Conference in 1958, and from shortly thereafter it had been coordinated and partly funded by arrangements under the Euratom treaty. Serious consideration had been given to charging CERN†, the European centre for high-energy

* Later to become the European Union after more countries had joined and the three original Communities (Coal and Steel, Economic and Euratom) had merged into one treaty.
† CERN has a wider membership than that of Euratom.

physics, with this task and the CERN Study Group on Fusion Problems under the chairmanship of the head of the proton-synchrotron division, John Adams, had already broken the ground, but when the CERN Council in 1959 decided against broadening its activities, it fell to Euratom to step in. Under the leadership of the Italian physicist Donato Palumbo, a network was established that had the objective of coordinating European fusion research and development until the commercial stage. Euratom had failed to achieve this for fission reactor development and was eager to rehabilitate itself in the new field, but its efforts were regarded with appreciable suspicion, particularly from the French side. Whereas CERN was a scientific collaboration that had acquired political support, Euratom was a political undertaking in search of scientific and technical content. Without Palumbo's scientific authority and diplomatic skill the venture might well have headed for another failure.

The United Kingdom and Denmark already had fusion programmes and so the European fusion 'club' was enlarged when these countries joined the European Communities in 1972. The first discussions on the construction of a big tokamak for Europe started as early as 1971. The worldwide tokamak boom that had followed publication of the results from the Russian T-3 experiment was well under way. New tokamaks were being planned and built in most of the major fusion laboratories inside and outside the Soviet Union. Moreover, the Americans had approved already the construction of the 1 MA PLT, which was due to start operation in 1975 [3].

As the Americans had a head start with PLT, the Europeans considered that it would be desirable to leapfrog the 1 MA size of tokamak and to construct an even larger device to bridge the gap to a future prototype fusion reactor. Such a project would be beyond the budget of a single national laboratory and so a European collaboration was proposed. A working group recommended that a detailed design for a 3 MA tokamak should be prepared and it was agreed to establish a multinational project team to design the new machine which became known as JET—the Joint European Torus. The United Kingdom offered to host the design team at Culham Laboratory near Oxford. The first members of the multinational team, which grew to about thirty scientists and engineers, began to arrive in September 1973. The team was headed by Paul-Henri Rebut, who previously had been responsible for the construction of the French tokamak TFR.

In 1973, when the main parameters of the design were fixed, the largest tokamak actually operating had a current of 400 kA and the largest in construction would have about 1 MA. The original specification for JET, a plasma current of 3 MA, already was quite ambitious and even this was extended during the design phase to 4.8 MA. In fact the machine was so well designed that all that was required later to reach 7 MA was to strengthen the coil supports and buy additional power supplies. The contrast with earlier

Figure 8.1. Schematic of the JET experiment in its initial configuration with limiters on the outer mid-plane. Heating systems and the many diagnostic systems are not shown.

fusion experiments that had failed to reach their design goals was striking and JET would set a new milestone in the standard of construction of fusion machines. Water-cooled copper coils were selected for the toroidal magnetic field and, in order to minimize the bending stress, they were made D shaped.* The vertically elongated vacuum chamber (figure 8.1) gave room to a plasma with a bigger volume and a larger plasma current than one with a circular cross-section. The horizontal plasma radius would be 1.25 m and the vertical minor radius 2.1 m, making the elongation about 1.7. The aspect ratio (the major radius divided by the horizontal minor radius) would be about 2.4—much smaller than in previous tokamaks. The toroidal field was specified initially to be 2.8 T, to be increased in a later 'enhanced performance' stage to 3.4 T. Further increases during the life of JET subsequently raised the field to almost 4 T.

* The straight part of the D is supported by a central column; the curved part has the shape of a flexible wire in which the $I \times B$ force is balanced by the tension. This requires that its radius of curvature is proportional to the toroidal field strength, i.e. inversely proportional to R.

The JET design team published its proposal in 1975 [2] and put forward a powerful case for the new European experiment, but the early momentum was lost by several years of indecision and delay at the political level. The choice of site proved to be a major impasse. This had been deliberately left open at the outset and the design was such that it could have been built on any one of several suitable sites. The initial contenders included sites in Italy, France, Germany and the UK and it took nearly three years to agree on the British site at Culham. This remarkable episode is recorded in books by Willson and Shaw [4]. JET was to be the principal experiment of the European Fusion Programme. The organizational structure, a Joint Undertaking under European law, was formally established for a duration of 12 years beginning on 1 June 1978.* The major part of the cost of JET, 80%, was to be charged to the European Community budget. The United Kingdom, which would benefit from local expenditure, would pay a special contribution of 10% and the remaining 10% would be shared between those countries which had fusion research programmes. These included Sweden† and Switzerland, which although not members of the European Communities had meanwhile become equal partners in the common fusion programme.

Hans-Otto Wüster, a German physicist who had demonstrated his managerial skill at CERN, was appointed as Director of the project and the construction, on the Culham site, of the JET buildings started in 1979. The final stage of assembly was completed in January 1983 and the original design mandate of 3 MA was achieved by the end of that year.

8.1.2 TFTR—the Tokamak Fusion Test Reactor

Europe's decision in 1973 to design a big tokamak had provoked a quick response from the Americans. Determined to maintain a leading position, they started a conceptual design study of a large experiment early in 1974 and this produced a documented design for the Tokamak Fusion Test Reactor (TFTR) two years later [5]. The stated objectives were to demonstrate fusion energy production from the burning, on a pulsed basis, of deuterium and tritium in a magnetically confined toroidal system, to study the physics of large tokamaks, and to gain experience in the solution of engineering problems associated with fusion systems approaching the size of experimental power reactors.

Construction of TFTR started in April 1976 at the Plasma Physics Laboratory of Princeton University, which had become the largest fusion laboratory in the US and, with PLT and PDX under construction, had

* This original agreement was extended several times and finally concluded at the end of 1999, after which JET continued under a new organizational structure.

† Sweden subsequently joined the European Communities.

abandoned stellarators to concentrate most of its efforts on the tokamak line (section 7.1). One of Princeton's strengths had always been that it had very good motor-generator power supplies. Combined with engineering skills and a philosophy to build robust fusion experiments without unnecessary frills, new experiments could be built relatively quickly and economically. Other sites had been considered for America's proposed big tokamak, but Princeton's reputation and experience with tokamaks won the day.

The TFTR design specified a plasma current of 2.5 MA in a circular cross-section vacuum torus with inner minor radius 1.1 m and major radius 2.65 m. The toroidal field of 4.86 T on the plasma axis would be produced by 20 water-cooled copper coils. The main heating would use neutral beams, but these could do more than just heat the bulk plasma. John Dawson and colleagues at Princeton [6] had recalled that the main contribution to the fusion reactivity came from a relatively small number of ions in the high velocity tail of a Maxwellian distribution; the bulk of the plasma contributed more to the pressure (which is subject to beta limits) than to the reactivity. Injection of 50–100 keV ions would enhance the population of high velocity ions and therefore provide a superreactive plasma, as long as no anomalous loss or slowing down would take place. The idea had been worked out in Princeton [7] and was incorporated in the design of TFTR, which aimed at energy break-even ($Q = 1$) in two-component operation. The two components could be a deuterium beam in a tritium plasma, or vice versa (box 8.1) The proposal required as much as 20 MW of 120 kV D^0 to be delivered into the plasma for a pulse length of 0.5 s. As we will see later, these parameters were all exceeded during TFTR's operational life. It was planned to operate TFTR in D–D and D–T plasmas but with respect to radiation shielding and remote handling a slightly different policy was adopted compared with JET. Whereas JET was to be built inside a large hall whose walls would be thick enough from the outset to provide all the radiation shielding eventually required for D–T operation, TFTR would be built in a hall with rather thinner walls (1.8 m compared with JET's 3 m), within which an additional shield close to the machine (an 'igloo') would be erected before the start of the work with tritium.

The intense efforts to build TFTR (figure 8.2) all came into focus on the afternoon of Christmas Eve 1982 when, for the first time, a puff of hydrogen was injected into the torus as the magnetic coils were energized forming a 51 kA plasma that lasted for about 50 ms. A few months later, after rigorous plasma and systems testing and the installation of the extensive plasma diagnostics, TFTR began an experimental programme with Ohmic heating prior to the installation of the first neutral beams.

8.1.3 JT-60

The third and last of the big tokamaks to start operation, in April 1985, was the Japanese machine JT-60. Although interest in fusion had been aroused in

Box 8.1 Beam–plasma reactions

When an energetic beam of charged particles is shot into a solid, liquid or gaseous target, most of the particles give up their energy in exciting electrons to higher quantum states or ionizing target atoms or molecules—the beam is stopped in the target. In this process, the direction of the beam particles undergoes small changes—straggling—which on average tend to cancel out. In rare cases, a particle will undergo an elastic collision with a target nucleus, resulting in large-angle scattering accompanied by an appreciable exchange of momentum and energy. Still less frequent are inelastic nuclear collisions, in which nuclear energy is gained or lost and new particles may be formed. But averaged over all the particles the gain is always much smaller than the loss of beam energy. This is why fusion requires a hot plasma.

In a thermonuclear plasma with $T_\text{i} \approx 10\,\text{keV}$, the ions contributing most to the reaction rate are those in the far tail of the Maxwellian velocity distribution. But when one fast ion loses its energy by elastic collisions with other ions, this is not a loss because the energy stays in the thermal reservoir from which the Maxwellian tail is continuously replenished by other elastic collisions. Yet, it may pay to enhance the tail of the velocity distribution and one way to accomplish this is called 'two-component operation'.

When a beam is injected into a completely ionized plasma target, the three competing energy transfer processes are elastic collisions with free electrons, elastic collisions with plasma ions and nuclear reactions. Their probabilities depend on the relative velocities of the colliding particles, which in most cases of thermonuclear interest are in decreasing order: the electron thermal velocity, the beam velocity and the ion thermal velocity. Because of the v^{-4} dependence of the Coulomb-scattering cross-section, the stopping power of the target plasma decreases as $T_\text{e}^{-3/2}$, which if the beam is to produce net energy sets a lower limit to the electron temperature. Since reaction energies are typically one hundred times greater than beam energies, roughly 1% of the beam needs to react for break-even. For injection of deuterium into a tritium plasma, or vice versa, the critical electron temperature is about $T_\text{e} = 3.5\,\text{keV}$. This somewhat alleviates the problem of electron energy confinement, when compared with what is required if $T_\text{e} = T_\text{i}$.

Japan by the 1955 and 1958 Geneva Conferences, it had been decided that research initially should be of a basic nature and reside under the Ministry of Education before the Japan Atomic Energy Research Institute (JAERI), under the Ministry of Trade and Industry, would become involved. So, the

Figure 8.2. Schematic of the TFTR experiment. Three neutral beam heating systems are shown.

first phase of the fusion programme at JAERI, which included the construction of the small JFT-2 tokamak, had not started until 1969. The second phase programme, which started in 1975, included plans for the construction of a big tokamak, JT-60, which was intended to reach conditions in deuterium that would be equivalent to break-even in D–T. Thus, Japan joined the big tokamak league with rather fewer intermediate-sized machines than the other countries.*

JT-60 was to be constructed on a new site at Naka close to the main JAERI laboratory and would be housed in a massive building with thick radiation shielding walls, although JT-60 was not intended to operate with tritium. The tokamak was specified to operate at 2.7 MA in limiter mode with a plasma of circular cross-section with minor radius of 0.95 m, major radius of 3.0 m and toroidal field at the plasma centre of 4.5 T. As with the other two big tokamaks, these design parameters were comfortably exceeded in operation.

JT-60 was the only one of the three big tokamaks to be designed at the outset with a divertor, in this case a set of three poloidal coils inside the vacuum vessel on the low field side of the tokamak. This divertor took up so much space that the plasma was small relative to the dimensions of the toroidal field coils; it proved effective at controlling impurities but not so successful in accessing the H-mode which seemed to prefer divertors at the top or bottom of a torus (the H-mode was not discovered until 1982 by

* Fusion research in Japan continued along parallel paths, with a broadly based programme in the universities and a programme focused on tokamaks at JAERI, until the beginning of 2001 when the two programmes merged into a single entity.

Figure 8.3. Interior of JT-60U after the vessel walls had been completely covered with carbon tiles. The open divertor region can be seen at the bottom of the torus. In later stages of the experiment this was replaced with a more closed divertor.

which time JT-60 was at an advanced stage of construction). After operating with this configuration for about four years, JT-60 was shut down for a radical upgrade in 1989, which included the installation of a divertor of more conventional design with a single null at the bottom of the torus. Taking advantage of the large dimensions of the toroidal field coils, this would allow a substantial increase in the plasma cross-section and current (figure 8.3) and the performance would match that of JET. The decision to rebuild JT-60 was a bold one and the speed of the reconstruction was impressive. The new version, renamed JT-60U, would come into operation in March 1991 and would immediately begin to play a major role.

8.2 Operation and results

8.2.1 Heating the big tokamaks

The principles of the additional heating methods were first demonstrated on smaller tokamaks in the early 1970s, but these techniques had not yet been applied on the large scale now required. In fact, when the design of the big tokamaks started there was virtually no experience of additional heating. It was assumed that the scaling of temperature with heating power would be close to linear and there was even some optimism that things might actually get better when tokamaks no longer relied on the plasma current for heating. The bad news that confinement actually degraded with heating power was still coming in when the big tokamaks came into operation a decade later. The heating requirements for all the big tokamaks had to be upgraded several times in the light of this experience. For example, the first specification for JET in 1973 had provision for only 3 MW of neutral beams, although it was noted that this might need to be enhanced to 10 or even 25 MW as heating techniques were refined and additional funding became available. In fact, by 1990, JET had installed 24 MW of neutral beams plus a nominal 32 MW of ion cyclotron heating and 12 MW of lower hybrid current drive. Likewise, TFTR's initial specification of 20 MW of neutral beams was increased to 40 MW plus 10 MW of ion cyclotron heating.

Following the first applications in mirror machines, the technology of neutral beam sources had been developed extensively. The early sources developed at Oak Ridge, Culham and Berkeley had evolved during the late 1970s to the multipolar type that later came into widespread use. This is a large volume ion source where a low temperature plasma produced by electrons from heated tungsten filaments is confined by a multipolar magnetic field. The ions are extracted from this source plasma and accelerated though a series of carefully aligned multi-aperture grids. The extracted ions are then neutralized in a neutral gas from which they pick up an electron with virtually no change of energy or momentum. Those ions that have not been neutralized are deflected magnetically towards dump plates while the neutral beam passes into the recipient tokamak plasma. The neutralization efficiency is high at energies in the 10–20 keV energy range, but falls at higher energies. Yet, the beam energies inevitably had to be increased from 15 keV, sufficient to penetrate to the cores of small tokamaks, to some 80 keV for hydrogen or 160 keV for deuterium, needed to penetrate the big tokamaks.

The three large tokamaks all started operation without any additional heating; this was added in stages and upgraded over a period of years to the values quoted in table 8.1. A common feature was the dependence on neutral-beam injection heating. This was particularly so for TFTR, which relied on powerful beams to create the conditions for optimum confinement that will be described later. The final specification for the TFTR beam system consisted of four beamlines, each with three sources with a maximum operating voltage of 120 kV. The JET system consisted of two beamlines each with eight sources capable of operating at 80 keV with hydrogen or 160 keV with deuterium or, in a later stage, with tritium. JT-60 had 14 beamlines each with two sources delivering a total of 28 MW and this was increased to 40 MW in JT-60U. Extrapolation to a reactor will require even higher beam energies, in the range of 1 MeV and this necessitates the

Table 8.1 Summary of main parameters* of the big tokamaks

Parameter	Symbol	Unit	JET	TFTR	JT-60U
Plasma major radius	R_0	m	2.96	2.62	3.4
Plasma minor radius (horizontal)	a	m	1.25	0.97	1.0
Plasma elongation ratio	κ		1.68	1.0	1.5
Toroidal field on axis	B_t	T	3.45–4.0	5.9	4.2
Plasma current (limiter mode)	I	MA	7.0	3.0	—
Plasma current (divertor mode)	I	MA	5.0	—	6.0
Neutral beam power	P_{nb}	MW	24	40	22
Ion cyclotron power	P_{icrh}	MW	32	10	6
Lower hybrid power	P_{lh}	MW	12	—	15
Total additional heating	P_{add}	MW	68	50	43

* These parameters are the maximum values achieved and were not necessarily used routinely or simultaneously. This applies in particular to the heating powers. Generally the optimum confinement was obtained with operation at less than maximum current.

development of a new type of neutral beam source. The efficiency of accelerating and neutralizing positive-ion beams, as used hitherto, becomes very low at energies above 200 keV. The technology of producing and accelerating negative-ion beams, which are much more efficient at higher energies, is being developed and the JT-60U team has achieved a beam-driven current of 1 MA with 3.75 MW beam power at 360 keV.

The three large tokamaks also had ion cyclotron heating, but only in JET was this pushed to power levels that matched the neutral beams. None of the three big tokamaks used electron cyclotron heating—a consequence of the difficulty of producing large powers at the relatively high frequencies needed to match toroidal fields exceeding 3 T. Lower hybrid current drive has been included in table 8.1, although in these experiments this is generally used for controlling the plasma current distribution rather than for plasma heating.

8.2.2 Keeping clean

For all of the large tokamaks it became an immediate and continuing task to reduce impurities to acceptably low concentrations. We have discussed the effects of impurities and the measures to reduce their impact that were developed on other tokamaks already in sections 6.5 and 7.4, but many of these lessons came too late in the day to influence the construction of the big machines and had to be painfully relearned. All three tokamaks were designed with metal torus walls and these had to be covered progressively with carbon tiles in order to reduce metallic impurities in the plasma. Impurities and density control were the main obstacles to improved performance, particularly as heating powers were raised, and an important part of the operation had to be devoted to these issues.

Although it had been recognized from the outset that the limiters (and the divertor in the case of JT-60) would be required to handle substantial heat loads, the initial choice of materials was found wanting. JET had installed a set of water-cooled nickel limiters, equally spaced around the midplane on the outside of the torus. These were designed to cope with the expected power loads but concerns about the effect of nickel impurities in the plasma caused them to be replaced with carbon limiters just before the first operation. However, these could not be cooled and were replaced later by two cooled, continuous belt limiters, equally spaced above and below the outer mid-plane. Both carbon and beryllium tiles were used on these belt limiters. TFTR combined a continuous toroidal 'bumper' limiter on the inner wall, poloidally extending 60° above and below the midplane to take most of the heat load, with a movable limiter to control the plasma size. Initially these were made of carbon coated with titanium carbide, but the coating was quickly eroded by plasma and was abandoned in favour of bare carbon (figure 8.4).

Both JET and TFTR experienced serious problems with massive bursts of carbon influxes at high heating powers—referred to as 'carbon blooms' on the American side of the Atlantic and as 'carbon catastrophes' on the European side. These influxes, which were caused by localized hot spots on exposed edges of limiter or divertor tiles, became a serious barrier to high performance. The problem was resolved eventually only by very careful

Figure 8.4. Interior of TFTR showing the carbon limiter on the high field side of the plasma (i.e. at small major radius) and the smaller limiter belt on the low field side.

attention to the design and alignment of individual tiles to eliminate all exposed edges and by replacement of amorphous carbon tiles in the most critical areas with others made of carbon fibre reinforced graphite that had better thermal and mechanical properties.

However, carbon absorbs large quantities of hydrogen, which would cause further problems in terms of recycling control in the big tokamaks. Following the introduction of a new carbon limiter with an area of $22\,m^2$ in TFTR in 1986, it appeared [8] that extensive discharge cleaning in deuterium was required to reduce oxygen to acceptable levels. But this left the limiter saturated with so much deuterium that it became very difficult to control the density and made further conditioning in helium necessary to remove the deuterium. The resulting Ohmic target plasmas with very low levels of recycling were found to be prerequisite for the enhanced confinement conditions with powerful neutral beam heating and fuelling—the so-called 'supershots' discovered in TFTR in June 1986 [9].

Wall conditioning was also found to be very important. Although JET could be baked to 500°C and operated with the walls hot, impurities remained a problem. To reduce oxygen levels, carbonization was attempted until beryllium gettering came into routine use around 1989. The penalty to be paid for the use of this toxic metal was that all work inside the torus had to be carried out in pressure suits. Initially TFTR had used chromium gettering, but this was abandoned due to fears of contamination of carbon limiters. TFTR later saw improvement after injecting lithium pellets into the plasma, which formed an active layer on the walls, and JT-60U employed boronization to good effect.

8.2.3 Pushing to higher performance

The results obtained in JET with Ohmic heating alone exceeded expectations. With a plasma density of about $4.2 \times 10^{19}\,m^{-3}$, the central ion and electron temperatures were 3 and 4 keV respectively and the energy confinement time was 0.8 s. The fusion triple product $n_D(0)\tau_E T_i(0)$ reached values of about $1 \times 10^{20}\,m^{-3}\,s\,keV$. Initial results with neutral beam and ion cyclotron heating raised the temperature but degraded the confinement time so that there was no significant improvement in the fusion product. Similar observations were made with Ohmic heating in TFTR, though with pellet fuelling the fusion product could be raised to $2 \times 10^{20}\,m^{-3}\,s\,keV$. It was clear that confinement would have to be improved substantially in order to push the big tokamaks closer to the value of $10^{21}\,m^{-3}\,s\,keV$ required for $Q \approx 1$.

Even before the big tokamaks started to operate, the discovery of the H-mode in ASDEX had raised the question of whether they could take advantage of this breakthrough in confinement. Only JT-60 had been designed from the outset with a divertor, but this—located on the outer mid-plane—did not in fact allow access to the H-mode. Although extensive

Box 8.2 From $n\tau$ to $n\tau T$

From what was discussed in section 1.2.3 and box 1.6, it follows that the ratio between power produced in and power lost from a steady-state thermonuclear plasma scales as

$$R \propto n_\mathrm{i}\tau_\mathrm{E}\langle\sigma v_\mathrm{i}\rangle/T_\mathrm{i} \propto \beta B^2\tau_\mathrm{E}\langle\sigma v_\mathrm{i}\rangle/T_\mathrm{i}^2.$$

So, in a β-limited system with given field strength and with optimized τ_E, the optimal temperature is that where the function $\langle\sigma v_\mathrm{i}\rangle/T_\mathrm{i}^2$ reaches its maximum. (We assume here that τ_E may depend on β but not explicitly on T_i, as it does in mirror confinement.) In the vicinity of this maximum, the function is approximately constant so that $\beta B^2\tau_\mathrm{E}$, that is to say the triple product $n_\mathrm{i}\tau_\mathrm{E}T_\mathrm{i}$ becomes a good measure for R.

For D–T fuel, the maximum lies in the range 10–15 keV, where

$$R \approx 10^{-21}n_\mathrm{i}\tau_\mathrm{E}T_\mathrm{i}.$$

When, in the 1980s, the big tokamaks routinely reached these temperatures, it was simpler to express their performance in this single number rather than as a point in an $n\tau$ versus T diagram, but one should note that for temperatures outside this range the triple product overestimates R. Obviously, a precise estimate of R requires knowledge of electron and ion temperature and impurity concentration, as well as temperature and density profiles.

Progress in fusion is usually stated in terms of the parameter Q which is defined as the ratio of the fusion power produced to the heating power applied externally, $Q = R/(1 - R)$.

research on LHCD had made it possible by 1990 to find a limiter H-mode [10], the long-term solution for JT-60 would lie in the rebuild as JT-60U with a single-null divertor and increased plasma cross-section. TFTR and JET had been designed with limiters. The circular cross-section of TFTR precluded any serious prospect of adding a divertor retrospectively and consequently TFTR concentrated on enhancing confinement by other means, very successfully as it would turn out.

Fortuitously, JET's D-shaped torus would allow a divertor and this was implemented in stages. It was realized that both single-null and double-null divertor-like configurations* could be produced in JET without any major modifications, simply by a suitable combination of currents in the existing

* In JET publications referred to as 'X-point configurations', an X-point being the point in the poloidal plane where the separatrix between open and closed magnetic surfaces crosses itself.

poloidal magnetic coils enhanced by the effect of magnetic saturation in the central transformer core. The single-null configuration was demonstrated in 1986 at currents up to 3 MA and the double-null up to 2.5 MA [11]. The plasma went into the H-mode for about 2 s when the neutral beam heating power was increased above about 5 MW. At a plasma temperature of about 6.5 keV and a density of $1 \times 10^{19}\,\mathrm{m}^{-3}$, this H-mode produced about $2 \times 10^{20}\,\mathrm{m}^{-3}\,\mathrm{s\,keV}$ for $n_{\mathrm{D}}(0)\tau_{\mathrm{E}}T_{\mathrm{i}}(0)$, twice as good as the best so far obtained in JET limiter plasmas and extrapolating to a fusion power exceeding 1 MW with tritium. The results achieved in JET's first three years of operation were widely acclaimed when announced [12] in November 1986 at the Kyoto Conference.

The TFTR team was achieving impressive results without a divertor. Powerful neutral beams were used to heat and refuel the core plasma and extensive helium conditioning of the wall and inner limiter reduced the hydrogen recycling and carbon influx [9]. This resulted in the so-called 'supershot' regime characterized by strongly peaked density profiles, broad electron temperature profiles and high ion temperatures, the central value even reaching 20 keV.

In the following few years the intellectual and material efforts invested in these two big tokamaks were amply rewarded with increasingly impressive scientific results. At the Washington Conference in 1990, the TFTR team reported [13] that plasma parameters in the supershot regime had been extended to $T_{\mathrm{i}}(0) \approx 35\,\mathrm{keV}$, $T_{\mathrm{e}}(0) \approx 12\,\mathrm{keV}$ and $n_{\mathrm{e}}(0) \approx 1.2 \times 10^{20}\,\mathrm{m}^{-3}$ producing D–D reaction rates of $8.8 \times 10^{16}\,\mathrm{s}^{-1}$. Values of the fusion parameter had reached $n_{\mathrm{e}}(0)\tau_{\mathrm{E}}T_{\mathrm{i}}(0) \approx 4.4 \times 10^{20}\,\mathrm{m}^{-3}\,\mathrm{s\,keV}$ in TFTR* and $n_{\mathrm{D}}(0)\tau_{\mathrm{E}}T_{\mathrm{i}}(0) \approx 8 \times 10^{20}\,\mathrm{m}^{-3}\,\mathrm{s\,keV}$ in JET [14], although the JET team added an important note of caution that these conditions could not be sustained in steady state. The best TFTR supershots in D–D had $Q_{\mathrm{DD}} \approx 1.9 \times 10^{-3}$, the equivalent of $Q_{\mathrm{DT}} \approx 0.3$ if the same conditions could be attained in a 50/50 D–T mixture. In JET the equivalent $Q_{\mathrm{DT}} \approx 0.8$ provided encouragement for the plans for actual experiments with tritium in 1991.

The final round of performance improvement in D–D plasmas in JET and TFTR before D–T operations was reported in 1992 at the Würzburg Conference. Further improvements to the supershot regime in TFTR with injection of lithium pellets and a massive 33 MW of NBI resulted in a D–D fusion reaction rate of $10^{17}\,\mathrm{s}^{-1}$. In JET the hot-ion H-mode had been optimized with 14.9 MW of NBI to reach $9 \times 10^{20}\,\mathrm{m}^{-3}\,\mathrm{s\,keV}$ in deuterium, equivalent to $Q_{\mathrm{DT}} = 1.14$. But from there on the JT-60U team took the lead in D–D performance and two years later announced in Seville [15] that $1.2 \times 10^{21}\,\mathrm{m}^{-3}\,\mathrm{s\,keV}$ had been reached. Thus in a period of about ten years, the three big tokamaks had increased the fusion product by an

* For a strict comparison, the TFTR result needs to be multiplied by $n_{\mathrm{D}}(0)/n_{\mathrm{e}}(0) \approx 0.75$.

order of magnitude from the values obtained when they first operated. This brought them within a factor of five of the value required for ignition, but this step was too large to be feasible with any realistic upgrade and would require a substantially bigger machine.

These results allowed a fairly accurate definition of the parameters of a fusion reactor. With a toroidal field up to 5 T, it would require linear dimensions two to three times larger than the present generation of big tokamaks and, with a plasma current of more than 20 MA, it would produce a fusion power output of several GW (thermal). Even without special means for driving the plasma current, the pulses would last at least one hour. As we will see in the next chapter, the world's major fusion programmes had meanwhile jointly undertaken to design, and hopefully construct, a test-reactor known as ITER, the International Tokamak Experimental Reactor. As the design of ITER evolved, it turned out that it would have so much in common with JET that the latter became increasingly important as a scale model for ITER. So, a good part of JET's programme was specifically directed to provide input for this new project.

8.2.4 Real fusion power at last

JET and TFTR had been designed from the outset to carry out experiments in D–T plasmas. Both tokamaks had substantial shielding against fusion neutrons and gamma rays as well as provisions for remote handling and maintenance. The original plans for both machines had included a period of several years' operation in deuterium to bring the machine systems and the plasma conditions up to the performance levels which would justify operating with tritium. After an extensive period of tritium operation, substantial repairs or modifications to the tokamaks, though feasible, would be difficult and time consuming. The natural place for extensive tritium operation was clearly at the end of their programmes, after which the tokamaks could be shut down or used for engineering studies to gain further experience of remote handling. However, once TFTR and JET started operating, it became clear that it would be more difficult to achieve their aims than had been envisaged, and that the process of optimization during the deuterium phase would take much longer. The starting dates for tritium therefore tended to get pushed farther back. Moreover, it was recognized that these big tokamaks were unique and powerful facilities with the capacity to carry out important physics studies that were needed for the final design of a reactor. Thus there was pressure, on the one hand to extend the physics programmes in deuterium by adding new elements, but on the other hand to give much-needed new momentum to these programmes by the demonstration of 'real' fusion.

JET resolved this dilemma by deferring its plans for the major tritium phase (which should have been completed on the original schedule before

Figure 8.5. Interior of JET. The detailed arrangement of limiters, divertors and wall tiles was changed several times during the life of JET. This view shows the arrangement in 1998 after the installation of the highly enclosed 'gas-box' divertor with its deep recesses at the bottom of the torus. By this stage almost all of the interior surface—in particular the inner wall—had been covered with carbon tiles. On the extreme right of the view (between two rails of the protective poloidal limiter) can be seen four ion cyclotron heating antennas. To the left of this group of antennas are the lower hybrid current drive antenna (surrounded by its own carbon protective frame) and the articulated arm of the remote manipulator that is inserted into the vacuum vessel during maintenance periods (as here) after the D–T campaign.

1990) and proposed instead an extension of the programme to include both the tritium experiments and a new divertor (figure 8.5). A very short preliminary series of experiments in tritium would be conducted in the second half of 1991 and the final round of tritium experiments would take place later, after the installation of the new divertor. A major constraint on the preliminary tritium experiments was the relatively low limit on neutron production set by the need to minimize activation of the machine structure in order to carry out engineering work for the new divertor. The target plasma was a hot-ion H-mode in the single-null configuration with a plasma current of 3 MA and toroidal field of 2.8 T. In order to minimize activation of the vessel, the initial optimization of this discharge was carried out in deuterium. This was followed by several plasma pulses with tritium concentrations as low as 0.1% to test the diagnostics. When everything was judged to be ready, tritium was introduced into the plasma by operating two of the 16 neutral beam sources in tritium. This produced a plasma mixture with about 10% tritium and 90% deuterium. The performance lay in the middle of the range of conditions produced during the optimization experiments in deuterium and the actual Q_{DT} of 0.15 would have been 0.5 in the optimum 50% tritium mixture. The fusion power peaked sharply for 2 s, reaching a maximum of 1.7 MW. Detailed analysis of the data provided important information on transport and confinement and the public image of fusion received a boost from the extensive press coverage.

But now it was TFTR's turn. Princeton conducted an extensive campaign of high-power D–T experiments between 1993 and 1997 [16], utilizing the high-performance plasma scenarios already developed in deuterium plasmas. With 39.5 MW of deuterium and tritium neutral beams having energies up to 115 keV, the fusion power output rose to 10.7 MW. The fusion power density in the plasma core reached about $2\,\mathrm{MW\,m^{-3}}$, a value comparable with that expected in a fusion reactor. The experiments in TFTR provided the first opportunity to study the effects of plasma heating by α-particles by comparing the electron temperature in D–T plasmas with that in otherwise similar D–D plasmas; they showed α-particle confinement to be consistent with theoretical predictions.

During 1997, JET returned to tritium operation and a broad-based series of D–T experiments was carried out, setting new records for fusion power and producing a great deal of scientific and technological data for future reactors. The highest fusion power in JET, establishing a new record value of 16 MW (figure 8.6), was obtained using the so-called 'ELM*-free H-mode', in which there is an energy transport barrier near the plasma edge. Careful control of the plasma density before the start of the H-mode allowed 20 MW of neutral beam heating power to be coupled efficiently to the ions—the ion temperature rose to nearly 30 keV while the electrons stayed around 10 keV. The ratio of fusion power to heating power, Q, was equal to 0.6 and would be equivalent to a value of 0.9 if corrected for the fact that the temperature was still rising. However, these favourable conditions could not be maintained for very long. The good particle confinement inherent in the H-mode caused the density to rise as well as the temperature and, after a couple of seconds, the density became so high that the H-mode terminated.

One way to prevent the density rise was to allow a mild instability at the edge of the plasma to develop. This so-called 'steady-state ELMy H-mode' had been foreseen as the most likely mode in which ITER would operate in order to avoid the transient nature inherent to the ELM-free H-mode. The ELM instabilities at the plasma edge would degrade the particle confinement sufficiently to prevent the uncontrolled increase in density. There was of course a price to pay in the form of some degradation also of the energy confinement. JET ELMy H-mode experiments in D–T had electron and

* Edge-localized mode.

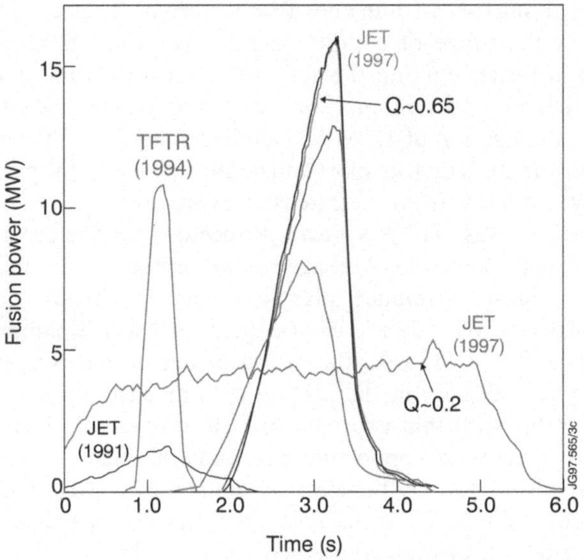

Figure 8.6. Fusion power with D–T operation of JET and TFTR. JET's first D–T campaign in 1991 produced a peak of about 1.7 MW lasting about 2 s. TFTR's campaign between 1993 and 1997 reached a new record of 10.7 MW. JET's second campaign in 1997 is illustrated by the record value of 16 MW in an ELM free H-mode and by the quasi-steady-state ELMy H-mode with about 4 MW lasting for nearly 5 s.

ion temperatures of about 8 keV and produced about 4 MW of fusion power using about 20 MW of heating power. This quasi-steady state was maintained for nearly 5 s, a limit imposed by JET's heating systems. These results put the extrapolation to reaching ignition in ITER on a much more reliable basis than hitherto. An important physics result was that the heating power needed to access the H-mode was lower in pure tritium than pure deuterium, approximately by the inverse of the average atomic mass. The significance for ITER would be a potential 33% reduction in the power needed to access conditions with high thermal insulation in a pure tritium plasma (for example, during start-up) and a 20% reduction in the power needed to maintain such conditions during high-fusion-power operation. A study of α-particle heating confirmed TFTR's finding that the α-particle confinement was as predicted by neoclassical theory, an important result as anomalous α losses would have important consequences on the fusion energy balance.

An important aspect of the tritium experiments in TFTR and JET was the technological experience gained in the safe handling of tritium in a fusion experiment. The first short series of experiments with tritium in JET had required only a relatively small amount of tritium, about 0.2 g, but much larger quantities were required for the later series. The 1997 JET experiments utilized a stock of about 20 g but this was recycled repeatedly though the plant, so that it was the equivalent of nearly 100 g. TFTR, which had no facility for recycling tritium on site, used about 100 g of tritium to produce over 1000 D–T discharges with a total of about 1.6 GJ of fusion energy. The quantities of deuterium and tritium actually consumed by the fusion reactions were very small, a total of about 1 mg of each for the whole series of JET experiments, but this produced as much energy by the fusion process, a total of 675 MJ, as burning about 50 kg of hydrogen in a chemical reaction.

8.2.5 The end of the era

By the end of the twentieth century, the big tokamaks had reached the targets that had been set for them some 25 years previously—the fusion triple product had been brought up to the break-even value and operation with tritium had demonstrated real fusion power. The difficulties had been much greater than could have been foreseen and the programmes had taken a decade or so longer than the early optimistic plans had allowed. But at last the long-sought goal of ignition was within sight, though just out of reach, for these experiments.

After completing the tritium campaign, TFTR was closed down in April 1997. This left the US fusion programme without its major facility. The aims and the budget for fusion research had been re-evaluated at the government level and, as we shall see in section 8.4, plans to construct a new major experiment at Princeton to replace TFTR had failed to win approval.

JET's second tritium campaign in 1997 was followed by further experiments in deuterium. JET had been scheduled to close at the end of 1999 to release funds for the anticipated European participation in ITER, but when it became clear that decisions on ITER would be delayed, it was decided to extend the life of JET. Under EFDA, the European Fusion Development Agreement, a new organization was set up to use the JET facilities until at least the end of 2002 with the possibility of enhancement of the machine performance and further operation beyond that date.

JT-60U also was threatened with closure in 2001 but received a reprieve pending decisions on plans for its replacement at JAERI with a new superconducting tokamak.

8.3 Improving the tokamak

During the lifetime of the big tokamaks, there was a steady improvement in the understanding of tokamak physics, albeit as much on a pragmatic and empirical basis as from any major advances in theory. Much important work was carried out on the big tokamaks in parallel to the 'parameter pushing' that determined their main programmes. Many valuable results also came from the smaller and medium-sized tokamaks and stellarators that operated alongside the big machines. The results from the big tokamaks combined with data from the small and medium-sized tokamaks operating through the 1990s allowed further refinement of the empirical scalings for energy confinement that had been evolving for the past two decades. An increasingly important aspect of this work was that international data bases were compiled which we include here, though they remain at a less-advanced stage than conventional tokamaks.

Relatively few new conventional tokamaks were built after 1983, but there was new-found enthusiasm for stellarators and for so-called spherical tokamaks. In the following we attempt to summarize some of the most important tokamak issues.

8.3.1 Reactor-relevant divertor physics

The results from the first generation of divertor experiments in the 1970s and early 1980s had shown that poloidal-field divertors offered a means for impurity control and, with the serendipitous discovery of the H-mode, for improved energy confinement. In the 1990s, the subject was followed up in DIII-D, JET, JT-60U, ASDEX U and Alcator C-mod. In fact, the main purpose for the construction or upgrade of this generation of tokamaks often was to incorporate advanced divertors. Not only did these experiments need the divertors to improve their own performance, they also had to face the tasks of integrating the divertor principle into a reactor-like magnetic coil geometry and of making the conditions for divertor operation compatible with those required for a reactor. In a fusion reactor, a divertor is required to perform three primary functions: to handle the power exhaust while keeping erosion of the target plates to a minimum, to control the impurity level of the core plasma and the recycling of the hydrogen fuel, and to pump the helium 'ash'. Work along these lines received additional impetus when the difficulties of designing a divertor for ITER to satisfy all of these requirements simultaneously became fully appreciated. Detailed modelling calculations [17] of the plasma in a divertor scrape-off layer guided the optimization of the divertor design, but experimental tests were required to validate and to benchmark these models.

The reactor studies had revealed that it would be difficult to design a divertor capable of handling by direct plasma–plate contact more than about 15% of the power deposited in the plasma by alphas and other heating. Hence, the plasma flowing into the divertor has to be cooled to a temperature of only a few eV before it contacts the plates. This requires that the bulk of the power flowing out of the plasma core is dissipated by radiation or charge-exchange processes in the scrape-off layer and the divertor chamber (figure 8.7). Model calculations and experiments showed that radiation

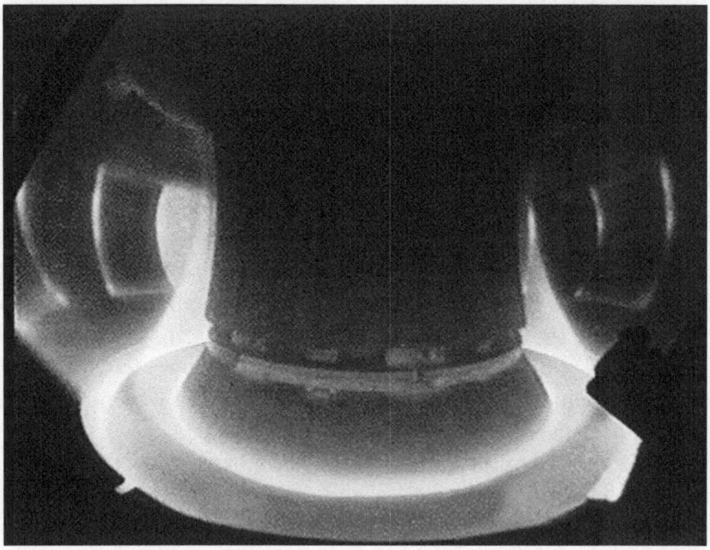

Figure 8.7. Plasma light emission showing the poloidal divertor in the ASDEX-U experiment.

cooling by the recycling hydrogen and intrinsic impurities would generally be insufficient so that the divertor plasma would require 'seeding' with additional impurities. But as the fraction of dissipated power increased towards 100%, plasma became detached from the divertor and a particular concern was that the detachment would be unstable—the region of radiation cooling might encroach into the core regions and degrade the fusion performance. It turned out however that such detached plasma conditions could be maintained in stable equilibrium, without disruptions [18].

8.3.2 Advanced tokamak scenarios

The ultimate viability of the tokamak as the basis for an economically attractive fusion power plant requires sufficient energy confinement time τ_E for ignition and sufficient volume-averaged toroidal beta β_t for adequate fusion power density. Both of these parameters are observed to increase approximately linearly with plasma current I, but the L-mode scaling for τ_E combined with the Troyon beta limit could lead to a minimum current for ignition as high as 30 MA with consequently large reactor dimensions and power output. Reductions in size, output and cost require improvements in τ_E and β_t, and at the same time the design must take full advantage of the bootstrap mechanism.

The discovery of the H-mode improved confinement in ASDEX had given hope that energy confinement in tokamaks could be lifted above the

L-mode value τ_L. The enhancement factor, now usually expressed as $H = \tau_E/\tau_L$, is generally taken as the benchmark for improved confinement. As we have seen, the H-mode is characterized by a steep gradient region near the plasma edge and by confinement enhancement with $H \approx 2$. Combined with normalized $\beta_N \approx 2.5$, close to the Troyon limit, this leads to a requirement of about 20 MA for ignition. It is interesting to compare this figure with earlier estimates. The demand-driven reactor studies in the 1970s and 1980s that will be discussed in the following chapter had focused on tokamak reactors with plasma currents around 10 MA. But in 1988, the results from heating experiments in JET led Rebut [19] to propose a 30 MA facility as the next step in tokamak research. The physics of the 1990s has shown that this figure can be brought down to 20 MA and the question is now what more Nature has in store in the way of improvement of the tokamak concept.

The study of tokamak operational scenarios with the potential to combine H values in the range 3–4 and β_N in the range 4–6 acquired the title 'advanced tokamak research'. Success would result in a significant reduction in size and cost of a fusion reactor. The key to these advanced regimes lies in optimizing the shape of plasma cross-sections and internal profiles of parameters such as the plasma current density—which determines the q profile—the plasma density and the poloidal rotation. Interest has focused on two regimes: (i) the high magnetic shear, or high internal inductance (l_i) regime in which the current profile is strongly peaked on axis and (ii) the regime with low or negative magnetic shear in the core of the plasma, in which the current flows primarily in the outer regions.

The TFTR, D III-D and JT-60U groups created transient high-l_i profiles both by rapidly reducing the plasma current and by increasing the plasma volume. The maximum achievable β was shown experimentally to increase with l_i as predicted by theory. A major difficulty with this scenario is the incompatibility with the natural distribution of the bootstrap current density which tends to be highest near to the plasma edge, thus tending to reduce rather than enhance l_i. Steady-state tokamak reactor scenarios will need a high fraction of bootstrap current to reduce the amount of external current drive. Bootstrap current is a natural consequence of high β and combines well with low central shear.

A variety of names—reverse shear, negative central magnetic shear and optimized shear—have been given by different groups to regimes that are essentially the same. The potential of this regime was demonstrated in the early 1990s in experiments in DIII-D and JET. High β plasmas were created in DIII-D by rapidly increasing the current and elongating the plasma during beam heating. Rather similar conditions—the so-called pellet enhanced performance or PEP mode—were found in JET with a combination of central density peaking by pellet injection and ion cyclotron heating. In both cases the q profile was not measured directly but was inferred from the radii at which low-order rational MHD modes occurred. Diagnostic advances (section 5.3) subsequently permitted a more systematic study of the production and formation of these regimes. The most widely used technique to produce these low central shear conditions is to apply additional heating during the current ramp-up phase. Created in this way, these are necessarily transient conditions, but pseudo-steady-state low central shear has been obtained in JET, Tore Supra, FTU and JT-60U by applying off-axis lower hybrid current drive [20].

An alternative regime that holds the prospect of combining high confinement at high density, exceeding the empirical Greenwald limit, with the major fraction of the plasma power radiated at the periphery is the so-called radiative I-mode developed on TEXTOR. Paradoxically, this improvement in confinement is obtained by adding impurities, usually neon, to plasmas that are heated by neutral beams and ion cyclotron waves. The physics mechanisms responsible for the improved performance are not fully understood.

8.3.3 Spherical tokamaks

In 1986, Peng and Strickler [21] proposed a tokamak device with extremely low aspect ratio, which they called the 'spherical tokamak'. It was foreseen that this would have the potential to operate at high β. Troyon's theoretical scaling (section 7.5) predicts that the highest β limit is obtained by combining low aspect ratio with high elongation and triangularity. DIII-D used this prescription to get to record values of β but it is difficult to squeeze the aspect ratio much below about 2.5 in a conventional tokamak because toroidal field coils take up space on the inside of the plasma. A very tight aspect ratio like 1.2 is achieved in spherical tokamaks by dispensing with the usual arrangement of toroidal field coils in favour of a single current-carrying rod down the vertical axis. The spherical tokamak benefits also from strong toroidal effects on the edge safety factor q_a and can operate at values of toroidal field that are typically an order of magnitude lower than in a conventional tokamak [22]. It may be seen as belonging to the alternative toroidal systems discussed in chapter 4, but for better understanding is introduced here as a recent offspring of the tokamak line.

The first studies of this configuration were made by inserting central rods into existing spheromak-type experiments (section 4.4). The first purpose-built spherical tokamak, START [23] at Culham, came into operation in 1991. START had striking results with Ohmic heating and when operated later with neutral beam injection heating [24] reached $\beta \approx 40\%$, more than three times the value obtained in a conventional tokamak (the previous highest value of 12.6% was achieved in DIII-D with an aspect ratio of 2.8). Values of the normalized β (or Troyon factor, section 7.5) $\beta_N = \beta / I_N$ as high as about 6 were reached, well in excess of the usual limit of about 3.5. Moreover, the operating limit in a spherical tokamak appears to be caused not by external kink or ballooning instabilities that would be expected at larger aspect ratio, but by internal plasma instabilities (the so-called internal reconnection events) that do not appear to be related to β. Internal reconnection events also appear at the limits to high-density and low-q operation. When START first came into operation, it was observed to be free from major disruptions, but disruptions at low q began to be observed later following the installation of divertor coils close to the plasma.

Smaller spherical tokamaks have been operated in the USA, Japan and Australia. This was one of the few remaining areas of tokamak physics where novel work could be done with modest facilities, and it became one of the most exciting fields of fusion research during the late 1990s. The potential of spherical tokamaks to work at high β would result in a high fusion power density and their apparent freedom from disruptions would also be an attractive feature for a fusion reactor. Energy confinement in spherical tokamaks appears to be consistent with the ITER scalings based on data from conventional tokamaks, but these scalings need to be tested over a bigger parameter range. A new generation of bigger experiments—the Mega Amp Spherical Tokamak (MAST) [25] at Culham (figure 8.8), the

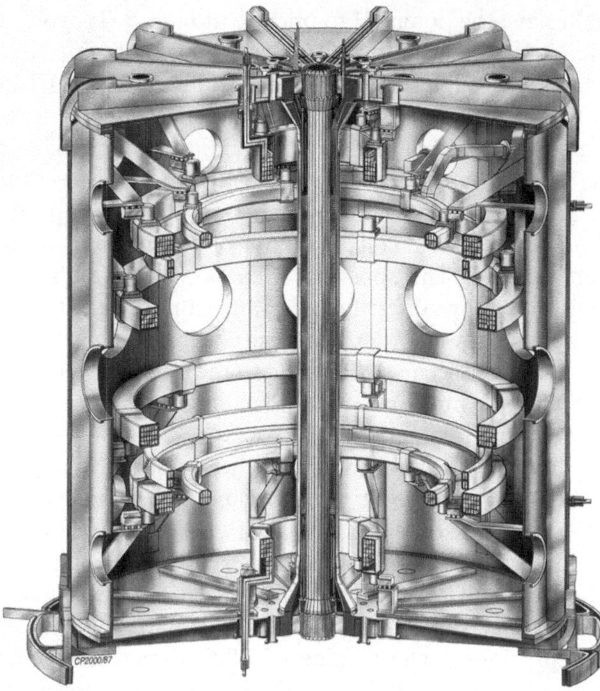

Figure 8.8. Schematic of the MAST experiment—first of a new generation of large spherical tokamaks.

National Spherical Torus Experiment (NSTX) at Princeton and Globus-M in St Petersburg—came into operation in 2000–2001 and will be an important step to establish whether the promising results of the smaller experiments extrapolate well with size.

8.4 Towards ignition

The long-term strategy for the development of fusion energy formulated in the 1970s required the attainment of three steps: scientific feasibility, technical feasibility and commercial feasibility. In the simplest scenario, these steps were sequential and each was associated with a specific generation of machine—JET, TFTR and JT-60 were there to establish scientific feasibility and would be followed by a 'next step' to establish technical feasibility ahead of a 'demo' that would demonstrate commercial feasibility. This has continued to serve as a useful roadmap for fusion even though it was clear that these issues are interdependent and there cannot be such a sharp distinction between the steps. Thus the JET generation of experiments demonstrated scientific feasibility as defined by $Q = 1$, but could go no further towards reaching ignition and understanding the physics of burning plasmas—so these tasks would fall to the 'next step' along with the demonstration of the technology needed for a reactor. This required that the 'next step' would be a large and expensive facility—a full size reactor except for the electricity generating and tritium breeding plant—and plans for it came to be developed through international collaboration in the INTOR and ITER reactor studies that will be discussed in the next chapter. But some thought this placed too big a burden on a single machine and they sought quicker and cheaper ways of approaching ignition and exploring the physics of burning plasmas by separating the ignition physics from the fusion technology.

The high-field, compact tokamak offered the most promising route towards ignition. The high densities and confinement times achieved in the Alcator experiments at MIT led Bruno Coppi to propose [26] in 1976 a series of compact experiments to which he gave the name Ignitor (**Ign**ited **Tor**us). The objective was to demonstrate α-particle heating earlier and at lower cost than was possible in bigger experiments. The original reference design for Ignitor envisaged that the assembly would be reduced to a basic minimum '*with little concession to diagnostic access ports*'. Heating would be primarily Ohmic, with the possibility of auxiliary heating if required to fine-tune the temperature—though it was not clear in the original proposal how this would be accomplished without concessions to ports. The initial proposal envisaged a plasma current up to 3 MA in a tokamak with major radius of 0.6 m and minor radius of 0.2 m (only marginally bigger than Alcator C). The toroidal field would be 15 T at the plasma centre and as high as 25 T at the coils. Later proposals revealed [27] a somewhat larger device with the major radius increased to 1 m and the plasma heated by toroidal compression to a final major radius of 0.7 m and plasma current of 4 MA. One of the most critical components would be the toroidal magnet for which the Ignitor designers envisaged a structure with composite plates of steel and copper, cooled to low temperatures (but not superconducting).

Soon after TFTR came into operation in 1982, design studies were started for a Compact Ignition Tokamak (CIT) to be built at Princeton specifically to study the physics of burning plasmas. Starting from the Ignitor concept, the CIT study investigated a broad range of engineering approaches. By now, a plasma current of 9 MA and a toroidal field of 10 T were judged to be necessary to reach ignition. But CIT's physical dimensions grew dramatically—from a major radius of 1.22 m in 1986 to 1.75 m in 1987 and to 2.14 m (with 11 MA) in 1989—reflecting the difficulty of building a high-field tokamak with realistic engineering structures that left enough space for a plasma. Fears that CIT would not ignite led to it being abandoned in 1990 and replaced by a new proposal—the Burning Plasma Experiment (BPX). This also increased in size (the major radius grew to 2.6 m) and cost. But by then the USA was collaborating internationally in design studies for ITER and the logic of pursuing both ventures in parallel was thrown into question. If ITER came to be built it would do all that BPX promised and more—but if BPX was really necessary to resolve important issues of physics ahead of ITER, then why go ahead with ITER?

The outcome of the squabble was that the US got neither BPX nor ITER. A review of the US fusion programme in 1991 concluded that BPX could not be supported within the foreseeable future budgets for magnetic fusion. These declining budgets soon forced the US programme to shift its focus from reactor development to plasma science and to withdraw from the ITER collaboration. But this policy is not without its critics, who point to the failure to profit from the considerable progress made by TFTR and the compelling need for an experiment to study plasmas dominated by α heating. This has led to FIRE (Fusion Ignition Research Experiment) [28]—a proposal for a compact high field (10 T) tokamak utilizing cooled copper alloy coils with major radius 2.1 m and plasma current 7.7 MA. The plasma would have a highly shaped cross-section and double-null divertor. These parameters may change as the design progresses and whether FIRE will be built remains uncertain. But the mood in Congress seems to swing back and there is a new interest in a burning-plasma experiment. Its approval would indicate that the US is returning to support fusion as an energy research programme.

Ignition experiments were also considered outside the US. Garching made a study of a high-field tokamak brought to ignition by ICRH, but this Zephyr project was abandoned when new scalings showed that ignition was out of reach. Proposals to build an Ignitor in Europe were reviewed several times but always failed to receive Euratom support because of the relatively high cost, demanding engineering problems and concern about it actually achieving ignition [29]. Nonetheless the Ignitor concept (now with toroidal field of about 13 T and plasma current of about 11 MA) continues in Italy without funding from the European programme and has reached a stage where prototype components are under development [30].

The Alcator experiments also inspired Boris Kadomtsev and Evgeniy Velikhov to propose around 1979 a high-field compact tokamak—TSP (*Tokamak Silnoye Pole* meaning 'with a strong field') that came to be built at the Troitsk laboratory* near to Moscow [31]. It was planned to operate TSP in tritium to demonstrate α-particle heating. The plan was to create a plasma with a major radius of 1 m, minor radius of 0.32 m and toroidal field of 2 T. With ion cyclotron or neutral beam heating, the temperature was expected to reach 2 to 3 keV and the plasma would be compressed adiabatically in both minor and major radii by rapidly increasing both the toroidal and vertical fields. The compression was estimated to be equivalent to a heating power of 50 to 100 MW and was expected to push temperatures to 10 keV. The key to the compression was a powerful and innovative inductive energy storage system driven by flywheel generators. A series of technical problems with this complex electrical system combined with reduced funding for research in Russia prevented satisfactory operation of TSP. In the meantime, as we have seen in section 7.6, the optimistic scaling derived from the early Alcator results has been reassessed and we must assume, on present knowledge, that the expected performance could not have been attained.

* At that time the Troitsk laboratory was a branch of the Kurchatov Institute for Atomic Energy, but around 1990 it became an independent institute known thereafter as TRINITI.

Chapter 9

Towards a fusion reactor

To discuss the field of fusion reactor design and technology in the same depth as we did the physics is beyond the scope of this book. But the subjects are inseparable and through the years reactor studies have fed back ever stronger into the physics experiments, to the point where they now determine the points to be placed on the scientific agenda.

9.1 First thoughts

Some early ideas of what a reactor might look like come surprisingly close to present-day reactor models (figure 9.1). Others now seem rather naïve.

The 1946 patent application submitted by Thomson and Blackman [1] proposed electrostatic confinement of ions in a torus of major radius $R_0 = 1.3\,\mathrm{m}$ and minor radius $a = 0.3\,\mathrm{m}$. The potential well would be created by a toroidal stream of electrons, driven by a travelling electromagnetic wave and confined by the pinch effect. The fuel would be deuterium and the total thermal energy yield of 9 MW would be exploited as heat or as neutrons for breeding fissile materials.

Immediately following Tamm's initial work in 1951, Sakharov considered a toroidal deuterium-fuelled reactor [2]. Assuming $\beta = 1$ and equating the charged-particle reaction yield with the collisional heat-conduction loss, he found as a minimum for a self-sustained reaction:

$$B_t a = 10\,\mathrm{T\,m}$$

where B_t is the magnetic field on axis. Taking $B_t = 5\,\mathrm{T}$, $a = 2\,\mathrm{m}$ and $R_0 = 12\,\mathrm{m}$, and choosing $n = 3 \times 10^{20}\,\mathrm{m}^{-3}$ to make $\beta = 1$ at $T = 100\,\mathrm{keV}$, he obtained a nuclear power production of 880 MW(th).

That the product $B_t a$ is a useful figure of merit, with respect to both single-particle confinement and collisional diffusion, was clear from elementary scalings. At a given temperature, the confinement time, τ, as determined by classical diffusion makes Lawson's product (boxes 1.6 and 9.1) scale as

$$n\tau \propto (B_t a)^2.$$

This would apply to cross-field diffusion in closed systems; in mirror devices one has to deal with diffusion in velocity space, which is independent of field strength and size. But the $B_t a$ product reappears if one considers the energy dissipated in resistive coils. In geometrically similar systems with equal plasma temperatures, the ratio between the thermonuclear power, P_{out}, and the power dissipated in the coils, P_{coil}, scales as

$$P_{\text{out}}/P_{\text{coil}} \propto (\beta B_t a)^2.$$

With 400 MW(e) required to energize the copper toroidal-field coils, Sakharov's 880 MW(th) reactor was about the smallest possible deuterium burning reactor. Clearly, then, this reactor had to be large in terms of $B_t a$, whether the concern was to overcome energy loss from the plasma or dissipation in the coil.

The argument is slightly different for pinches. Since $rB_p = 0.2I(r)$, where $I(r)$ is the current (MA) within the radius r, and 3.5 MeV α-particles have $B\rho = 0.27$ T m, self-heating by fusion reactions requires a pinch current of several MA. At the Geneva Conference in 1958, Rosenbluth presented similarity rules, based on the classical theory of diffusion in linear systems, for the unstabilized toroidal pinch and suggested a current of 3 MA, induced in a $R = 30$ cm, $a = 6$ cm torus, the plasma yielding 30 MJ of fusion energy per pulse.

In Princeton, Spitzer foresaw a succession of stellarators named A, B, C and D of which the D stellarator with a tube diameter of 90 cm would be a full-scale model for power production [3]. The last one actually built was C; we met this device first as a stellarator and later as a tokamak experiment.

The first reactor-feasibility study for mirror machines was carried out in 1960 by Post. The reactor was thought of as having a 50–100 m long cryogenic solenoid with short mirror sections; the vacuum vessel had a radius of 50–100 cm and the β-values were up to 0.1. Assuming efficiencies of 0.9 and 0.5 for injection and for thermal-to-electricity conversion, respectively, Post arrived at power plants with electrical output in the range of 100–200 MW, and acceptable circulating-power levels. In the original version, the coils would be high-purity sodium at 10 K, but the discovery of hard superconductors in 1961 made the outlook much better, as shown by a revised version of the paper [4]. Even so, end losses left only a narrow margin; mirror reactor studies have always needed to assume extremely efficient energy conversion, in particular recovery of charged-particle energy losses.

Box 9.1 Lawson revisited

The significance of the $n\tau$ product was implicit in the earliest work, but was expressed most clearly in Lawson's 1957 paper, which specified $n\tau > 10^{22}\,\mathrm{m^3\,s^{-1}}$ for power gain from a D–D reactor and $n\tau > 10^{20}\,\mathrm{m^3\,s^{-1}}$ for D–T (box 1.6). Lawson assumed a pulsed system, in which fuel is made to react and all of the fusion energy, together with that of the unburned fuel, is extracted from the reactor as heat, converted into electricity at an efficiency $\eta \approx 33\%$ and returned in part to heat fresh fuel. Self-heating by α-particles played no role in this model.

When thoughts turned towards steady-state systems with injection of fuel into and removal of helium from the burning plasma, it became customary to define an energy-multiplication factor, Q, equal to the ratio between reaction power output and external power injected into the plasma. If one fifth of the D–T reaction energy, that is the fraction carried by the α-particles, is deposited in the plasma, the relation between Q and the quantity R, defined in box 8.2, is:

$$Q = 5R/(5 - R), \qquad R = 5Q/(5 + Q).$$

In a fully ignited D–T system the plasma is heated internally by the α-particles and Q is infinite. A D–T reactor (figure 9.1) at the optimum temperature of 12 keV requires $n_i\tau_E > 1.5 \times 10^{20}\,\mathrm{m^{-3}}$ for ignition. Toroidal reactors can in principle operate in an ignited mode, but it is likely that some external particle and power input will be required to control the density, temperature and current profiles of the burning plasma, as well as the power output. Recirculating a few per cent of the output ($Q \simeq 50$) would, in view of the efficiencies of NB and RF heating—perhaps 50%—not seriously impair the economy of the system. Anyway, the confinement quality needed for net power output is nearly the same as that for ignition and either one is commonly referred to as the Lawson criterion.

Mirror reactors by themselves can at best be energy multipliers with $Q \approx 5$. Tritium-burning mirror reactors are therefore sometimes thought to be equipped with energy-multiplying fission blankets. The alternative is a direct-conversion scheme which collects the energy of charged particles escaping through the mirrors, but this requires a reaction like D–^{3}He, which deposits most of its energy in charged particles. (For this reason, most advocates of neutron-free or neutron-lean reactions favour open systems, whereas those who contend with D–T opt for closed systems.)

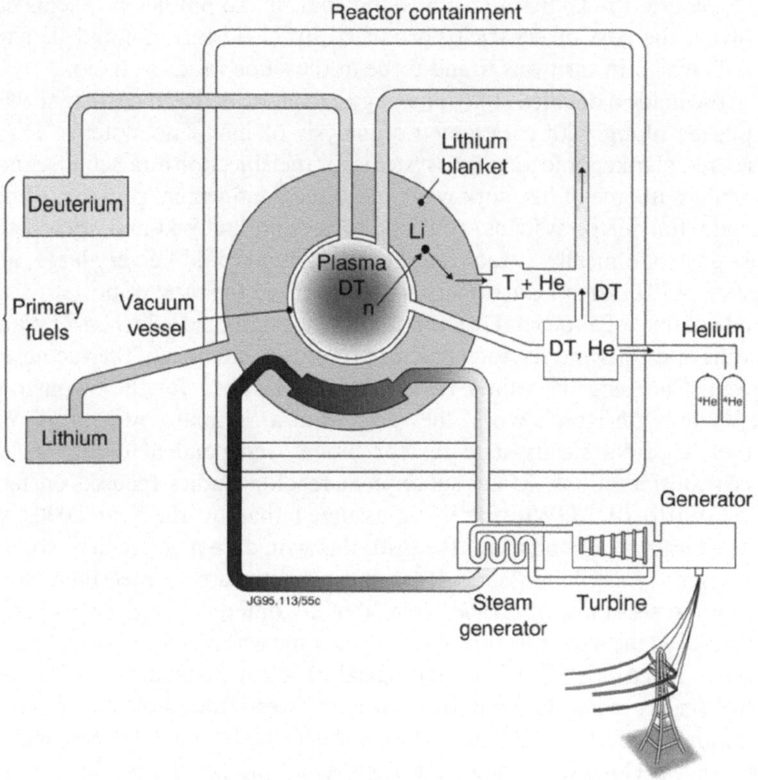

Figure 9.1. Schematic of a fusion reactor.

9.2 Second thoughts

During most of the 1960s, talking about reactors was not really fashionable among plasma physicists. The Lawson criterion had set the goal for confinement, but Bohm diffusion would make it impossible to reach the promised land. The first priority was, therefore, to place the relevant plasma physics on a more solid foundation and to develop a better understanding of instabilities and of methods to suppress their growth. But towards the end of the decade, voices were heard again reminding the plasma physicists that really they were expected to produce the design of a viable reactor. Among the first to reconsider the problem were Mills [5] from Princeton, Ribe [6] from Los Alamos, Carruthers [7] from Culham and Rose [8] from Oak Ridge and MIT. The Culham study focused on a steady-state toroidal D–T reactor with 5 GW(th) output. It argued that open geometries required systems to convert charged particles into energy that were beyond current technology and that the cost of energy storage in capacitor banks for pulsed systems would be prohibitive. While noting that a D–D reactor required higher $n_i \tau_E$

and T_i values, the Culham group found that, if the physics problems could be solved, the cost of electricity produced by D–D fusion could be similar to D–T, which in turn was found to be in the same range as fission.

Rose made a detailed study of energy exchange between particle species in the plasma along with a parametric analysis of magnetic systems. He also considered blankets and auxiliary systems for fuel injection and helium exhaust. The main outcome of his work was that, if the confinement problem (difficult especially for mirror systems) could be solved, both closed and open systems appeared economically attractive, but the reactor would be very large, in the range of 5–20 GW(th). Carruthers *et al.* optimized the ratio of power to 'engineered volume' of a toroidal reactor, taking the combined thickness of blanket, shielding and coils and the total reactor power per unit area of the vacuum wall as given. Choosing the values 1.75 m and 13 MW m^{-2} for these constraints, based mainly on Rose's work, they arrived at a thermal power of 5 GW(th) for their toroidal steady-state reactor model, independent of any plasma-physics considerations. Many subsequent reactor studies focused on figures like 5 GW(th) or 2 GW(e), it being assumed that by the year 2000, when fusion energy might enter into the grid, this would be a convenient size for a power plant of any kind. When Rose and Carruthers presented their work in an informal session at the 1968 Novosibirsk Conference, it drew as large an attendance as the main meeting. Clearly, the time was ripe for a formal meeting on fusion reactors and their technological problems. Culham was the venue; Carruthers, Rose and Golovin from Moscow were among the organizers and the lecturers. The 1969 Culham meeting was to be for the engineers what the 1958 Geneva Conference had been for the plasma physicists: the seed from which grew an international community of specialists.

Post [9] discussed mirror reactors and stressed that these would require either direct energy conversion or energy multiplication in a fission blanket. His direct-conversion scheme would employ four steps (figure 9.2): expansion of the plasma in a static magnetic field, separation of ions from electrons, deceleration and energy-selective collection, and combination of the collected currents at a common potential by passing them through voltage inverters; all this to be done with an overall efficiency between 90 and 95%. Designs were also presented for short-pulse reactors based on theta pinches or toroidal pinches [10].

Golovin [11] had expanded on Sakharov's original idea of a gas-blanket tokamak reactor, assuming neoclassical transport loss and, of course, now using superconducting toroidal-field coils. A typical reactor would have $R_0 = 5.2$ m, $a = 1.8$ m, $B_t = 4$ T and a power output of 5 GW(th). He also sketched a laboratory tokamak capable of ignition, which had parameters very much like the later T-10 tokamak, as well as a 'final laboratory unit' with $R_0 = 3.2$ m, $a = 0.65$ m, $B_t = 7$ T and 1.5 GW(th) power output, which became known as T-20. (Some years later, Kadomtsev [12] suggested $R_0 = 5$ m, $a = 2$ m, $I = 5$ MA for the power-producing T-20.)

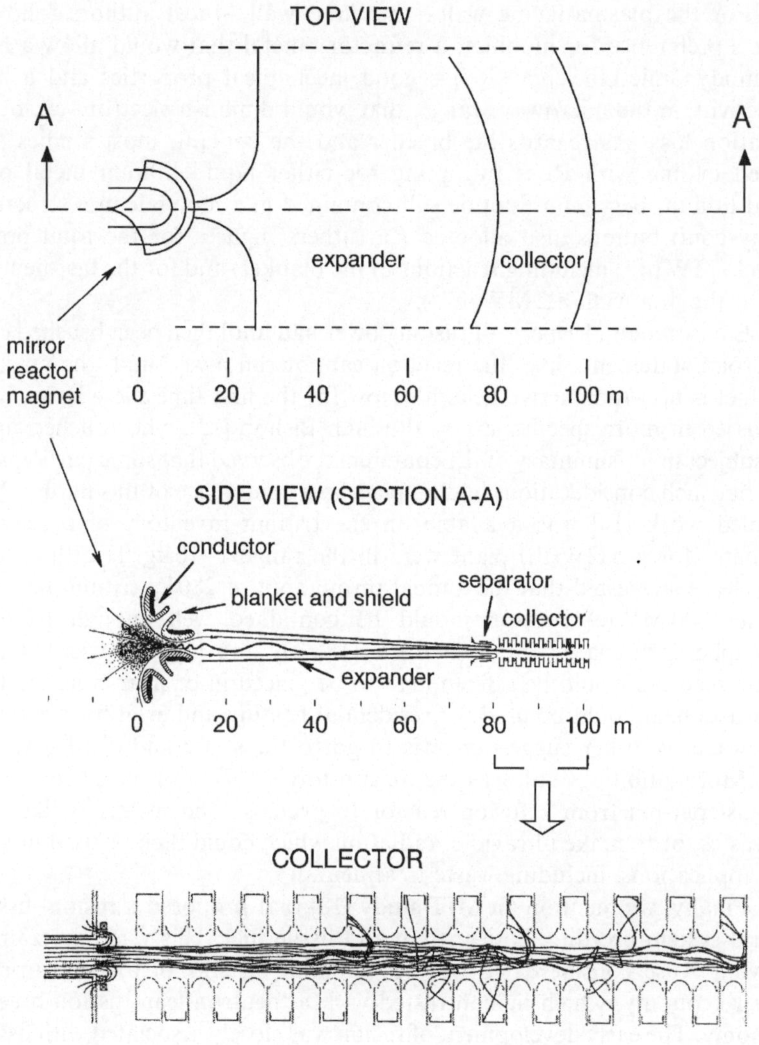

Figure 9.2. Energy recovery in mirror reactor. Top: the quadrupole mirror field fans out over 180°, horizontally to the right and vertically to the left (compare figure 3.4); only a truncated view of the right-hand side is shown here. Centre: side view of right-hand side collector. Bottom: when the ions arrive in the low-field region, 80–100 m away from the mirror, they can, after electrostatic deceleration to recover their energy, cross the field lines and arrive at a collector. The electrodes are at outwardly-increasing positive potentials; each ion is meant to penetrate just deep enough into the collector to be collected at the electrode whose potential corresponds with its original kinetic energy.

For the plasma-facing wall—the 'first' wall—most authors followed Rose's preference for niobium, a refractory metal that would allow a high thermodynamic efficiency. It has good mechanical properties and a high reflectivity in the microwave range, that would diminish electron–cyclotron radiation loss. As regards the breeder and the coolant, most studies also stayed in line with Rose by opting for either liquid lithium metal or a fused lithium beryllium fluoride salt contained in a molybdenum structure. Many contributions also adopted Carruthers' figures for the total power flux (13 MW m^{-2} including reactions in the blanket) and for the fast-neutron flux at the first wall (8.2 MW m^{-2}).

Environmental aspects of fusion power had until then been brushed over by broad statements like 'the reaction cannot run away' and 'the reaction product is non-radioactive helium'. Now for the first time these issues were discussed in more specific terms, although Bishop [13], who touched upon the subject in his summary of the conference, observed that some participants felt that such considerations had no place in a conference of this nature. Yet, detailed work [14] was available on the tritium inventory of a reactor; estimates for a 5 GW(th) plant were in the range 1–10 kg. Health criteria for release suggested that the annual throughput of 280 kg tritium per year for a 1000 MW(e) reactor should be contained well enough to limit atmospheric discharge to about 0.1% [15]. By and large, it was felt that fusion reactors would be safe enough to be placed in or near cities, so that the waste heat could be used for residential heating and even for treatment of sewage. Another suggestion was to go to the shore and distil seawater [16]. More ambitious still was the 'fusion torch' [17], a proposal to use the exhaust plasma from a fusion reactor to process solid materials like ores or wastes, or to make ultraviolet radiation which could then be used in various applications, including waste treatment.

Finally, we mention an MIT study [18] that proposed a fusion–fission hybrid system breeding both tritium and fissile fuel. This was one example of what Rose considered to be the potential benefits of a 'neutron-rich' power economy, which he contrasted with a 'neutron-lean' fission-breeder economy. The early development of fusion was closely associated with fission energy. When, not many years later, fission came under attack, the fusion community was ill prepared to argue its case as having the potential to be developed into a vastly more safe and clean source of energy than fission. We return to this subject in section 9.4.

9.3 Pioneering studies

The strongest efforts in the field of fusion-reactor studies developed in the US, where both Congress and the AEC were getting impatient with the physicists' emphasis on basic plasma physics and wanted to see to what sort of application these studies might eventually lead. Under the influence of the impending energy crisis and the growing concern about pollution of the environment, the question of fusion's possible role in the world's future energy supply took on increasing significance [19]. Under the leadership of R L Hirsch, who joined the AEC (later to become the Department of Energy) in 1968 and became its director of controlled thermonuclear research in 1972, the laboratories were pressed to concentrate on systems with reactor potential and to make fusion reactors competitive with fission breeders [20]. In this climate not only the national fusion laboratories but also some of the universities began to take an interest in fusion reactor engineering. In 1970, a Symposium on Thermonuclear Fusion Reactor Design was held at Texas Tech University in Lubbock, and from 1974 the American Nuclear Society held a biennial series of Topical Meetings on the Technology of Controlled Thermonuclear Fusion. Also the biennial IEEE Symposium on the Engineering Problems of Thermonuclear Research, which had hitherto been concerned mainly with experimental fusion devices, began to attract papers on reactors.

Some fairly detailed design studies on tokamak reactors were carried out at Oak Ridge [21] and Princeton [22]. Mirror devices were treated at Livermore [23] and theta pinches at Los Alamos jointly with Argonne National Laboratory [24]. Preliminary results of these studies formed the basis of a report 'Fusion Power, an Assessment of Ultimate Potential', issued in February 1973 by the AEC. The report stressed the importance of the potential advantages of fusion, as compared with fission, in matters of safety and environmental impact, and urged fusion research and development to concentrate on this potential.

The approach in the Soviet Union was different. The country had committed itself to an ambitious fission energy programme and saw fusion primarily as a source of neutrons for breeding fissile materials. The goal of the fusion laboratories was, therefore, to develop the fusion core for a hybrid system, not the pure-fusion reactor favoured elsewhere.

In Europe, where the French *Commissariat à l'Energie Atomique* was not interested in developing a near-term competitor to its fission-breeder programme, emphasis had to be on scientific, rather than technological aspects of fusion. There could be no place, therefore, in the European Communities' fusion programme for reactor design studies of the kind that sprang up in the US, and all work on specific technical problems had to be disguised as technology for the next generation of experiments. In Britain, which did not join the European Communities until 1973, Culham's fusion programme retained a strong technology component. The laboratory had been subjected to a severe budget cut in the late 1960s and their reactor studies served the dual purpose of demonstrating the feasibility and usefulness of fusion power as well as identifying areas for further technological development.

Japan responded to the energy crisis of 1973 by greatly expanding work on alternative energy sources, notably solar and fusion. Its reactor studies emphasized features related to safety and environment.

The most comprehensive study to appear in these years came from the University of Wisconsin [25]. This UWMAK-I study was conservative in that it chose conventional stainless-steel metallurgy and liquid-lithium cooling—building upon the liquid-sodium technology developed for fast breeder reactors. It proposed niobium–titanium superconducting coils and assumed a neutron flux at the first wall as low as $1.25\,\mathrm{MW\,m^{-2}}$. The study considered various forms of radiation damage like erosion of the first wall, swelling of structural materials and, above all, embrittlement, and the first estimate of the Wisconsin group was that these effects would limit the lifetime of the first wall to two years. This study also drew attention to environmental aspects, in particular to the quantities of radioactive wastes to be disposed of. Moreover, it laid bare the principal weaknesses of the tokamak reactor—its large size and its complexity. Periodic replacement by remote handling of its innermost components would restrict availability and the combination of cyclic stresses (UWMAK-I would have operated with 100-minute pulses) with swelling and embrittlement augured ill for its reliability.

The American reactor studies were reviewed at an IAEA workshop held at Culham in 1974, along with less elaborate conceptual design studies carried out by laboratories in other countries. JAERI [26] had been among the first to respond to risk analyses, which had pointed to a lithium fire as the most serious accident, by proposing a gas-cooled tokamak with a solid breeder. A group at Nagoya [27] had considered an open system with RF fields to plug the mirrors. Culham [28], had studied a set of mirror systems linked end-to-end to make a toroidal device and had updated their earlier RFP study. The Frascati–Ispra–Napoli (FINTOR) collaboration in Italy [29] presented preliminary results of a study aimed at finding the minimum size of a tokamak reactor designed to demonstrate the technical feasibility of D–T burning. The parameters of early tokamak reactor studies are listed in table 9.1.

Among the technical problems, tritium containment and first-wall damage emerged as the most critical issues. Circulating energy remained a serious problem for pulsed devices like the theta pinch and the plasma energy balance was considered to offer little hope for steady-state mirror machines unless in the form of toroidally linked mirrors or bumpy tori. Thus, attention focused on toroidal devices, particularly on the tokamak in which plasma confinement was far better than in pinches and stellarators. Environmental aspects, such as release of tritium in normal operation, production of radioactive wastes and accidental dispersion of radio-isotopes, were—contrary to five years earlier—firmly on the agenda and choices like niobium as the material for the first wall and liquid lithium for the breeder

Table 9.1

	Major radius, R_0 (m)	Plasma radius, a (m)	Vertical elongation κ	B_t on axis (T)	Plasma current I (MA)	Thermal power (GW)	Neutron wall loading (MW m^{-2})	First wall	Breeder	Coolant
Culham MK-I 1972	12.5	2.5	1	9.5	9.7	5.8	3.9	Nb-Zr	Molten Li	Molten Li
Wisconsin UWMAK 1973	13	5	1	4	20.7	5.3	1.25	316SS	Molten Li	Molten Li
Princeton PRD 1973	10.5	2.8	1	4	10	1	0.44	Nb–Zr	Molten Li	Molten Li
Oak Ridge 1974	10.5	3.25	1	6	14.6	5.3	1.76	PE-16	Fused FLiBe	He
JAERI 1974	10	2	1	6	8	2	1.25	Incoloy-800	Solid Li_2O	He
FINTOR 1974	11.25	2.25	1	4.5	5	0.17	0.1	316SS		He
Wisconsin UWMAK III 1976	8	2.5	2	4.0	15.8	5	2.5	Mo(TZM)	Molten Li	He/Li
Culham MK-II 1976	7.4	2.1	1.75	4.1	11.6	5.8	5.7		Molten Li	Molten Li
Jülich 1979	6.9	1.8	1.7	3.7	8.8	5	7.3	Mo	Molten Li	He
Oak Ridge DEMO 1977	6.2	1.55	1.6	3.4	4.0	2.4	2.7	316SS	Molten Li	Molten salt
Wisconsin NUWMAK 1978	5.2	1.15	1.6	6.0	7.2	2.1	4.3	Ti	Li62Pb38	Boiling water
MIT HFCTR 1978	6.0	1.2	1.5	7.4	6.7	2.5	3.4	Mo(TZM)	Molten Li	FLiBe
Culham MK-IIB 1978	6.7	1.9	1.75	3.2	10.2	3.4	4.5		Molten Li	He
USA Starfire 1980	7.0	1.9	1.6	5.8	10.1	4.0	3.6	Austenitic SS	Li compound	Pressurized water

and coolant were being criticized on these grounds (the former because it has long-living activation products).

9.4 Drawing fire

The early fusion reactor design studies were not only useful in outlining crucial physical and technical problems and encouraging in suggesting that consistent solutions might be found, they also brought to light vulnerable spots like energy recovery in mirror machines and theta pinches and, above all, the size and complexity of tokamak reactors. Power companies in the US were not enthusiastic about putting multi-gigawatt plants of unproven design and high investment cost into their grids. Nor did they, or the higher levels in the AEC, want to spread the notion that fusion might be an early competitor to their fission breeder development.

In 1971, the R&D Task Force of the American utilities' Electrical Research Council had classified fusion research as a first priority, requiring equal support with the fission breeder. Demonstration of scientific feasibility was foreseen between 1976 and 1979 and commercial sales before 2000 [30]. But the suggestion that fusion might enter the market before the turn of the century and become a serious competitor to the fission breeder was not left unchallenged. The Electric Power Research Institute (EPRI), established in 1972, adopted a different line, attacking the claim that fusion might have significant advantages over fission with respect to safety, radioactive wastes and non-proliferation of nuclear weapons [31]. Comparing a D–T reactor with niobium as a structural material and liquid lithium as a tritium breeder on the one hand against a sodium-cooled plutonium breeder on the other, they denied that fusion had any decisive advantage over fission.

The 1970s were a time of controversy and confusion about, among other things, resources (energy and materials) and environment. The oil crisis in 1973 and the Reports to the Club of Rome from 1972 and 1974 had accentuated the world's dependence on fossil fuels and this, against the background of the emerging debate about the global climate, stimulated the search for new energy sources. Yet, fusion did not attract unequivocal political support, because on one side there was a tendency to deny the need for an alternative to the fossil fuels, while at the other extreme there was a desire to return to a world without any large-scale production and distribution of electricity—the sector of the energy market where fusion was looking for its share.

The fusion reactor studies of the late 1970s and the 1980s paid much attention to the size and cost of the reactors, which was to be minimized, and the choice of materials, which had to be viewed in the light of their availability and their liability as radioactive wastes. The materials problems had been brought forward in particular by the UWMAC-I reactor study, which had gone to great lengths to estimate the quantities of materials required for a fusion economy and to compare these with known reserves. They found that beryllium (specified in some studies as a neutron multiplier) was in very short supply, and that a mature fusion economy would also cause supply problems in the US—though not throughout the world—for niobium, as well as chromium and nickel if conventional stainless steel were to be used for the first wall and the blanket.

The UWMAC study set the tone for the ensuing discussion on the activation of materials in fusion power plants by comparing them with the fission products of an equivalent uranium reactor. The comparison is complicated in that both produce a variety of isotopes with different decay times, so that the ratio of their total activities is a function of the time after the shut-down of the reactor. Moreover, the activities as well as the chemical properties of the active materials are of different kinds. The actual hazard suffered by humans, integrated over time, may be expressed in terms of a 'Biological Hazard Potential', but when discussing this one still has to distinguish the risks of ingestion and inhalation. Ingestion is to be considered in connection with waste products that could enter the food chain via ground water, while both inhalation and ingestion can contribute to the risk in case of an accident. The UWMAC study made it clear that, although the volume of radioactive waste would be large, its BHP would be markedly lower than that of fission products—by one to three orders of magnitude. Niobium came out worst and a vanadium–titanium alloy best, with stainless steel in between.

As we saw, these conclusions were not immediately accepted by the fission-breeder community. And indeed, when activities were measured in other units like the volume to be disposed of, the picture looked somewhat different. The discussion on these questions brought together a group of fission and fusion specialists who carried out a detailed comparison under the auspices of the International Institute for Applied Systems Analysis (IIASA) in Austria. This study did confirm that the intrinsic flexibility in fusion technology '*offers the possibility for D–T fusion to be quantitatively superior to fission in important environmental respects, despite qualitative similarities*'. It stressed, however, that these advantages would not materialize automatically and that priority had to be given to environmental characteristics from the earliest stages of designing fusion systems [32], in other words that niobium and liquid lithium were not the last words in the quest for clean and safe fusion.

The IIASA study highlighted one dilemma confronting fusion research. On the one hand, there was the desire to produce first a burning plasma and next a working reactor as soon as possible, to show that fusion was not just a mirage. On the other hand, there was the fear [33] that if the wrong kind of reactor were to be chosen for early commercialization, the development would risk being frozen, so that it would become impossible to return to a track leading to environmentally and socially more attractive reactors.

These concerns related as much to the choice of materials and to the fuel, as to the size of the reactor. Whereas most early reactor studies focused on D–T fuel, liquid lithium breeder and niobium or stainless steel structure, the alternative solid breeders and structural materials like vanadium, and titanium, and ultimately the 'advanced fuels' like D–^{3}He and H–^{11}B, were advocated as more benign alternatives.

Whether it was for threats to the environment—real or perceived—or for the threat that it might one day deliver abundant energy, fusion never gained much sympathy from environmental activists. Their attitude was most clearly expressed by Amory Lovins [34]:

> *Assuming (which is still not certain) that controlled nuclear fusion works, it will almost certain be more difficult, complex and costly—though safer and perhaps more permanently fueled—than fast breeder reactors* [35]. *But for three reasons we ought not to pursue fusion. First, it generally produces copious fast neutrons that can and probably would be used to make bomb materials. Second, if it turns out to be rather 'dirty', as most fusion experts expect, we shall probably use it anyway, whereas if it is clean, we shall so overuse it that the resulting heat release will alter global climate: we should prefer energy sources that give us enough for our needs while denying us the excesses of concentrated energy with which we might do mischief to the earth or to each other. Third, fusion is a clever way to do something we don't really want to do, namely to find yet another complex, costly, large-scale, centralized, high-technology way to produce electricity—all of which goes in the wrong direction.*

Obviously, nuclear safeguarding would be simpler in a fusion than in a fission economy. Further, it is not heat release but the greenhouse effect that threatens the climate. And energy will never be cheap, so there will always be an incentive to avoid wasteful consumption. What remains is the argument that fusion appears more like a threat than a salvation to those like Lovins in 1976, who hold that:

> *For all these reasons, if nuclear power were clean, safe, economic, assured of ample fuel, and socially benign per se, it would still be unattractive because of the political implications of the kind of energy economy it would lock us into.*

Lovins' paper carried the title '*The Road Not Taken*' and indeed, when by the end of the 1970s the dust had settled, it was clear that when facing a dilemma between economic and environmental interests, societies generally opt for economic growth. Fusion was broadly recognized as a desirable alternative, provided that the economic barriers that stood in its way could be overcome, and fusion research emerged from the debate with much clearer objectives than it had before. But attempts to accelerate the development failed. A DOE study that indicated that it would be possible to have a demonstration reactor working in 20 years' time was shelved and the legislation that required the DEMO to be in operation by the turn of the century was not followed up by appropriate funding. And in Europe, a similar study in 1981 that arrived at a period of 25 years was suppressed by the European Commission when political consultations indicated that it had no chance of being accepted.

The trends in tokamak reactor design through the 1970s were discussed by Conn [36]. He compared twelve studies, emphasizing the reduction in size from $R_0 \approx 12\,\mathrm{m}$ in the early designs to $R_0 \approx 6\,\mathrm{m}$ in the later ones. Concurrently, the toroidal current went down from 21 MA in UWMAK-I to 10 MA or less in the later ones. The reason for this reduction in size is not of a scientific nature, it is rather that one had to admit that the earlier models were too big to be attractive to potential customers, so that one had to incorporate what were euphemistically called 'reasonable extrapolations from known plasma physics'. This tension between what science had to offer and what economy demanded has haunted reactor design studies ever since. Our table 9.1 is mainly drawn from Conn's more comprehensive tables XXV and XXVI [36], but includes figures for STARFIRE [37].

The STARFIRE design was based upon an elaborate study carried out by a collaboration between Argonne National Laboratory and industrial partners. It was influenced by the various criticisms levelled against earlier designs and represented what manufacturers and potential customers considered to be an acceptable product. The study received strong contributions from manufacturing industries, and was supervised by a Utility Advisory Committee, which helped to bridge the gap between what the reactor designers could propose with some confidence and what the electrical power industry might be interested in buying. No exotic materials were used in the design, the cost of electricity appeared quite acceptable, much attention had been paid to safety and environmental issues and the physics assumed seemed reasonably close to the state of the art. It came at the time, not long after PLT and Alcator had made their big jumps forward (section 6.1.1), when the US Congress passed the Magnetic Fusion Energy Engineering Act of 1980. All this helped to strengthen confidence in the tokamak concept as the basis for a viable reactor and paved the way for further steps towards actual construction of experimental tokamak reactors.

With the construction of the big tokamaks well under way and with credible reactor concepts at hand, the sailing was smooth—for a while. But the physics assumptions underlying the STARFIRE design would not stand up in later years. New clouds, called confinement degradation, beta limits and density limits, would rise at the horizon and the wind would turn against ambitious plans for the next generation of fusion devices that would have lighted the thermonuclear fire. When we come to the international reactor projects, INTOR and ITER, in section 9.6, we shall see that the prospects for an INTOR-size tokamak reactor still depend on whether the ongoing research into advanced geometries and operating regimes will lead to some form of breakthrough. But first we round off our discussion of the economic and social aspects of fusion.

9.5 Economic and social aspects of fusion

Early studies of tokamak reactors had identified engineering problems, technology requirements and their physics implications. But the second round of studies, when attempting to address these problems, had arrived at reactor designs that either were not attractive commercially or, in order to make them appear so, demanded too much of the engineering and physics. The goal of building a reactor that could compete with fossil fuels went out of reach and the task of the reactor studies in the 1990s became to focus on improving the economy, safety features and design simplicity of fusion reactors. An important feature of these reactor studies, exemplified by the work of the ARIES team in the USA, has been the introduction of much more rigorous and systematic methodologies to take proper account of conflicts between the physical, technological and commercial constraints.

The ARIES studies have covered several different tokamak regimes including pulsed and steady state operation and various advanced modes with improved energy confinement properties. They also investigated different technologies for the structural and blanket materials. The broad conclusion is that a fusion reactor based on the 'conventional' tokamak needs to incorporate both advanced tokamak physics and advanced technology in order to arrive at a fully cost-competitive position. The debate between the fusion community and the utilities about the optimum size for a commercial plant has continued without reaching a clear conclusion. Some scenarios favour small units that could be sited within a distributed network anywhere there is a demand; others support the construction of large 'energy parks'. Fusion appears better suited to the second scenario—the present understanding of the economics of advanced tokamak reactors pointing to unit sizes of at least 1 GW(e). Economies of scale might favour even larger unit sizes—3 GW(e) or more—possibly combining electricity generation with the production of hydrogen fuel [38].

ARIES and other studies have considered reactor designs based on low aspect ratio tokamaks—the so-called spherical tokamaks. These studies have to make a bigger extrapolation in size from present-day experiments. The best indications are that a low aspect ratio tokamak reactor would be comparable in size and cost with one based on conventional tokamaks but may not necessarily lead to a more compact reactor [39]. Advocates of the low aspect ratio tokamak argue that a series of modest size machines producing significant fusion power could provide a lower cost development path to a full size reactor [40] and they point to potential advantages in terms of avoiding disruptions. On the other hand, the central current-carrying rod in a spherical tokamak would have to be replaced frequently and the very large plasma current (typically 30 to 40 MA) requires a bootstrap current that is precisely aligned to the equilibrium current profile but even so the current drive power could be several hundreds of MW [41].

In the mid-1990s the ARIES Team conducted the Starlite study [42] to determine the requirements and goals of a fusion power plant. Consultation with American electric utilities established a set of criteria that a fusion reactor should meet to be attractive commercially and a similar set of criteria was established by an EPRI working group. These criteria fall into three general categories: (i) cost, (ii) safety and environmental issues and (iii) operational features such as reliability, maintainability and availability. We shall, without restricting ourselves to the American scene, concentrate on the first two of these points.

Cost projections using standard energy forecasting models show that fusion could be competitive with other fuels around 2050. Of course these models contain assumptions about the likely growth of energy demand, the cost and scarcity of other fuels, the stringency of future environmental standards and the cost of meeting them, and these issues introduce uncertainties at least as large as those about the cost of building a fusion reactor. Extreme positions can be taken on all of these factors, but the prevailing view seems to be that the world demand for energy will double by 2050 and double again by the end of the century, and it is broadly recognized that fossil fuels will not be able to meet all of this demand due to a combination of rising costs and environmental constraints. But how this will affect the future market for fusion energy is impossible to predict, as long as the impact of climate change remains controversial. If a political consensus should develop that current measures to reduce CO_2 emissions will fail to produce the necessary result, and that the world should urgently work towards a future in which all electricity is produced without CO_2 emission, the market would be different. The options would be few and the solution would have to be sought in a combination of all available means: energy savings, hydropower, small-scale renewable sources, sequestration of CO_2 and nuclear energy—preferably fusion. In such a market fusion might become competitive much sooner than is currently assumed and what is now a long-term aim of energy research would turn into a top priority. And the effect on CO_2 emissions would be enhanced because the availability of an environmentally acceptable and abundant source of electrical energy would encourage its use in areas where this now has to be discouraged.

Recent studies of the safety and environmental aspects of fusion have been more rigorous than the early ones and have generally confirmed the inherent advantages of fusion reactors. An assessment carried out for the European Commission in 1995 [43] considered two alternative conceptual designs for a 3000 MW power plant broadly based on the ITER-EDA design. One concept was based on conventional technology with reduced-activation martensitic steel, lithium–lead alloy breeder and water cooling. The other assumed advanced technology with vanadium alloy structures, lithium ceramic tritium breeder and helium cooling, emphasizing low activation materials, avoidance of chemical reactions between coolant and reactor materials and improved energy conversion with vanadium alloy components.

The European study concluded that fusion plants would be inherently safe with no possibility of uncontrolled power runaway and that the plasma would switch off passively in the event of a total loss of active cooling. Radiation doses to the public and operators were well below safety limits both in normal operation and in the most severe accidents. Over their lifetimes, as a result of component replacement and decommissioning, fusion reactors would generate activated material similar in volume to that of fission reactors, but qualitatively different in that the long-term radiotoxicity would be very much lower. Low-activation materials offer potential for further reduction. After about a hundred years, radiotoxicity would fall to levels comparable with the ashes from coal-fired power plants. Similar conclusions were reached in American and Japanese studies. To what extent these factors will simplify the licensing procedures for a fusion power plant and allow fusion to avoid the problems of public acceptability now confronting fission, will depend on how societies perceive the need for fusion energy.

As well as having economic and social advantages compared with other options for generating electricity, any new technology such as fusion must demonstrate that it can achieve the necessary degree of reliability and availability before it will be taken up by the utilities. This will require the construction and operation of a DEMO type reactor based on technology that is as close as possible to that of the future commercial plants, although DEMO, being the first of a kind, will inevitably cost more to build and operate. How the responsibility for long-term energy research and development is shared between government, utilities and manufacturing industry varies from one country to another, but the world-wide trend towards privatization and liberalization of the electricity supply industry is generally accompanied by a decline of the funding for projects with such long payback times [44]. It has often been remarked that it would make the introduction of fusion on the energy market much easier if this could be done with small production units. In the absence of a viable concept for this, there is no choice but to climb over the barrier and try to reach the slope where the cost of energy goes down with further increase in size [38].

9.6 Joining forces for the 'next step'

International collaboration in fusion research has a long history, which started with the secret wartime Cockroft–Libby agreement between Britain and the USA and entered the public domain when the Soviet leaders visiting Britain in 1956 made their first overtures in this direction. Through bilateral and multilateral arrangements, the latter mostly under the IAEA, fusion physicists developed a close world-wide network. Since the early 1970s, nuclear engineers have joined their ranks and produced a variety of reactor design studies with different degrees of credibility. In August 1977, David Rose from MIT invited a group of senior engineers from different countries to discuss how the programmes could be better integrated. There were two international agencies that could possibly play a role, the 'four letter agency', IAEA, under the auspices of the United Nations, and the 'three letter agency' IEA. The latter had been set up by the OECD in the wake of the Middle-East oil crisis, and aimed at strengthening energy research and development in the non-communist world. It had already made an inroad into fusion by sponsoring some small international collaborations and so could perhaps be given a greater task. The fusion community, however, was not inclined to leave the Soviet colleagues out and was more than relieved when their chief, Evgeniy Velikhov, and at the political level their minister of foreign affairs, proposed to set up a workgroup under the IAEA to study the modalities of a common effort. The outcome was that the IAEA would host an international team, charged with designing an **int**ernational **to**kamak **r**eactor (INTOR).

The specific aim of INTOR was to be a test facility for the plasma engineering and tritium-breeding blanket technology required for the demonstration reactor, DEMO, that would follow it. Thus, INTOR would fill the role of the 'next step' experiment foreseen in the progression from the JET, TFTR, JT-60 generation to DEMO. There was hope that at least some of the partners would have operated burning plasma experiments before INTOR came on line, though as we saw in section 8.4 this did not occur. Although INTOR would not be required to breed its own tritium, the original objectives specified demonstration of both tritium breeding and electricity production in blanket modules. The total neutron fluence available for materials testing was to be $5\,\mathrm{MW\,yr\,m^{-2}}$ (later reduced to $3\,\mathrm{MW\,yr\,m^{-2}}$).

INTOR studies took the form of a network of general and specialist 'workshop' meetings supported by homework in each of the participating countries and served to unify tokamak reactor studies in two ways. National design teams compared their studies and incorporated what they learned from each other in their own models, to the extent that comparative studies showed their products to be almost indistinguishable, given equal boundary conditions. Also INTOR brought physicists and engineers together, not just

Table 9.2 INTOR and national engineering test reactors

Name	INTOR	NET	FER	TIBER	OTR
Origin	IAEA	EC	Japan	USA	USSR
Major radius, R_0 (m)	5.0	5.18	4.42	3.0	6.3
Plasma radius, a (m)	1.2	1.35	1.25	0.83	1.5
Elongation, κ	1.6	2.05	1.7	2.4	1.5
Toroidal field, B_t (T)	5.5	5.0	4.61	5.55	5.8
Plasma current, I (MA)	8.0	10.8	8.74	10	8.0
Safety factor, q	1.8	2.1	1.8	2.2	2.1
Average density, $\langle n_e \rangle$ (10^{20} m^{-3})	1.6	1.7	1.14	1.06	1.7
Energy confinement τ_E (s)	1.4	1.9	1.7	0.44	1.7
Average beta (%)	4.9	5.6	5.3	6.0	3.2
Fusion power, P_{th} (MW)	585	650	406	314	500
Average neutron wall load (MW m^{-2})	1.3	1	1	1	0.8
Neutron fluence (MW a m^{-2})	3	0.8*	0.3	3.2	5
Energy multiplication, Q	Ignited	Ignited	$Q > 20$	$Q > 5$	$Q > 5$
Burn pulse (s)	150	>200	800	Steady state	600

* 0.08 in first (physics) phase.

on an *ad hoc* basis as in so many previous incidental reactor studies, but now as representatives of the four major fusion programmes, backed by strong home bases. Although INTOR (and its successor ITER) were studies specifically focused on the 'next step' rather than on a commercial reactor, they largely took the place of conceptual reactor studies and their development reflects progress in both tokamak physics and engineering over the last two decades of the twentieth century.

In a series of reports [45] INTOR assembled a database of the relevant scientific and engineering information, developed a conceptual design for a reactor that would demonstrate the required plasma performance and establish the technological feasibility of fusion power generation, performed a critical assessment of the developing database, and analysed critical issues as well as possible innovative features. In parallel with the INTOR study, the four partners developed their own concepts of 'next step' devices which could take the place of the common project if this was not to be built. In its final sessions [46], the INTOR workshop analysed and compared the characteristics of these 'national' designs (the European Community is referred to as a nation in this context). Our table 9.2, extracted from the final INTOR report [47], lists their vital statistics.

In 1979, INTOR had reviewed the literature on existing empirical scalings for energy confinement and, mainly on the basis of experiments with Ohmic heated plasmas, had proposed what became known as the INTOR scaling:

$$\tau_E = 5 \times 10^{-21} n_e a^2.$$

This served both as the basis for the INTOR design and as a benchmark against which the results of on-going experiments could be assessed. However, as we have noted earlier, tokamak physics in the early 1980s ran into difficulties with strong additional heating. Goldston's scaling for the confinement time with its $P^{-1/2}$ dependence and Troyon's beta limit fell below the assumptions underlying the INTOR conceptual design. Though the situation improved somewhat when the H-mode was discovered in ASDEX and confirmed in a variety of other machines, extrapolation to the INTOR parameters still failed to give assurance that the required performance in terms of plasma pressure and energy confinement time would be reached. The discrepancy is highlighted by the fact that INTOR and the parallel national studies assumed that ignition could be achieved with plasma currents in the range 8–10 MA, but, by this time, JET had operated already at 7 MA (albeit with a limiter so that confinement was in the L-mode) and was clearly much too small to ignite.

The penultimate INTOR Workshop Report [48] admitted this discrepancy between physics assumptions and results and (by way of exercise to the reader) gave recommendations for updating the conceptual design. Design studies had often started from what would be acceptable and, when the science failed to perform accordingly, had called for further research and development with the expectation that one day this would resolve the problem. But INTOR was not to have been merely a reconnaissance mission; the workshops had set out to design a reactor that, on current knowledge, could be built. Yet, it ran into a deadlock when physics turned against it and the partners, particularly the US side, felt that they could not admit defeat and come home with a realistic design. The INTOR database assessment and conceptual design, along with the national studies, helped to focus tokamak physics and fusion technology research into areas of reactor relevance, but clearly the time was not ripe for a joint venture—the four partners had different perceptions of the need for a test reactor and what precisely they expected from it.

9.7 ITER

In the mid-1980s, General Secretary Gorbachov of the Soviet Union in a meeting with President Mitterand of France promoted the idea of an East–West collaboration to establish the feasibility of fusion energy. Walking in Igor Kurchatov's and Lev Artsimovich's footsteps, Evgeniy Velikhov, Director of the Kurchatov Institute and a close advisor to Gorbachov, had been a strong supporter of the INTOR project. The new initiative was taken up in 1985 at the first Summit Meeting in Geneva between the American and Soviet leaders Reagan and Gorbachov. Their final communiqué announced that their countries, joined by others, would work together to establish the feasibility of fusion energy for '*the benefit of all mankind*'. This led to the signing in 1987 of the ITER agreement between the United States, the Soviet Union, the European Community and Japan. The name ITER (usually pronounced as 'eater') has a dual significance, being a Latin word meaning 'the way' as well as an acronym for the International Thermonuclear Experimental Reactor.

It is reported that although President Reagan thought that he had agreed to the construction of ITER [49], concern in the US Department of Defense about possible transfer of sensitive software and hardware capability to the USSR led to approval only of the first phase—the Conceptual Design Activities (CDA). This started in 1988 with a team of about 40 scientists and engineers from the four partners meeting for several months each year in Garching. The West German government was particularly supportive and there was talk of a site to construct ITER somewhere near to the border with East Germany. However, a few years later the unexpected break-up of the Soviet Union* and the reunification of Germany brought good news for world peace and bad news for fusion—the Russians were no longer able to support a big international project and neither was a Germany faced with the costs of reunification.

Whatever the political realism might be, ITER was regarded by those working on it as a machine that would be built. A realistic design had to be developed and difficult problems had to be solved, not simply deferred in the hope that they could be resolved later. A serious reappraisal of the minimum size that would be required to reach ignition indicated a current closer to 20 MA than the 8 MA that had been assumed for INTOR. The parameters of INTOR are compared in table 9.3 with three different stages of the ITER design, including the most recent version of ITER(2001) which we will discuss in section 9.7.4. In comparing the values of parameters in this table it should be noted that INTOR, ITER-CDA and ITER(1998) had the target of ignition, whereas ITER(2001) has reduced technical objectives, in particular $Q \approx 10$. The substantial increase in plasma current from INTOR to ITER-CDA resulted mainly from the updated physics data base, the detailed changes in machine dimensions from ITER-CDA to ITER(1998) reflect the refinements of translating a conceptual study into a detailed engineering design, and the size reductions for ITER(2001) are a consequence of the reduced technical objectives and reduced cost target.

9.7.1 The ITER EDA design

The conceptual design phase was followed by an agreement, signed in 1992 by the four parties, to proceed with a six-year programme of Engineering Design Activities (EDA). This had the objective of delivering a detailed

* The Russian Federation replaced the Soviet Union as party to the ITER agreement.

Table 9.3 Comparison of INTOR and ITER parameters

	INTOR 1986	ITER CDA 1990	ITER 1998	ITER 2001
Reference	[50]	[51]	[52]	[53]
Plasma current, I (MA)	8	22	21	15 (17)*
Major radius, R_0 (m)	5.0	6.0	8.14	6.2
Plasma radius, a (m)	1.2	2.15	2.8	2.0
Elongation, κ	1.6	2.0	$\sim$1.6	1.7–1.85
Toroidal field on axis, B_t (T)	5.5	4.85	5.68	5.3
Safety factor, q	1.8	3.1	3.0	3.0
Average $\langle n_e \rangle$ ($10^{20}\,m^{-3}$)	1.6	1.23	0.98	1.0
Average $\langle T_i \rangle$ (keV)	10	10	12.9	8.1
Average $\langle \beta \rangle$ (%)	4.9	4.2	2.2–3.0	2.5
Confinement time, τ_E (s)	1.4	3.6	5.9	3.7
Energy multiplication, Q	Ignition	Ignition	Ignition	10
Nominal fusion power, P_{th} (MW)	600	1000	1500	500 (700)
Neutron wall load ($MW\,m^{-2}$)	1.3	1.0	1.0	0.57 (0.8)

* The design has capability for plasma current up to 17 MA with the parameters shown in parentheses at some reduction in other parameters, notably the pulse length.

design that would provide a basis for a decision to proceed to the construction phase, either as an international collaboration or individually by one of the four partners. The EDA was carried out with a dedicated 'Joint Central Team' of about 170 scientists and engineers working full time for ITER, supported by many specialists working part time in the so-called 'Home Teams'. It proved impossible to agree on a single site for the Joint Central Team—the Europeans had assumed that it would naturally succeed the CDA at Garching but the Americans favoured a move to San Diego and the Japanese proposed Naka. The compromise decision was to split the team between all three sites—an ominous early warning of future indecision and procrastination that would henceforth plague the project at the political level. Nonetheless, the work went ahead with considerable enthusiasm. Paul-Henri Rebut left JET to be appointed overall Director of ITER* at the San Diego site, which (with Yasuo Shimomura from Japan and Valerij Chuyanov from the Russian Federation as deputy directors) was to be responsible for overall coordination, physics and safety. The Garching site, working under an American deputy director, Ron Parker from MIT, had the responsibility for the vacuum vessel and everything inside it. The Naka site, with a

* In 1994, Rebut was replaced by another Frenchman, Robert Aymar, who had previously been responsible for the construction of the superconducting tokamak Tore-Supra.

European deputy director, Michel Huguet from JET, was to be responsible for everything outside the vacuum vessel, including the magnets, buildings and power supplies.

The goal for the ITER programme was to demonstrate the scientific and technical feasibility of fusion energy for power generation. The detailed technical objective of the EDA was to design a machine that would demonstrate controlled ignition and burn, initially with an inductively driven current with a pulse duration of about 1000 s. A longer-term objective was to demonstrate steady-state operation using non-inductive current drive. ITER was to test and demonstrate the technologies essential for a fusion reactor including superconducting magnets and tritium breeding—but it would not have to be self-sufficient in tritium and would not generate electricity.

The EDA team focused their attention on a tokamak with a single null poloidal divertor, with similar D-shaped plasma (but more than twice the linear dimensions) as JET. The toroidal and poloidal fields would be produced by superconducting coils* with a maximum toroidal field at the plasma centre of about 6 T. Empirical scaling from JET and other tokamak experiments indicated that ITER would require a plasma current of at least 20 MA to reach ignition. These basic requirements determined the dimensions: the plasma major radius would be about 8 m, the minor radius 2.8 m and the nominal fusion power 1.5 GW. The ITER magnetic fields would be provided by a combination of 20 toroidal and nine poloidal superconducting coils. These magnet systems, with a combined weight of conductor and structural support in the region of 25 000 t, would have a major impact on the overall cost and construction schedule.

ITER differed from previous tokamaks in terms of the exposure of machine components to the radiation and heat fluxes associated with the average neutron load at the plasma edge of 1 MW m^{-2}. The principal shielding requirements were to reduce the heat flux deposited by neutrons in the superconducting magnet coils and to reduce the radiation damage of the vacuum vessel wall to a level that would allow repair by welding should the need arise. The plasma was to be surrounded by a radiation shield structure consisting of replaceable water cooled steel modules attached to a massive toroidal steel shell suspended inside the vacuum torus. The shield would have to remove heat received directly on the front surface by plasma radiation (about 300 MW) in addition to the bulk heating by neutrons (about 1200 MW). Under start-up and fault conditions the localized surface heat loads on some sections of the wall could be very much higher. Moreover, this massive structure had to be designed with the ability to withstand electromechanical forces from a disruption.

* The requirement that the engineering had to be 'reactor relevant' ruled out any serious consideration of copper coils.

The axisymmetric divertor at the bottom of the torus, whose function is to exhaust both power and particles, also presents serious design problems. The divertor target could handle by direct plasma contact only about 50 MW compared with the nominal 300 MW of fusion power deposited in the plasma by α heating—radiative cooling of the edge and divertor plasmas was crucial. The design of the ITER divertor was supported by a programme of research on divertor physics that combined computational work with experiments on other tokamaks [54] (section 8.3.1).

ITER would require additional heating in order to make the transition into the H-mode, to heat the plasma to ignition, to drive non-inductive currents, to provide a means of control of plasma equilibrium and fusion burn and possibly to suppress instabilities. Estimates of the amount of power required for each of these tasks fell in the range 100 to 200 MW. Three different systems—neutral beams, ion cyclotron and electron cyclotron—were evaluated in detail with reference designs of about 50 MW for each system. As well as heating the plasma, all three methods could be used to drive non-inductive currents and a lower hybrid system was kept in reserve as a fourth option for current drive. In view of the substantial ongoing technical developments in all of these heating techniques, the choice was kept open with the intention to make the final selection at the time of construction.

9.7.2 The physics basis

The physics basis for ITER [55] was established by 'Expert Groups' involving physicists working on the leading experiments in the Home Teams. This helped to ensure that ITER was firmly based on state of the art tokamak physics and also helped to push these working tokamaks into studies aimed at clarifying points at issue for ITER. In particular, ITER-specific confinement data bases were compiled and scalings refined to a much greater degree than hitherto.

To combine good energy confinement with quasi-steady-state control of plasma density, the ELMy H-mode regime was selected as the preferred mode of operation. A fundamental issue that stimulated a great deal of work on existing tokamaks was the prediction of the energy confinement. An extensive database was developed for the ELMy H-mode regime based on measurements in eleven tokamaks. The ITER scaling IPB98(y,2) (figure 9.3)

$$\tau = 0.0365 I^{0.97} B_{\mathrm{t}}^{0.08} P^{-0.63} n^{0.41} M^{0.20} R_0^{1.93} \varepsilon^{0.23} \kappa^{0.67}$$

has broadly similar dependence on plasma current, major radius and power to the well known Goldston scaling dating from the 1980s, but with more precise and reliable values of the exponents. There is also explicit dependence on the inverse aspect ratio ε, elongation κ and average ion mass M.

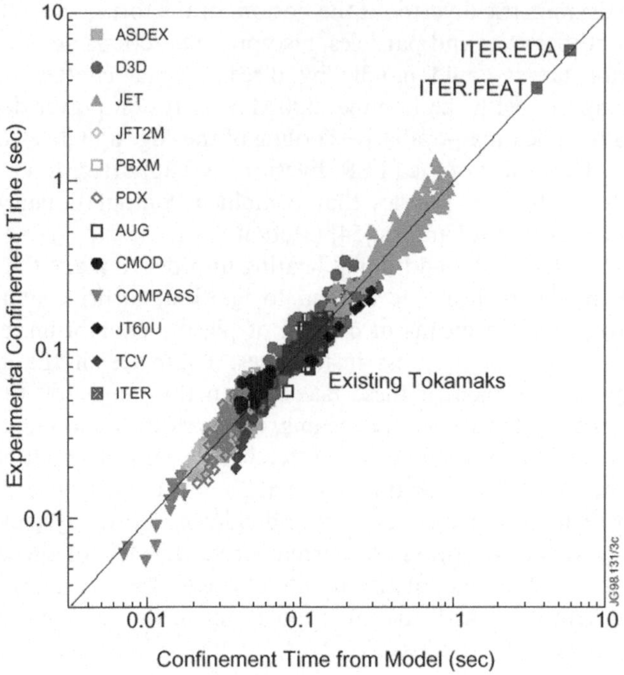

Figure 9.3. Scaling to ITER. The experimental energy confinement time measured in a wide range of tokamaks from Compass ($R_0 = 0.56\,\text{m}$, $I \approx 0.3\,\text{MA}$) to JET ($R_0 = 3\,\text{m}$, $I \approx 5\,\text{MA}$) plotted against the ITER scaling formula given in the text. The extrapolation to ITER is indicated by two points: ITER.EDA is the 1998 design ($R_0 = 8.14\,\text{m}$, $I \approx 21\,\text{MA}$) and ITER.FEAT is the 2001 design ($R_0 = 6.2\,\text{m}$, $I \approx 15\,\text{MA}$).

An important issue was to predict accurately the heating power required for the transition to the H-mode. In deuterium plasmas the threshold power scaled with electron density, toroidal magnetic field and plasma major radius. The extrapolation to ITER predicts that at a density of $5 \times 10^{19}\,\text{m}^{-3}$, the H-mode threshold power would lie in the range 40 to 170 MW; the uncertainty is mainly in the dependence on R_0. An important result from the JET tritium experiments in 1997 is that there appears to be a favourable dependence on ion mass, which if confirmed by further experiments would reduce the power threshold in ITER by 20–30%.

In terms of the operating limits for plasma stability, the ITER study adopted a fairly conservative policy but there were still some areas of uncertainty. The maximum plasma current of about 20 MA was consistent with the usual $q_a = 3$ limit. Theoretical models had suggested that ITER plasmas would be stable to ideal MHD modes at values of the normalized β sufficient to maintain ignition and operation at the nominal 1.5 GW. However, various effects that reduce the ideal limit had been observed in some experiments and this raised concern that the effective β limit for long pulse operation might be less than that predicted theoretically. The nominal operating density would have to exceed the Greenwald limit by a small margin in order to achieve ignition. On the one hand there were grounds for optimism since this limit had been exceeded by quite substantial margins in plasmas refuelled by means other than gas puffing, for example by firing pellets of frozen deuterium into the core; on the other hand there were concerns because high density operation was often accompanied by a substantial degradation in confinement.

9.7.3 Decision and indecision

The 1998, then called 'Final Design Report' for ITER made impressive reading. For the first time, here was a proposal for an experimental fusion reactor that was firmly based on the current generation of experiments, with a detailed engineering design and conscientious cost estimates. Moreover, prototypes of some of the key components, including the superconducting coils and the vacuum vessel, were being built and tested already. Construction was planned to take about ten years and would be followed by a programme of operation divided into two phases, a physics phase followed by a technology phase, each of about ten years' duration. The work of the ITER Team had been monitored and guided by the four partners so there were few surprises. Moreover the Detailed Design Report, published a year earlier, had been subjected to independent review by all the partners and endorsed as technically and scientifically sound. The American review panel's report [56] to the Department of Energy had '*re-affirmed the importance of the key elements of ITER's mission—burning-plasma physics, steady-state operation, and technology testing*' and went on to state that it had '*great confidence that ITER will be able to make crucial contributions in each of these areas.*'

All seemed very positive, but ITER failed to receive approval. Instead of the smooth progression to decisions on site selection and construction that had been hoped for, there was a compromise extension of the design phase by a further three years with instructions to the team to go back and design a device with reduced technical objectives that would meet the same programme objectives but cost less to build. Worse was to come. The Americans vacillated about their continued support for ITER and, after some further months of cliff hanging, were instructed by Congress to withdraw completely. The American team members at Garching and Naka were recalled and the San Diego headquarters closed down early in 1999. Their government even took the unprecedented measure of forbidding American scientists to participate further, even as observers, in any ITER activities and meetings.

The factors that led to this situation are complex and have as much to do with events outside fusion as with anything within the programme. National policies as well as international relations had changed dramatically during the 1990s—mostly to ITER's disadvantage. After the Soviet Union collapsed, ITER could no longer be promoted as a vehicle of East–West collaboration. Clearly, it always had to stand on its own feet as leading to a new energy source, but in spite of long-term concerns about energy resources and global warming, there was no sense of crisis—indeed budgets for energy research, including fusion, had tightened in many countries.

The 1998 Final Design Report had put the estimated cost of building ITER in the range of \$5.2 to \$5.7 billion—well within the anticipated target of \$6 billion. The average annual cost over the planned ten-year construction period, with peripheral costs included, would be about \$750 million and annual operating costs thereafter would be around \$400 million. For comparison,* total world spending on fusion research in 1998 was about \$1400 million†—so in principle ITER was affordable. But this existing funding could not be redirected so easily—it came from many different budgets in different countries and was committed already to existing projects. Building ITER would be possible only with a substantial increase in funding, especially from the host country that would be expected to pay for the site and much of the infrastructure, and even then would put a significant squeeze on the existing programmes. No longer could the construction costs be shared equally between the four partners. When ITER had been conceived in the mid-1980s, the fusion efforts in Europe, Japan, the Soviet Union and the US were roughly comparable. A decade later the US fusion effort had become considerably smaller than its counterparts in Europe and Japan and the Russian programme was but a shadow of what it once had been.

Within the American fusion community there had always been reservations about ITER, which was acceptable to many only if accompanied by research aimed at finding either an alternative to the tokamak or a way to make it more compact. But by 1995 the US fusion budget had fallen to about one half of its value in real terms compared with 1977—forcing the closure of many alternative lines and putting renewed pressure on the few that remained. Competition between the demands of ITER and the basic fusion programme resulted in an erosion of support for ITER among the US fusion community and pressure from the Chairman of the Senate Appropriations Committee to focus only on building ITER backfired and fuelled further negative reaction. The DOE had asked for a substantial

* ITER cost estimates are referenced to year 1989 values and need to be inflated by about 30% for comparison with year 1998 values—though this factor varies slightly depending on where ITER would be located.

† Overall annual expenditure on magnetic fusion research in Europe in 1998 was equivalent to about \$500 million (40% from Community funds, the rest from national research budgets), Japan's programme was of similar size to that of Europe, America's was about about \$230 million and there were smaller programmes in other countries including Australia, Brazil, Canada, China, India and Korea.

annual increase in the fusion budget to reverse this trend in order to maintain a balanced domestic programme alongside participation in the construction of ITER, but the fusion budget, along with those of other scientific research programmes, was further reduced by a Congress that was determined to control public spending.* To stay consistent with the reduced budget, the US fusion programme thereafter adopted a revised strategy which focused on the underlying science—a substantial departure from the previous strategy [57], endorsed as recently as 1991, of a demonstration fusion power plant by 2025 leading to the operation of commercial fusion power plants by 2040. Clearly, participation in ITER could not be reconciled with the new constraints. The fusion research community became deeply divided and statements undermining ITER's credibility received widespread media coverage, whatever their scientific value, making the project an easy prey for a hostile Congress.

Attitudes to ITER were generally much more positive in the three other partners. Support for fusion and a determination to build an ITER-like machine appeared particularly strong in Japan. But Japan's economy was going through a difficult phase and its government ruled that no commitment could be made for the time being. In Europe, participation in ITER was said to be the keystone in the future programme. The 1996 evaluation of the European Fusion Programme [58] had concluded that

> *Fusion has now reached a stage where it is scientifically and technically possible to proceed with the construction of the first experimental reactor, and this is the only realistic way forward. Starting the construction of ITER is therefore recommended as the first priority of the Community Fusion Programme under the Fifth Framework Programme.*†

By closing down JET after 1999 as foreseen and redirecting that funding to ITER, Europe could have maintained its other fusion research programmes and participated in an ITER built outside Europe with only a modest (~10%) increase in the centrally funded part of the fusion budget. The Review had also recommended that

> *ITER should be built in Europe, as this would maintain Europe's position as world leader in fusion and would be of great advantage to European industry and laboratories.*‡

* The DOE had asked for an increase from about $370 million in 1995 to about $860 million in 2002 (with an average of $645 million per year between 1995 and 2005) but it was reduced by Congress to $230 million.

† 1998–2002.

‡ These conclusions were reaffirmed in 2000 when the next review panel reported on the sixth framework programme.

But construction of ITER in Europe would require a 50% increase in the Euratom fusion budget as well as a substantial cost to the host country for the site and infrastructure. This became the barrier—none of the European countries came forward with an offer. Germany, which had long been the strongest partner in the European fusion programme and its most effective supporter at the political level, had shown strong interest in acting as host. Not long after the reunification, however, it began to show signs of hesitation, which became even stronger after a new, anti-nuclear Government took power in 1998. And the attitude of the 'Green' faction in the European Parliament has always been negative.

These uncertainties allowed issues to surface again that had been debated throughout the history of fusion every time an increase in scale was contemplated. Even within the fusion community some felt that, despite the impressive results that had been obtained in the big tokamaks, so much still remained to be understood about the behaviour of the plasma in these devices that it would be foolhardy to proceed with such an expensive step as ITER until all these uncertainties were resolved. The protagonists for ITER argued that the scientific basis was much more reliable and secure than had been the case 25 years previously when the big three tokamaks had been designed. One had learned how to optimize and enhance the performance of these machines and to push forward in ways that could not be foreseen when they were conceived. Certainly, fusion research would never have progressed this far had these big tokamaks not been built. Many of the physics issues and most of the technological issues confronting fusion can be resolved only on a reactor-grade plasma. Experience gained on smaller plasmas is valuable, but the physics often changes in subtle but significant ways in going to a new scale of experiment.

9.7.4 Back to the drawing board

With the realization that the ITER Parties could not proceed to the construction of the 1998 design, the new directive was to find a set of reduced technical objectives that would still satisfy the overall programme objectives of demonstrating the scientific and technical feasibility of fusion power—but at a cost of approximately 50% of the previous target. Two alternative strategies were compared by a special working group: an ITER-like machine capable of addressing both scientific and technological issues in an integrated fashion or a number of complementary lower cost experiments each of which would focus on specific issues. It was concluded that the complex non-linear interactions between α-particle heating, confinement barriers, pressure and profile control and their compatibility with a divertor could be addressed only in an integrated experiment capable of long pulses with the α-particles as the dominant source of heating. The working group expressed the unanimous opinion that the world programme was '*scientifically and*

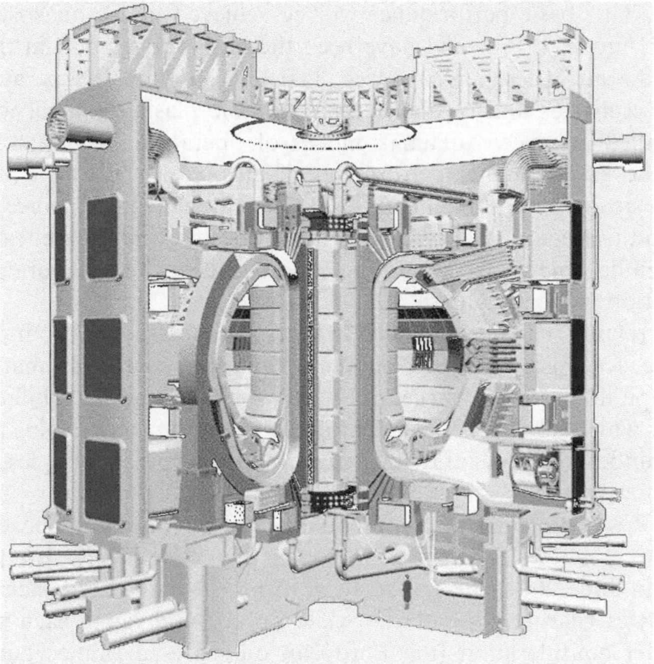

Figure 9.4. Schematic of ITER in the reduced-size version proposed in 2001. The cut-away shows the interior of the vacuum torus with the internal neutron and thermal shield and the divertor regions. The superconducting toroidal and poloidal magnetic field coils lie outside the vacuum vessel and are contained in a secondary vacuum within a cryostat. The vacuum torus and cryostat are penetrated by various ducts (or ports) for cooling pipes, plasma heating systems, diagnostics systems etc. These systems and other ancillaries outside the secondary vacuum are not shown. The scale of the machine is indicated by the figure at the bottom.

technically ready to take the important ITER step'. Clearly the construction cost could be reduced significantly only by making ITER smaller. To operate with α-particles as the dominant source of plasma heating requires $Q > 10$, which determined the reduced size. As a consequence, the average neutron flux available for component testing fell to $0.5\,\mathrm{MW\,m^{-2}}$ and the average fluence to $0.3\,\mathrm{MWa\,m^{-2}}$. Notwithstanding this technological sacrifice, $Q = 10$ became the new boundary condition for the design*.

A new Final Design Report, presented in 2001, proposed a tokamak with major and minor radii as well as plasma current roughly three quarters of the previous values (table 9.3 and figure 9.4). Fusion power will be about

* Initially the new design was known as ITER-FEAT, which supposedly stood for Fusion Energy Amplifier Tokamak, but was never officially recognized by the ITER Council, nor even unambiguously explained. Its official name later became simply ITER; we refer to it as ITER-2001 to distinguish the new design from the previous one.

500 MW. This basic performance will be achieved in the quasi-stationary 'ELMy H-mode' that would have been the basis for ignition in the larger version. As an experimental device, ITER is required to operate over a range of scenarios and configurations. Flexible plasma control will allow the so-called advanced tokamak modes of operation (section 8.3.2) that will be the key to higher performance if means can be found to sustain them in a steady state. To this end ITER is now designed with increased elongation and triangularity. There is no doubt that to exploit these features the device will become much more of an experiment in which elaborate physics optimization will be required.

The revised schedule allows eight years for construction from receipt of the licence to construct to first plasma operation. The cost estimate for the direct capital cost of constructing ITER is now about $3.8 billion (2000 value) of which about 53% is taken up by the tokamak machine core and the remainder by ancilliaries including buidings, power suppies, heating and cooling systems etc.

ITER was designed so that it could be built in any of the participating countries, with the capability for conducting experiments from remote centres. In June 2001, the Canadian government announced their offer to site ITER at Clarington, Ontario, while Europe and Japan each have two sites under consideration (the European ones are in France and Spain) from which they have yet to make their choice. Decisions on whether and where to build ITER will have far-reaching consequences on world-wide fusion research and development. If ITER is built it will require commitment of significant resources, both in terms of money and manpower, and it will become the cornerstone of the programme—the test bed on which the feasibility of fusion power is to be established. The general consensus is that fusion is scientifically and technically ready to take this step—the barriers are financial and political.

Chapter 10

Epilogue

Looking back

The quest for fusion, which started in earnest in the late 1940s, has experienced several cycles of optimism and pessimism. Much ground was covered already in the first decade, though at that time most of the work was done in secrecy. The first reports reached the public in 1956 and, by the time all shrouds were finally lifted at the 'Atoms for Peace' Conference in Geneva in 1958, one could see most of the magnetic confinement configurations that have since been investigated, along with indications of the difficulties with stability and anomalous transport that this research was going to meet. Several groups had claimed to have seen thermonuclear neutrons, but the toroidal pinch had already made a painful retreat before the Conference and only for the theta pinch could a convincing case be made. Much work had already been done on new diagnostic methods and in most cases these revealed rapid recycling and insufficient heating of the plasma. Thus, the spectre of 'pump-out' or Bohm diffusion rose to haunt fusion research for much of the second decade. Theoreticians proposed new and more threatening instabilities and experimentalists retreated from plans to build bigger successors to experiments like Zeta and C-stellarator to concentrate on attempts to understand the physics of hot plasmas in magnetic fields. But in the second half of the 1960s the experimental results began to improve. The pessimism began to dissipate at the IAEA fusion Conference at Culham in 1965 and turned into optimism at Novosibirsk three years later.

The Novosibirsk Conference in 1968 was to prove a major milestone in fusion research. Since Ioffe had surprised the Salzburg Conference with his minimum-*B* configuration, seven years earlier, the open-ended systems had not produced new results of a comparable impact. Most laboratories were therefore turning towards closed systems and, although it was not the only one defying Bohm, the tokamak emerged as the clear leader. Thereafter tokamaks proliferated world-wide while stellarators and reversed-field pinches took second place—kept alive with the hope that they might have long-term advantages as reactor systems in areas where the tokamak was weak and that in the meantime they could contribute to understanding the physics of toroidal confinement.

Although high-beta linear confinement systems such as the z-pinch and the theta-pinch had provided relatively easy access to high temperatures, their potential as reactor systems was put into question by end-losses and instabilities. Attempts to avoid end-losses in the theta pinch by making it toroidal failed because of instabilities, but the toroidal z-pinch continued to be studied extensively. This finally achieved a measure of success by Novosibirsk, when it was realized that the essential ingredient was field reversal, which had been seen earlier but not properly recognized. It would take several more years to understand that this was a consequence of the plasma attaining a minimum energy state consistent with conservation of helicity, but by this time even more impressive confinement results were being obtained with the much stronger stabilizing toroidal field of the tokamak.

The only open-ended system with reactor potential that remained after Novosibirsk was the low-beta mirror configuration, but even this was in serious difficulty until the mid 1970s when long-standing barriers of burnout and stability were finally overcome and new proposals for tandem mirrors offered a way to control end-losses. This line of research flourished for another decade until it was all but abandoned in the mid 1980s as the performance of the big tokamaks moved ahead and these machines claimed an increasing share of the tightening budgets. Mirrors had little margin to spare as reactor systems and they fell out of the competition when results failed to meet the required targets.

Throughout these developments, some systems were more attractive with a view towards a possible reactor, while others were investigated mainly because they were more suitable for studying particular plasma phenomena. These performance-oriented and physics-oriented, or one might say holistic and reductionistic, approaches each had their place in fusion research. But as the science matured and the questions to be answered became more specifically related to reactor plasmas, the large devices that were built principally to achieve better performance in terms of plasma density, temperature and energy confinement became also the places where much of the relevant physics was discovered. So, whereas progress initially depended on studies in a wide variety of plasma containers, research from the mid-1970s on gradually focussed on the physics of reactor-grade plasmas in reactor-like devices. In historical perspective, the merger of the performance-oriented and physics-oriented streams in fusion research is a sign of health and accomplishment, even if particular steps in the process, such as the untimely termination of the US mirror programme, were rather a sign of poverty.

In the three decades after Novosibirsk, the increasing emphasis on the tokamak was rewarded by steady progress, and the encouraging performance of the newly built machines added more fuel to the tokamak fire. In the aftermath of acute oil shortages resulting from conflict in the Middle East, fusion budgets increased in the 1970s. Although the short-term effects of the oil crisis were soon overcome, they accentuated long-term concerns about energy reserves and industrial nations readily supported fusion research, if only as an insurance premium against the risk of future energy shortages. In this climate it was not too difficult to decide upon the construction of three very large tokamaks—JET in Europe, TFTR in the United States and JT-60 in Japan. The Soviet Union, while remaining very strong in plasma theory, was falling behind in the construction of experimental fusion facilities and in data-handling technology, having chosen to concentrate on other priorities such as space research.

Eventually, the large tokamaks would produce the most substantial advances in fusion research. They explored new regimes of improved confinement, both with limiters and with divertors, extended the database for the empirical scaling relations for energy confinement that form the basis for extrapolation to ITER, found ways to reduce the impurity content of their plasmas, observed the onset of alpha particle heating, brought the fusion triple product, which since 1958 had steadily increased by one or two orders of magnitude per decade, up to the break-even value and produced nearly 20 MW of real fusion power when operating with tritium. Scientific progress in understanding and predicting plasma behaviour flourished in the big tokamaks and the smaller machines that operated alongside them. Topics like equilibrium and stability, heating, current drive, edge and divertor physics have reached new levels of understanding—although much remains to be done, both in theory and experiment, to explain energy and particle transport in the presence of microturbulence, and more generally to solidify the scientific basis under our empirical knowledge.

Meanwhile, the four major parties in fusion research, the USSR, the USA, Japan and the European Community, had started to think about plans for their next steps and in parallel considered ways to pool their expertise and budgets to make the largest possible step forward. In 1977 the IAEA initiated the INTOR study with the aim to produce a design of an experimental tokamak reactor. The initial optimism generated by the tokamaks of the 1970s, however, could not be sustained when strong heating proved to lead to deteriorating heat insulation and the plasma pressure turned out to be more severely restricted than had been anticipated. Although the next decade brought regimes of better confinement, the scaling relations and operating limits underlying the INTOR design gave way to less optimistic ones, which showed that the proposed 8 MA tokamak would not ignite. The disappointment was felt most strongly in the USA, where the conviction had grown that a much bigger reactor would have no chance of finding a place in the market.

However, in meetings with his European and American counterparts, Soviet General Secretary—later President—Gorbachov took a new initiative towards a fusion reactor. When he reached an agreement with US President Reagan, Japan and Europe were invited to join and in 1987 the four-party ITER agreement started work on a new design for a test reactor. Physicists and engineers of the world fusion community were brought together once more, this time with the clear mandate to design an experimental tokamak reactor that would work. Extrapolation to ITER from the database provided by the three big tokamaks and the host of smaller ones indicated that a plasma current of at least 20 MA would be required to reach ignition. The choice fell upon a shape similar to that of JET but with more than twice the linear dimensions. The nominal fusion power would be 1500 MW. After an outline design had been assessed and approved, the engineering design phase started in 1992. This produced a detailed design for a fusion reactor that was firmly based on a confinement scaling validated in the current generation of experiments. Prototypes of some of the key components, including the superconducting coils and the vacuum vessel, were built and tested. Construction was planned to take about ten years and the detailed cost estimate stayed within the target of 6 billion dollars at 1989 values.

But when the first 'ITER Final Design' Report was delivered in 1998, the political tide had turned against it. The global energy situation was no longer seen as demanding the rapid development of a new resource, despite concerns about environmental issues with fossil fuels and growing opposition against fission power. A decade earlier ITER had gained support as a high-profile demonstration of international collaboration involving both communist and capitalist countries, but after the end of the cold war the thrust of this argument was exhausted. ITER by itself could conceivably have been squeezed out of the existing budgets, but particularly in the US, where a new Republican Congress had already slashed the budget for magnetic fusion, this would have necessitated closing down all other lines of development. And since the three other partners had their own reasons for postponing a major investment, it was decided to lower the previously agreed cost target and to pursue a new design with reduced objectives. Even so, the Americans withdrew from the collaboration leaving its former partners to work out the reduced-cost option for which a new Final Design Report emerged in July 2001.

We will return to ITER, but first let us review some of the choices that had to be made in the drive towards building a tokamak test reactor based on the D–T reaction—a process of research and evaluation spanning half a century. None of these choices is final or irrevocable; none of the alternative options that have been under consideration has been definitively proven to be unfeasible, they have simply yielded to more promising alternatives. And each retains its advocates, who may well be able to challenge the intellectual grounds on which far-reaching decisions were taken. When major decisions must be made, science assumes a political dimension and whatever course is taken there is likely to be an opposition, but to find a consensus for an alternative course is a different matter.

For some time, the fusion-fission hybrid reactor figured prominently in the discussion. The idea was advanced already in the early 1950s, primarily with a view towards supplying fissile material for a rapidly growing thermal-reactor market, and became the accepted policy in the Soviet Union. The need for it would depend on the inventory doubling time of fission breeders (with thermal or fast neutrons), which might turn out to be too long—pessimistic estimates ran as high as 50 years—to support a rapidly expanding network of fission reactors. But the dwindling market for nuclear reactors took the pressure off breeder development; indeed the problem has rather become how to dispose of plutonium stocks. But the hybrid may take many different forms; for one, a fission blanket could serve as a neutron and energy multiplier, in case pure fusion failed to breed its own tritium fuel or to generate enough energy. Such proposals lost much of their attraction when environmental considerations gained prominence and fusion research concentrated on the areas where it could promise decisive advantages over fission. So, fusion research eventually settled on pure fusion. But the idea that fusion reactions could serve as a source of fast protons and neutrons competing with accelerators for the purpose of burning unwanted fission products has not entirely been abandoned.

There is a broad agreement that the first generation of fusion reactors should be based on D–T rather than on the inexhaustible D–D or the 'advanced' fuels, the neutron-lean D–^{3}He and the neutron-free H–^{11}B. The latter are advocated because they would reduce or eliminate problems with radioactivity, and the absence of a tritium-breeding blanket—and a neutron shield in the case of H–^{11}B—would simplify the construction of the reactor. On the other hand, they require a plasma pressure and a confinement time far exceeding the proven capabilities of any known confinement scheme. If there is a future for these fuels, it will lie beyond that of the deuterium–tritium mixture although pure-deuterium fusion would seem not so far away once the technology of D–T fusion has been mastered.

Magnetic confinement is further advanced than inertial confinement for use as an energy source. The notion is being entertained that this is a consequence of the former having had an earlier start, as if inertial confinement research commenced only with the invention of the laser, but the story is really more complicated. Both approaches were discussed already in the first years after the Second World War and the first thermonuclear device was brought to explosion in 1952, at a time when the physics of magnetic confinement was still in its infancy. Much thought was spent in the following years on ways to miniaturize the hydrogen bomb, and when the laser appeared on the scene, in 1960, there was a well developed conceptual framework in which it almost immediately fell into place. Plasma physics and computational methods had made great progress in the preceding decade and there was no lack of funding in the military laboratories, where ignition with lasers or particle beams attracted support because small explosions can serve as benchmarks for computational simulations of larger ones, and can simulate their radiation effects. Thus, in the USA, inertial confinement has always been promoted mainly as a military programme with possible civilian applications as a welcome by-product and it is now able to attract more support than magnetic confinement as a purely civilian programme. The technical and economic problems of energy generation by micro-explosions are formidable however and there are no indications that this method offers an easier route to an energy-producing reactor.

It is clear that the tokamak has advanced so much further than other magnetic confinement systems that to wait for one of these to catch up would be unproductive. At various times it was thought that stellarators, mirror machines or reversed-field pinches had intrinsic advantages, and that work on the tokamak might profitably be delayed on their behalf; and on looking back it is clear that such considerations have indeed slowed the main course. But not many critical observers have been convinced by the contention that if only more time and money had been spent on the alternatives, they would automatically have come as far as the tokamak. The tokamak was supported because it performed better than the others; it is difficult to argue the reverse. Nor does the argument that once the first tokamak reactor enters into service it will be impossible for any alternative reactor type to reach the development stage, carry much weight in a worldwide market where every industrial power will fight for its own niche.

Among the different shapes the tokamak may assume, the D-shaped cross-section with a poloidal divertor exemplified by JET, JT-60U, DIII-D etc. and on which the ITER design is based, has come out as the favourite one in virtually all tokamak reactor design studies since the mid 1970s. There exists a reliable database and there is extensive operational experience for this configuration, giving confidence in the relatively small extrapolations in size from where we stand to ITER and from there to a commercial reactor. Other variants of the tokamak line have their advocates but it is not clear that they lead to a cheaper and more compact reactor. High-field tokamaks might offer an alternative route to a burning plasma experiment but would not extrapolate to a commercial reactor. Spherical tokamaks show promise in relatively small-scale experiments, but the database needs to be expanded and extended to much larger devices before it could support the construction of a test reactor. Stellarators might offer advantages in terms of steady-state operation, but in terms of confinement and beta they appear to be roughly equal to tokamaks.

Nonetheless, at the height of the crisis into which the rejection of the original ITER proposal had thrown the fusion programme in 1998, one had to ask if any of these earlier choices needed to be reconsidered. Almost universally, the answer has been that none of the alternatives presents a clear path to a better solution. The research communities in Europe, Japan and Russia were generally inclined to proceed in the chosen direction of ITER, even if this could not be done at the originally envisaged pace. Attitudes differed most notably in the USA which adopted the policy of trying to optimize the concept by concentrating on fusion science before contemplating the construction of a test reactor, but this may turn out to have been a temporary deviation and the US shows signs of returning to the main stream.

It is widely held that the time has come for fusion-reactor technology to enter the phase in which consistent sets of reactor-relevant solutions for the problems of plasma-facing components, mechanical construction, tritium breeding, cooling, shielding and superconducting magnets are tested in a fusion-reactor environment. This is why the next step has mostly been envisioned as a means for not just studying the ignited plasma, but also serving as a test bed for fusion technology. In other words, it would combine a definitive step in confinement physics with a far-reaching reconnaissance in reactor technology. It is particularly on the last point that ITER may not go as far as had been envisaged and that the need for a dedicated materials testing facility is felt more strongly than ever.

It has been considered to separate the study of fusion technology from the demonstration of ignition. There have been proposals to build compact tokamaks of the smallest possible size and the highest possible field strength, that is to say with copper coils. Surely, if one of these machines had been built and the plasma had ignited, the world's fusion programme would now have been in a healthier state. But it would have been a far throw: the original proposals in this direction were based on scaling relations that meanwhile have lost their credibility. Since the engineering would have to be pushed to the very limits in these machines, they would also on this account risk to fail, and this is precisely the opposite of what has characterized the most successful experiments so far: conservative engineering and experimental flexibility which has allowed them to work in regimes that were not envisaged when they were designed. Moreover, the largest extrapolation from present-day experiments to a reactor is not primarily related to merely reaching ignition, but to the behaviour of the burning plasma and its material surroundings over thousands of energy confinement times.

Looking forward

At the time of writing, fusion stands at a critical point—perhaps the most critical point in its history. The decisions that have to be taken will determine if fusion is to progress as an energy technology or to take a slower course as a basic scientific research programme. For fusion to pass from the research stage into reactor development undoubtedly requires a substantial increase in funding and this will certainly not become available without strong pressure from within societies. But if fusion were to be regarded as a basic research programme, it is doubtful that funding could be sustained at a viable level to address the crucial issues.

The ITER project is the next essential step forward. Meetings that will determine if, where and how ITER is to be built began in June 2001 between Canada, the European Union, Japan and the Russian Federation. The choice of site, how the costs will be shared and who will provide the various ITER components and systems are among the many critical issues facing the negotiators. Above all, it will be important to find an organizational framework within which the ITER project can flourish—an international undertaking of this kind requires a degree of stability in policies, commitments, relations and structures that in the real world cannot be taken for granted. Formal signature of an agreement to build ITER as an international venture will require approval at the highest political level and the negotiations are geared to the assumption that this step could take place in 2003.

Already Canada has put forward an offer to host ITER at a site in Clarington, Ontario. A review by the Canadian Nuclear Safety Commission as a necessary step in the licensing procedure has started. Other site offers under consideration by the European Union and Japan are expected to join the Canadian offer in the negotiations during the course of 2002. In Europe, a site at Cadarache near Aix en Provence, where site studies have been under way for two years, has the strong support of the French government. The possibility of a site in Spain is also under consideration. The Japanese Council on Science and Technology Policy and its Minister for Science and Technology Policy have expressed the view that participation in ITER is desirable and Japan has sites at Naka and at Rokkasho under consideration. The Russian Federation will not put forward a site but has stated that participation in ITER is one of its highest priorities.

The USA, as an original ITER party, has the option to rejoin the present negotiations at any time and other countries (and the USA if it were to defer its decision to a later date) could join with the unanimous agreement of the negotiating parties. Political circumstances in the USA have steered its fusion programme in recent years into a science-oriented course, but there are signs that the US is rethinking its fusion strategy. Senior government scientific advisors and congressional committees have expressed their support for participation in ITER. The question remains, however, if the US fusion community will come together behind ITER—it is preparing to congregate in July 2002 at Snowmass, Colorado, to discuss options for burning plasma experiments and to make a comparative assessment. A US decision to rejoin ITER would be an important endorsement for fusion as an option for energy production.

These discussions take place against the background of the timescale for fusion energy, which presently foresees DEMO achieving net electricity production about 35 years after the decision to construct ITER and the beginning of large-scale electricity production after about 50 years. This is in contrast with studies performed in the US and Europe in the late 1970s, which signalled possibilities to reach that goal in 20 to 25 years. A recent European study [2] has again examined a possible faster track towards demonstrating the technical feasibility of fusion power on a 20–30 year timescale by fully exploiting the inherent flexibility of the ITER design. Tests of tritium-breeding and energy-extraction blanket modules would be advanced in ITER and a high-energy, high-intensity neutron source such as the International Fusion Material Irradiation Facility (IFMIF) to test and verify material performance would be built. The fusion scientists have—rightly or wrongly—often been accused of promising too much, too soon. In reaction, they have become perhaps overly cautious in pressing for more support. It is interesting to note that there now seems to be building up the kind of pressure from outside their community for which they have long been waiting.

We can now begin to answer the question posed in the opening of this review: '*Is a fusion reactor an interesting proposition, from a scientific, technical, social and economic point of view?*'. Yes, the D–T burning reactor appears to be feasible scientifically and technically and attractive socially, but the economic prospects depend to a large extent on the premium for reduction of the release of carbon into the atmosphere. The burning of fossil fuels may run up against increasing environmental constraints, long before an actual shortage arises, and if the worst fears of climate change come true there will be no cheap solution for maintaining our large-scale electricity supply. A future generation will only have the option to choose fusion if its predecessors have laid the scientific and technological foundation.

The solution for burning the inexhaustible fuel, deuterium, appears to lie in a further increase in scale of the reactor. Neutron-lean fusion is even farther away, not to speak of neutron-free fusion; reverse-field configurations and perhaps tandem mirror systems are among the long-term prospects for further reduction of the radioactivity associated with fusion. Thus, it would seem that the best way to learn about what may be speculated about as 'the ultimate potential of fusion' is to put the next-to-ultimate possibilities into practice first.

Recent history has reminded us of the obvious—that there will be no fusion until society asks for it. But if society calls for this new energy source, it must still allow some decades for the request to be answered. Only if no time is wasted can we say—and on much firmer ground than Lev Artsimovich used to say it—'*Fusion will be there when society needs it*' [3].

References

References to IAEA Conferences are abbreviated as follows:

1961 Salzburg — 1961 *Plasma Physics and Controlled Nuclear Fusion Research,* Salzburg, *Nucl. Fusion Suppl.* 3 vols (Vienna: IAEA 1962); Russian–English translation: AEC-tr-5589 2 vols (Washington: USAEC 1963)

1965 Culham — 1965 *2nd Conf. Plasma Physics and Controlled Nuclear Fusion Research,* Culham, *Proceedings Series,* 2 vols (Vienna: IAEA 1966), Russian–English translation: AEC tr-6760 (Washington: USAEC 1966)

1968 Novosibirsk — 1968 *3rd Conf. Plasma Physics and Controlled Nuclear Fusion Research,* Novosibirsk, *Proceedings Series,* 2 vols (Vienna: IAEA 1969); Russian–English translation: *Nucl. Fusion Suppl.* 1969

1971 Madison — 1971 *4th Conf. Plasma Physics and Controlled Nuclear Fusion Research,* Madison, *Proceedings Series,* 3 vols (Vienna: IAEA 1972); Russian–English translation: *Nucl. Fusion Suppl.* 1972

1974 Tokyo — 1974 *5th Conf. Plasma Physics and Controlled Nuclear Fusion Research,* Tokyo, *Proceedings Series,* 3 vols (Vienna: IAEA 1975); Russian–English translation: *Nucl. Fusion Suppl.* 1975

1976 Berchtesgaden — 1976 *6th Conf. Plasma Physics and Controlled Nuclear Fusion Research,* Berchtesgaden, *Nucl. Fusion Suppl.* 3 vols (Vienna: IAEA 1977)

1978 Innsbruck — 1978 *7th Conf. Plasma Physics and Controlled Nuclear Fusion Research,* Innsbruck, *Nucl. Fusion Suppl.* 3 vols (Vienna: IAEA 1979)

1980 Brussels — 1980 *8th Conf. Plasma Physics and Controlled Nuclear Fusion Research,* Brussels, *Nucl. Fusion Suppl.* 2 vols (Vienna: IAEA 1981)

1982 Baltimore — 1982 *9th Conf. Plasma Physics and Controlled Nuclear Fusion Research,* Baltimore, *Nucl. Fusion Suppl.* 3 vols (Vienna: IAEA 1983)

1984 London — 1984 *10th Conf. Plasma Physics and Controlled Nuclear Fusion Research,* London, *Nucl. Fusion Suppl.* 3 vols (Vienna: IAEA 1985)

1986 Kyoto — 1986 *11th Conf. Plasma Physics and Controlled Nuclear Fusion Research,* Kyoto, *Nucl. Fusion Suppl.* 3 vols (Vienna: IAEA 1987)

1988 Nice — 1998 *12th Conf. Plasma Physics and Controlled Nuclear Fusion Research,* Nice, *Nucl. Fusion Suppl.* 3 vols (Vienna: IAEA 1989)

1990 Washington — 1990 *13th Conf. Plasma Physics and Controlled Nuclear Fusion Research,* Washington, *Nucl. Fusion Suppl.* 3 vols (Vienna: IAEA 1991)

1992 Würzburg — 1992 *14th Conf. Plasma Physics and Controlled Nuclear Fusion Research,* Würzburg, *Proceedings Series* 3 vols (Vienna: IAEA 1993)

1994 Seville	1994 *15th Conf. Plasma Physics and Controlled Nuclear Fusion Research,* Seville, *Proceedings Series*, 3 vols (Vienna: IAEA 1995)
1996 Montreal	1996 *16th Conf. Fusion Energy, Montreal, Proceedings Series*, 3 vols (Vienna: IAEA 1997)
1998 Yokohama	1998 *17th Conf. Fusion Energy,* Yokohama, *Proceedings*, 4 vols (Vienna: IAEA 1999) also on CD as *Fusion Energy 1998 (2001 Edition)* IAEA-CSP-8/C (Vienna: IAEA 2001)
2000 Sorrento	2000 *18th Conf. Fusion Energy,* Sorrento, Proceedings on CD as *Fusion Energy 2000* IAEA-CSP-8/C (Vienna: IAEA 2001)

References to other conferences are abbreviated as follows:

1957 Venice	1957 *3rd Int. Conf. Ionization Phenomena in Gases*, Venice (Milan: Societa Italiana di Fisica)
1958 Geneva	1958 *Second United Nations Conference on the Peaceful Uses of Atomic Energy*, Geneva (Geneva: United Nations 1959); also 1959 *Prog. Nucl. En.* Series XI, volume **1**, *Plasma Phys. Thermon. Res.* (London: Pergamon)
1958 Leontovich	1958 M A Leontovich (ed.) *Plasma Physics and the Problem of Controlled Thermonuclear Reactions* (English translation: Pergamon Press 1961)
1969 Culham	1969 *British Nuclear Energy Society Conference on Nuclear Fusion Reactors* Culham (Culham: BNES)
1974 Culham	1974 *Fusion Reactor Design Problems* Culham *Proceedings of an IAEA Workshop* (Vienna: IAEA)

Chapter 1

[1] For historical reviews of the debate, see:
Badash L 1989 *Sci. Am.* August 78
Hallam A 1983 *Great Geological Controversies* (Oxford: University Press)
Burchfield J D 1975 *Lord Kelvin and the Age of the Earth* (New York: Science History Publications)

[2] The discussions that took place in astronomical circles in 1917–1920 have been reviewed in:
Hufbauer K 1981 *Hist. Studies Phys. Sci.* **11** 277. See in particular:
Eddington A S 1919 *Observatory* **42** 371
Eddington A S 1920 *Brit. Ass. Report* 34
Perrin J 1920 *Revue du Mois* **21** 113
Perrin J 1921 *Scientia* **30** 355

[3] Eddington A S 1926 *The Internal Constitution of the Stars* (Cambridge: University Press)

[4] Atkinson R d'E and Houtermans F G 1929 *Z. Phys.* **54** 656

[5] For a contemporary review, see:
Bethe H A 1937 *Revs. Mod. Phys.* **9** 69
Livingston M S and Bethe H A 1937 *Revs. Mod. Phys.* **9** 245
For a historical account see:
Hendry J 1987 *Annals of Science* **44** 143

[6] Atkinson R d'E 1936 *Astrophys. J.* **84** 73
Bethe H A and Critchfield C L 1938 *Phys. Rev.* **54** 248
[7] Weizsäcker C F von 1938 *Phys. Z.* **38** 176
Bethe H A 1939 *Phys. Rev.* **55** 103
[8] Bethe H A 1939 *Phys. Rev.* **55** 434
[9] Rolfs C, Trautvetter H P and Rodney W S 1987 *Rep. Prog. Phys.* **50** 233
Bahcall J N *et al.* 1982 *Rev. Mod. Phys.* **54** 767
[10] For a summary of von Weizsäcker's work, see:
Chandrasekhar S 1939 *An Introduction to the Study of Stellar Structure* (Chicago: University Press; also 1957 New York: Dover)
[11] Burbidge E M, Burbidge G R, Fowler W A and Hoyle F 1957 *Rev. Mod. Phys.* **29** 547
Cameron A G W 1957 *At. En. of Canada* CRL-41, also 1958 *Ann. Rev. Nucl. Sci.* **8** 299
[12] Eddington A S 1920 *Brit. Ass. Report* 34
[13] Oliphant M L E, Harteck P and Rutherford E 1934 *Proc. Roy. Soc.* **A144** 692
Oppenheimer J R and Philips M 1935 *Phys. Rev.* **48** 500
[14] Post R F 1956 *Rev. Mod. Phys.* **28** 338
Artsimovich L A 1958 Geneva **31** 6
[15] See e.g. Bhabha H 1958 Geneva **31** 40
[16] Smyth H D 1946 *Atomic Energy for Military Purposes* (Princeton: University Press),
Rhodes R 1988 *The Making of the Atomic Bomb* (New York: Simon & Schuster). For early fusion research, see also:
Bishop A S 1958 *Project Sherwood* (Reading, MA: Addison-Wesley)
Sakharov A D 1990 *Memoirs* (New York: Knopf)
Lawson J D and Hendry J 1993 *Fusion Research in the UK 1945–1960* (Culham: AEA Technology)
[17] Langmuir I 1928 *Proc. Natl Acad. Sci. USA* **14** 627
[18] Rogoff G L 1991 *IEEE Trans. Plasma Science* **19** 989
[19] See e.g. the books by
Thomson J J 1903 *Conduction of Electricity through Gases* (Cambridge: University Press)
Townsend J S 1915 *Electricity in Gases* (Oxford: University Press)
Thomson J J and Thomson G P 1928/1932 *The Conduction of Electricity through Gases* 3rd edition (Cambridge: University Press)
Engel A von and Steenbeck M 1932/1934 *Elektrische Gasentladungen, ihre Physik und Technik*, vol 2 (Berlin: Springer)
Langmuir I and Compton H T 1930 *Rev. Mod. Phys.* **2** 123
Langmuir I and Compton H T 1931 *Rev. Mod. Phys.* **3** 191
Druyvestein M J and Penning F M 1940 *Rev. Mod. Phys.* **12** 87
[20] Langmuir I 1928 *Proc. Natl Acad. Sci. USA* **14** 627; see also:
Tonks L and Langmuir I 1929 *Phys. Rev.* **33** 195
[21] Penning F M 1926 *Physica* **6** 241
[22] Ridenour L N (ed.) 1948 *MIT Radiation Laboratory Series* (New York: McGraw-Hill)
[23] Mitra S K 1947 *The Upper Atmosphere* (Calcutta: Royal Asiatic Society of Bengal)
[24] Bennett W H 1934 *Phys. Rev.* **45** 890
[25] Tonks L 1939 *Phys. Rev.* **56** 360
[26] The motion of a charged particle in a magnetic field was first treated by:
Riecke E 1881 *Ann. de Phys.* **13** 19

but because this author was primarily interested in the shape of the orbits he eliminated the time variable in his equations and so did not come to writing out the cyclotron frequency. For the history of electro-magnetic theory, see:
Whittaker E T 1951 *A History of Theories of Aether and Electricity* (London: Nelson)

[27] Birkeland K 1901 *Videnskapssels Rabelts Skrifter Mat-nat. kl.* No. 1 (Oslo: Kristiani)
Störmer C 1907 *Arch. de Sci. Phys.* IV **24** 317
Störmer C 1955 *The Polar Aurora* (Oxford: University Press)

[28] Alfvén H 1939 *A Theory of Magnetic Storms and of the Aurorae*; *Kungl. Svenska Vetens. Hand.* **18**(3)
The constancy of the magnetic moment was studied in depth by:
Hellwig G 1955 *Z. Naturf.* **10a** 508
Kruskal M D 1957 Venice 562

[29] Boot H A H and Shersby-Harvey R B R 1957 *Nature* **180** 1187
Boot H A H, Self S A and Shersby-Harvey R B R 1958 *J. Electr. Control* **4** 434
Gaponov A V and Miller M A 1958 *Sov. Phys. JETP* **34** 242 (Engl. transl. 168)
Weibel E S 1958 in *The Plasma in a Magnetic Field* (Stanford: University Press)

[30] Landau L D 1936 *Phys. Zs. Sovjet* **10** 154

[31] Vlasov A A 1938 *Zh. Eksp. Teor. Fyz.* **8** 291

[32] Landau L 1946 *Sov. Phys. JETP* **10** 25

[33] Chapman S and Cowling T G 1939 *The Mathematical Theory of Non-Uniform Gases* (Cambridge: University Press); see also:
Aleksandrov A F and Rukhadze A A 1997 *Plasma Phys. Reports* **23** 442 and comment in 2001 *Plasma Phys. Reports* **24** 192

[34] Alfvén H 1950 *Cosmical Electrodynamics* (Oxford: Clarendon Press)

[35] Precursors can already be found in:
Cowling T G 1934 *Month. Not. Roy. Astron. Soc.* **94** 39
A law of isorotation was formulated by:
Ferraro V C A 1937 *Month. Not. Roy. Astron. Soc.* **97** 458

[36] de Hoffmann F and Teller E 1949 *Phys. Rev.* **75** 1851 and 1864

[37] Schlüter A 1950 *Z. Naturf.* **5a** 72
Schlüter A 1951 *Z. Naturf.* **6a** 73

[38] Spitzer L 1956 *Physics of Fully Ionized Gasses* (New York: Interscience), see also references given there

[39] Post RF 1956 *Rev. Mod. Phys.* **28** 338

[40] Fermi E 1949 *Phys. Rev.* **75** 1169
Fermi E 1954 *Astroph. J.* **119** 1

[41] Post R F 1995 *Plasma Physics in the Twentieth Century*, chapter 22 in *Twentieth Century Physics* (Bristol: IOPP, New York: AIP) **3** 1617; see also in the same volume:
Longair M S, chapter 23, *Astrophysics and Cosmology*
Brush S G and Gillmor C S, chapter 26, *Geophysics*

[42] Rutherford E 1933 *Nature* **132** 433

[43] Paneth F and Peters K 1926 *Ber. Deutschen Chem. Ges.* **59** 2039
Paneth F, Peters K and Günther P 1927 *Ber. Deutschen Chem. Ges.* **60B** 808

[44] Wilner B 1989 *Nature* **339** 180

[45] The case for cold fusion was defended most strongly in:
Mallove E F 1991 *Fire from Ice: Searching for the Truth Behind the Cold Fusion Furor* (New York: Wiley)

It is dealt with critically, but not without sympathy, in:

Hoffman N 1995 *A Dialog on Chemically Induced Nuclear Effects: A Guide for the Perplexed about Cold Fusion* (La Grange Park: American Nuclear Society)

but it is conclusively demolished in:

Close F E 1990 *Too Hot to Handle: The Race for Cold Fusion* (London: Allen; Princeton: University Press 1991)

Huizenga J R 1992 *Cold Fusion: The Scientific Fiasco of the Century* (Rochester, New York)

Taubes G 1993 *Bad Science: The Short Life and Weird Times of Cold Fusion* (New York: Random House)

[46] Heppenheimer T A 1984 *The Man Made Sun: The Quest for Fusion Power* (Boston: Little, Brown)

Kantrowitz A 1992 Private communication

[47] Thonemann P C 1958 Geneva **31** 31

Lawson J D and Hendry J 1993 *Fusion Research in the UK 1945–1960* (Culham: AEA Technology)

[48] Ware A A 1951 *Phil. Trans. Roy. Soc.* **A243** 863

Cousins S W and Ware A A 1951 *Proc. Phys. Soc.* **B64** 159

Thonemann P C and Cowhig W T 1951 *Proc. Phys. Soc.* **B64** 345

[49] Bromberg J L 1982 *Fusion; Science, Politics, and the Invention of a New Energy Source* (Cambridge, MA: MIT Press) 28

Close F E 1991 *Too Hot to Handle: The Race for Cold Fusion* (Princeton: University Press)

Herman R 1990 *Fusion: The Search for Endless Energy* (Cambridge: University Press)

[50] Bishop A S 1958 *Project Sherwood: The U.S. Program in Controlled Fusion* (Reading, MA: Addison-Wesley)

Teller E 1958 Geneva **31** 27

[51] Sakharov A D see [16]

[52] Tamm I E 1951 in 1958 Leontovich **1** 35

Sakharov A D 1951 in 1958 Leontovich **1** 21

[53] Budker G I 1954 in 1959 Leontovich **3** 1

[54] Artsimovich LA 1961 *Controlled Thermonuclear Reactions* (Engl. transl. Edinburgh: Oliver & Boyd, 1964)

[55] For examples of early Japanese work, see:

Miyamoto G, Iwata G, Mori S and Inoue K 1957 *J. Phys. Soc. Japan* **12** 438

Miyamoto G, Kihara T, Iwata G, Mori S, Ohkawa T and Yoshikawa M 1958 Geneva **32** 308

Uo K 1961 *J. Phys. Soc. Japan* **16** 1380

The early history of fusion research in Japan was reviewed by:

Sekiguchi T 1983 *J. Fusion Energy* **3** 313

Oikawa A 1990 *Fusion Technology* **17** 231

[56] Sakharov A D 1951 in 1958 Leontovich **1** 21

Teller E 1954 *Sherwood Conference, Princeton;* see J L Bromberg, l.c. p 52

[57] Kruskal M and Schwartzschild M 1954 *Proc. Roy. Soc.* **A223** 348

[58] Kurchatov I V 1956 *At. Energ.* **1** 65; Engl. transl. in *J. Nucl. Energy* **4** 193

[59] The first neutrons had already been observed in July 1952, but after a brief wave of excitement Artsimovich had drawn the right conclusion that they had to be caused by an instability. See e.g.:

Sakharov A D see [16]

[60] Thonemann P C 1956 *Atomic Energy* August p. 169

[61] Kurchatov I V *et al.* 1956 *At. Energ.* **1** 76 (Engl. transl. 1957 *J. Nucl. Energy* **4** 203

Leontovich M A and Osovets S M 1956 *At. Energ.* **1** 81 (Engl. transl. 1957 *J. Nucl. Energy* **4** 209)

Lukyanov S Yu and Podgornyi I M 1956 *At. Energ.* **1** 79 (Engl. transl. 1957 *J. Nucl. Energy* **4** 224)

[62] Artsimovich L A 1956 *IAU Symposium on Electromagnetic Phenomena in Cosmical Physics,* Stockholm (Cambridge: Cambridge University Press) 451

Bezbatchenko A L *et al.* 1956 *IAU Symposium on Electromagnetic Phenomena in Cosmical Physics,* Stockholm 464

Shafranov V D 1956 *IAU Symposium on Electromagnetic Phenomena in Cosmical Physics,* Stockholm 491

[63] Pease R S 1956 *IAU Symposium on Electromagnetic Phenomena in Cosmical Physics,* Stockholm 99

[64] Lawson J D 1957 *Proc. Phys. Soc. London* B **70** 6

Pease R S 1957 *Proc. Phys. Soc. London* B **70** 11

Allen J E 1957 *Proc. Phys. Soc. London* B **70** 24

Tayler R J 1957 *Proc. Phys. Soc. London* B **70** 31, 1049, *Phil. Mag.* **2** 33

Carruthers R and Davenport P A 1957 *Proc. Phys. Soc. London* B **70** 49

Thompson W B 1957 *Nature* **179** 886

[65] Lundquist S 1951 *Phys. Rev.* **83** 307

[66] A review of early work on pinch stability is included in:

Bateman G 1978 *MHD Instabilities* (Cambridge, MA: MIT Press)

With respect to the pinch, notice in particular:

Normal-mode analysis:

Shafranov V D 1956 Stockholm 491

Shafranov V D 1957 *J. Nucl. En.* **2** 86

Braginskii S I and Shafranov V D 1958 Geneva **31** 43

Rosenbluth M N 1958 Geneva **31** 85

Tayler R J 1958 Geneva **31** 160

Arbitrary perturbations:

Tayler R J 1957 *Proc. Phys. Soc. London* B **70** 31

Trapped axial field:

Kruskal M and Tuck J 1958 *Proc. Roy. Soc.* **A245** 222

Trapped axial field and conducting walls:

Rosenbluth M 1957 Venice 903

Tayler R J 1957 *Proc. Phys. Soc.* London B **70** 1047

Shafranov V D 1957 *J. Nucl. En.* **2** 86

[67] Rosenbluth M N and Longmire C 1957 *Ann. Phys. (NY)* **1** 120

[68] Bernstein I B, Frieman E A, Kruskal M D and Kulsrud R M 1958 *Proc. Roy. Soc.* **A244** 17

Kadomtsev B B 1957 in 1958 Leontovich **4** 17

Kadomtsev B B 1958 in Leontovich **4** 450

Hain K and Lüst R 1958 *Z. Naturf.* **13a** 936

Hain K, Lüst R and Schlüter A 1957 Venice 458

Berkowitz J *et al.* 1958 Geneva **31** 171

[69] Various authors: 1957 *3rd Int. Conf. Ionization Phenomena in Gases, Venice* (Milan: Società Italiana di Fisica)

[70] Biermann L and Schlüter A 1957 *Z. Naturf.* **12a** 805
Biermann L, Hain K, Jörgens K and Lüst R 1957 *Z. Naturf.* **12a** 826

[71] Allen N L *et al.* 1958 *Nature* **181** 222
Honsaker J, Karr H, Osher J, Phillips J and Tuck J 1958 *Nature* **181** 231
Burkhardt L C and Lovberg R H 1958 *Nature* **181** 228
Hagerman D C and Mather J W 1958 *Nature* **181** 226
Spitzer L 1958 *Nature* **181** 221

[72] Bromberg J L 1982 *Fusion; Science, Politics, and the Invention of a New Energy Source* (Cambridge, MA: MIT Press) 28
Lawson J D and Hendry J 1993 *Fusion Research in the UK 1945–1960* (Culham: AEA Technology)

[73] Anderson O A *et al.* 1958 *Phys. Rev.* **110** 1375

[74] Rose B, Taylor A E and Wood E 1958 *Nature* **181** 1630

[75] *Nucleonics* May 1958

[76] Glasstone S and Lovberg R H 1960 *Controlled Thermonuclear Reactions* (Princeton: van Nostrand); for updated values, see:
McNally J R Jr 1979 *Oak Ridge Report* ORNL-TM-6914

[77] Bohm D 1949, chapter 1 in Guthrie A and Wakerling R K *The Characteristics of Electrical Discharges in Magnetic Fields* (New York: McGraw-Hill)
see also chapter 2 with Burhop E H S and Massey H S W and chapter 9 with these and Williams R W

[78] Simon A 1955 *Phys. Rev.* **98** 317; Simon A and Neidigh R V 1955 *Oak Ridge Report* ORNL-1890

Chapter 2

[1] Andrianov A M *et al.* 1958 Geneva **31** 348

[2] Petrov D P, Filippov N V, Filippova T I and Khrabrov V A 1954 and 1957 in 1958 Leontovich **4** 198

[3] Artsimovich L A 1958 Geneva **31** 6; the paper was read by E I Dobrokhotov

[4] Andrianov A M *et al.* 1958 Geneva **31** 348

[5] Anderson O A *et al.* Geneva 1958 **32** 150

[6] Post R F 1958 Geneva **32** 245
Post R F 1987 *Nucl. Fusion* **27** 1579
Bishop A S 1958 *Project Sherwood: The U.S. Program in Controlled Fusion* (Reading, MA: Addison-Wesley)
Bromberg J L 1982 *Fusion: Science, Politics, and the Invention of a New Energy Source* (Cambridge, MA: MIT Press)
Bishop credits Livermore's director, York, for the invention, but Bromberg and Fowler (private communication, 1988) confirm that Post proposed the mirror geometry.

[7] Budker G I 1954 in 1958 Leontovich **3** 1

[8] Bromberg *l.c.* Bishop *l.c.*

[9] Sinelnikov K D *et al.* 1961 Salzburg **3** 1147

[10] Barnett C F *et al.* 1958 Geneva **31** 298

[11] See footnote in 1958 Geneva **31** 40

Artsimovich L A 1968 private communication
Golovin 1992 private communication
[12] Kurchatov I V 1959 *Prog. Nucl. En.* Series XI: *Plasma Phys. Thermon. Res.* (London: Pergamon) **1** 42
[13] Barnett C F *et al.* 1958 Geneva **31** 298
[14] Yushmanov E E 1957 in 1958 Leontovich **4** 277
[15] Anderson O A *et al.* 1958 Geneva **32** 155
Boyer K *et al.* 1958 Geneva **31** 319
[16] Coensgen F H, Ford F C and Ellis R E 1958 Geneva **32** 266
[17] Kolb A C 1958 Geneva **31** 328
[18] Colgate S A 1958 Geneva **32** 123
[19] Elmore W C, Little E M and Quinn W E 1958 Geneva **32** 337, 339
[20] Kadomtsev B B 1992 *Tokamak Plasma* (Bristol: Institute of Physics Publishing) pp 4, 5
[21] Johnson J L, Oberman C R, Kulsrud R M and Frieman E A 1958 Geneva **31** 198
[22] Kruskal M D 1954 *USAEC Report* NYO-6045 (PM-S-12)
Kruskal M D *et al.* 1958 *Phys. Fluids* **1** 421
Shafranov V D 1953 in 1958 Leontovich **2** 197
[23] Bernstein W *et al.* Geneva 1958 **32** 210
[24] Glasstone S and Lovberg R H 1960 *Controlled Thermonuclear Reactions* (Princeton: van Nostrand) p. 305
Rose D J and Clarke M 1961 *Plasmas and Controlled Fusion* (Cambridge, MA: MIT Press) p. 435
Artsimovich LA 1961 *Controlled Thermonuclear Reactions* (English transl. Edinburgh: Oliver & Boyd, 1964) p. 241
The concept of helicity is discussed in:
Moffat H K 1978 *Magnetic Field Generation in Electrically Conducting Fluids* (Cambridge: University Press)
Berger M A 1999 *26th EPS Conf. Contr. Fusion and Plasma Phys. Maastricht; Proceedings in* 1999 *Plasma Phys. Contr. Fusion* **41** B167
[25] Spitzer L 1958 Geneva **31** 40; see also:
Artsimovich L A 1959 *Prog. Nucl. En.* Series XI: *Plasma Phys. Thermon. Res.* (London: Pergamon) **1** 6
[26] Rosenbluth M N 1958 Geneva **31** 85
[27] Suydam B R 1958 Geneva **31** 157
[28] Colgate S A, Ferguson J P and Furth H 1958 Geneva **32** 129
Tuck J L 1958 Geneva **32** 3
[29] Butt E P *et al.* 1958 Geneva **32** 42
[30] Dolgov-Saveliev G G *et al.* 1958 Geneva **32** 82
[31] Vedenov A A *et al.* 1958 Geneva **32** 239
Butler J W, Hatch A J and Alrich A J 1958 Geneva **32** 324
Clauser M U and Weibel E S 1958 Geneva **32** 161
[32] Stix T H 1958 Geneva **31** 125
Sagdeyev R S and Shafranov V D 1958 Geneva **31** 118
Akhiezer A I *et al.* 1958 Geneva **31** 99
Dubuvoi L V *et al.* 1958 Geneva **31** 288
Sinelnikov K D *et al.* 1958 Geneva **31** 292
[33] Wharton C B, Howard J C and Heinz O 1958 Geneva **32** 388
Harding G N *et al.* 1958 Geneva **32** 365

[34] Brown S C 1958 Geneva **32** 394
Drummond J E 1958 Geneva **32** 379
[35] Kadomtsev B B and Nedospasov A V 1960 *J. Nucl. En.* C **1** 230
[36] Barsanti G *et al.* 1958 Geneva **31** 238

Chapter 3

[1] Post R F 1958 Geneva **32** 245; see also:
Budker G I 1954 in 1958 Leontovich **3** 1
[2] Fowler T K 1969 *Nucl. Fusion* **9** 3
[3] Kelley G G 1967 *Plasma Phys. Contr. Fusion* **9** 503
[4] Baldwin D E 1977 *Comments Plasma Phys. Contr. Fusion* **2** 173
Lawrentiev O A 1963 *Ukr. Fys. J.* **8** 440
Vedenov A A *et al.* 1958 Geneva **32** 239
Consoli T 1967 *Proc. Int. Conf. Interaction RF Fields and Plasma*, Saclay, France 215
[5] Budker G I 1954 in 1958 Leontovich **3** 1
Post R F 1958 Geneva **32** 245
[6] Werner R W *et al.* 1974 Culham 171
[7] Sweetman D R 1961 Salzburg **1** 279
Post R F 1961 Salzburg **1** 99
Gourdon C and Prévot F 1961 Salzburg **1** 265
[8] Simon A 1958 *Phys. Fluids* **1** 495
Simon A 1960 *J. Nucl. Energy* Part C **1** 215
Simon A 1960 *Int. Summer Course in Plasma Physics, Risö Laboratory Report* 18 (Roskilde: Danish AEC) p 493
Barnett C F *et al.* 1958 Geneva **31** 298
[9] Dunlap J L *et al.* 1961 Salzburg **1** 233
[10] Bell P R *et al.* 1965 Culham **2** 77
[11] Ard W B *et al.* 1965 Culham **2** 153
[12] Artemenkov P I 1965 Culham **2** 45
[13] Brevnov N N, Romanovskii M K and Tomashchuck Yu F 1961 Salzburg **1** 289 (Engl. transl. **1** 97)
[14] Prévot F *et al.* 1965 Culham **2** 135
[15] Bernstein W *et al.* 1965 Culham **2** 23
Thompson E, Cordey J G and Sweetman D R 1971 Madison **2** 689
[16] Damm C C *et al.* 1968 Novosibirsk **2** 253
Berkner K H *et al.* 1971 Madison **2** 707
Anderson O A *et al.* 1974 Tokyo **1** 379
Baker W R *et al.* 1974 Tokyo **1** 329
[17] Gott Y V *et al.* 1974 Tokyo **2** 355 Engl. transl. 55
Zhil'tsov V A *et al.* 1974 Tokyo **2** 341 Engl. transl. 63
[18] Bardet R *et al.* 1974 Tokyo **2** 371
[19] Coensgen F H *et al.* 1976 Berchtesgaden **3** 135
[20] Post R F *et al.* 1960 *Phys. Rev. Lett.* **4** 166
Post R F 1961 Salzburg **1** 99
[21] Gott Y V, Ioffe M S and Tel'kovsky V G 1961 Salzburg **3** 1045
[22] Berkowitz J *et al.* 1958 Geneva **31** 171

Braginskii S I and Kadomtsev B B 1957 in 1958 Leontovich **3** 356
[23] Baiborodov Y T, Ioffe M S and Sobolev R I 1963 *Int Symp. on Mirrors, Fontenay-aux-Roses*
Gott Y V, Ioffe M S and Yushmanov E E 1965 Culham **1** 35 (Engl. transl. 1)
[24] Post R F 1961 Salzburg **1** 99
Post R F and Rosenbluth M N 1966 *Phys. Fluids* **9** 730
[25] Futch A H *et al.* 1965 Culham **2** 3
[26] For a note on the history of this scheme, see Hiskes J R 1967 *Phys. Today* **20** 9
[27] Harris E G 1959 *Phys. Rev. Lett.* **2** 34
[28] Bogdanov G F *et al.* 1961 Salzburg **1** 215 (Engl. transl. **1** 29)
Dunlap J L *et al.* 1961 Salzburg **1** 233
[29] Pistunovich V I and Shafranov V D 1961 *Nucl. Fusion* **1** 189
Postma H et al. 1966 *Phys. Rev. Lett.* **16** 265
[30] The field has been reviewed by:
Fowler T K 1969 *Nucl. Fusion* **9** 3
Post R F 1987 *Nucl. Fusion* **27** 1579
[31] Kadomtsev B B and Pogutse O P 1967 *Sov. Phys. JETP* **24** 1172
Post R F and Rosenbluth M N 1966 *Phys. Fluids* **9** 730
[32] Berk H L *et al.* 1976 Berchtesgaden **3** 147
[33] Baiborodov Yu T *et al.* 1973 Moscow **2** 122
[34] Baldwin D E 1977 *Comments Plasma Phys. Contr. Fusion* **2** 173
[35] Dimov G I, Zakajdakov V V and Kishinevskij M E 1976 Berchtesgaden **3** 177
[36] Fowler T K and Logan B G 1977 *Comments Plasma Phys. Contr. Fusion* **2** 167; see also:
Heppenheimer T A 1984 *The Man Made Sun: The Quest for Fusion Power* (Boston: Little, Brown) chapter 4
Fowler T K 1997 *The Fusion Quest* (Baltimore: Johns Hopkins University Press)
[37] Baldwin D E *et al.* 1978 Innsbruck **2** 427
Kishinevskij M E *et al.* 1978 Innsbruck **2** 411
Miyoshi S *et al.* 1978 Innsbruck **2** 437
[38] Baldwin D E and Logan B G 1979 *Phys. Rev. Lett.* **43** 1318
[39] Kesner J 1973 *J. Plasma Phys.* **15** 577
Simonen T C *et al.* 1983 *Phys. Rev. Lett.* **50**
[40] Kesner J, Post R S and Smith D K 1982 *Nucl. Fusion* **22** 577
Cohen R H 1983 *Phys. Fluids* **26** 2774
[41] Ryutov D D and Stupakov G V 1978 *Fiz. Plasmy* **4** 501
Ryutov D D and Stupakov G V 1980 Brussels **1** 119
[42] See the review article:
Hershkowitz N, Miyoshi S and Ryutov D D 1990 *Nucl. Fusion* **30** 1761
[43] Simonen T C *et al.* 1986 Kyoto **2** 231
[44] Pastukhov V P 1974 *Nucl. Fusion* **14** 3
[45] Cho T *et al.* 1986 Kyoto **2** 243
Cho T *et al.* 1988 Nice **2** 501
Miyoshi T *et al.* 1990 Washington **2** 539
Inutake M *et al.* 1992 Würzburg **2** 651
Tamano T *et al.* 1994 Seville **2** 399
Saito T *et al.* 1996 Würzburg **2** 105
[46] Kesner J *et al.* 1982 *Nucl. Fusion* **22** 549

Lane B G *et al.* 1987 *Varenna Course on Physics of Mirrors, Reversed Field Pinches and Compact Tori* **2** 873

[47] Baldwin D E 1983 *Varenna Course on Mirrors-Based and Field-Reversed Approaches to Magnetic Fusion* **1** 209

[48] Post R S *et al.* 1988 Nice **2** 493

[49] Ryutov D D 1987 *Varenna Course on Physics of Mirrors, Reversed Field Pinches and Compact Tori* **2** 791

[50] Hershkowitz N *et al.* 1988 Nice **3** 487
Breun R A *et al.* 1986 Kyoto **2** 263, 1988 Nice **2** 475

[51] Okamura S *et al.* 1984 London **2** 337

[52] Sato T et *al.* 1986 Kyoto **2** 343
Adati K *et al.* 1988 Nice **2** 677

[53] Itatani R, Takeno H and Yasaka Y 1986 Kyoto **2** 273

[54] Dimov G I 1987 *Varenna Course on Physics of Mirrors, Reversed Field Pinches and Compact Tori* **2** 703

[55] Andreoletti J 1963 *C. R. Hebd. Séances Acad. Sci.* **257** 1235
Furth H P 1963 *Phys. Rev. Lett.* **11** 308

[56] Ryutov D D 1988 *Sov. Phys. Usp.* **31** 300

[57] Post R F 1987 *Nucl. Fusion* **27** 1579, 1703

[58] Cohen B I (ed.) 1980 *Lawrence Livermore Laboratory Report UCAR-10049-80*
Baldwin D E 1977 *Rev. Mod. Phys.* **49** 317; see also [47]
For the history of MFTF-B, see also Booth W 1987 *Science* **238** 152

[59] Kanaev B I 1979 *Nucl. Fusion* **19** 347

[60] Filippov N V, Filippova T I and Vinogradov V P 1961 Salzburg **2** 577 (Engl. transl. **2** 311)

[61] Marshall J 1958 Geneva **31** 341
Marshall J 1960 *Phys. Fluids* **3** 134
Marshall J 1961 Salzburg **2** 663
Alfvén H 1958 Geneva **31** 3
Bostick W H 1963 *Prog. Nucl. En.* Series XI: *Plasma Phys. Thermon. Res. 292* (London: Pergamon)
Mather J W 1965 *Phys. Fluids* **8** 366.
Mather J W and Williams A H 1966 *Phys. Fluids* **9** 2080
Mather J W and Williams A H 1958 Geneva **32** 26

[62] For a detailed review of the experiments in Los Alamos, we refer to:
Mather J W 1971 *Methods of Experimental Physics* vol 9 ed. H R Griem and R H Lovberg (New York, London: Academic) chapter 15

[63] Morozov A I 1968 *Sov. Phys. Tech. Phys.* **12** 1580; see also 1968 Novosibirsk **2** 3

[64] Maisonnier Ch *et al.* 1972 *5th Euro. Conf. Contr. Fusion Plasma Phys.* Grenoble **2** 183

[65] Folkierski A, Frayne P G and Latham R 1961 Salzburg **2** 627
Bostick W H *et al.* 1966 *Phys. Fluids* **9** 2078
Kvartskhava I F *et al.* 1968 Novosibirsk **1** 237 (Engl. transl. 43)
Vasileva R P, Pergament M I and Yaroslavsky A I 1968 Novosibirsk **2** 39 (Engl. transl. 129)

[66] Filippov N V and Filippova T I 1965 Culham **2** 405 (Engl. transl. 270)

[67] For an example, see:

Potter D E 1970 *4th European Conference on Controlled Fusion and Plasma Physics*, Rome (Rome: CNEN) 116
[68] Bernard A *et al.* 1974 Tokyo **3** 83
[69] Maisonnier Ch *et al.* 1974 Tokyo **3** 99
[70] Filippov N V *et al.* 1974 Tokyo **3** 123, 133 (Engl. transl. 163, 169)
[71] Bostick W H, Nardi V and Prior W 1974 Tokyo **3** 109
[72] Michel L, Schönbach K H and Fischer H 1974 *Appl. Phys. Lett.* **24** 57
[73] Maisonnier Ch *et al.* 1974 Tokyo **3** 99
[74] Bernard A *et al.* 1978 Innsbruck **2** 159
[75] Ivanov V D *et al.* 1980 Brussels **2** 161
[76] Maisonnier Ch, Haegi M and Linhart J G 1965 Culham **2** 345
Maisonnier Ch *et al.* 1968 Novosibirsk **2** 345
[77] Haines M G and Marsh F 1982 Baltimore **2** 531
[78] Choi P *et al.* 1978 Innsbruck **2** 69
[79] Haines M G *et al.* 1990 Washington **2** 769
Haines M G *et al.* 1996 Montreal **2** 275
[80] Bakshaev Yu L *et al.* 1998 Yokohama **3** 879
[81] Sethian J D *et al.* 1984 London **2** 611
For other work on high-density *Z*-pinches, we refer to papers in: 1990 *Varenna Workshop on Physics of Alternative Magnetic Confinement Systems*
[82] Furth H P, quoted in Kolb A C, Dobbie C B and Griem H R 1959 *Phys. Rev. Lett.* **3** 5
[83] Tuck J 1959 *4th Int. Conf. Ionization Phenomena in Gases* Uppsala 2 vols (Amsterdam: North-Holland) **2** 920
Ribe F L 1959 *4th Int. Conf. Ionization Phenomena in Gases* Uppsala 2 vols (Amsterdam: North-Holland) **2** 1032
Boyer K *et al.* 1960 *Phys. Rev.* **119** 831
Nagle D E *et al.* 1960 *Phys. Rev.* **119** 857
[84] Kolb A C, Dobbie C B and Griem H R 1959 *Phys. Rev. Lett.* **3** 5
[85] Little E M, Quinn W E and Ribe F L 1961 *Phys. Fluids* **4** 711; see also 1961 Salzburg **2** 497
[86] (see e.g. 1961 Salzburg **2** 645)
[87] For a review of early theta-pinch research, we refer to:
Quinn W 1963 *Prog. Nucl. En.* Series XI: *Plasma Phys. Thermon. Res.* **150** (London: Pergamon)
[88] Goldman L M, Pollock H C, Reynolds J A and Westendorp W F 1962 *Phys. Rev. Lett.* **9** 9
Goldman L M, Kilb R W and Pollock H C 1964 *Phys. Fluids* **7** 1005
Goldman L M, Kilb R W, Pollock H C and Reynolds J A 1965 *Phys. Fluids* **8** 522
[89] Bodin H A B, McCartan J, Newton A A and Wolf G H 1968 Novosibirsk **2** 533
[90] Andelfinger C *et al.* 1965 Culham **1** 249
[91] McKenna K F *et al.* 1978 Innsbruck **2** 229
[92] Quinn W E and Siemon R E 1981 Linear magnetic fusion systems *Fusion* **1B** ed. E Teller (New York: Academic) ch 8
[93] Lawrentiev O A 1963 *Ukr. Fyz. J.* **8** 440
Lawrentiev O A 1963 *Ukr. Fyz. J.* **9** 1963
Lawrentiev O A, Potapenko V A and Stepanenko I A 1976 *Sov. Phys. Tech. Phys.* **21** 63

[94] Piterskii V V *et al.* 1988 *Sov. J. Plasma Phys.* **14** 94
[95] See e.g. Ware A A and Faulkner J E 1969 *Nucl. Fusion* **9** 353
[96] Consoli T 1968 Novosibirsk **2** 361
[97] Budker G I, Mirnow V V and and Ryutov D D 1971 *JETP Lett.* **14** 320
Bagryanskij *et al.* 1986 Kyoto **3** 467
[98] Early work on rotating plasmas is reviewed by:
Lehnert B 1971 *Nucl. Fusion* **11** 485
The work at Novosibirsk is reviewed by:
Bekhtenev A A *et al.* 1980 *Nucl. Fusion* **20** 579
Ryutov D D 1988 *Sov. Phys. Usp.* **31** 300
Pastukhov V V 1987 *Reviews of Plasma Physics* ed. B B Kadomtsev **13** 203
[99] Post R F 1967 *Phys. Rev. Lett.* **18** 232
Budker G I 1973 *6th Eur. Conf. Contr. Fusion and Plasma Phys.* Moscow **2** 136
[100] Logan B G *et al.* 1974 *Phys. Fluids* **17** 1302
[101] Sakharov A D 1951 in 1958 Leontovich **1** 21
[102] Budker l.c
Chebotaev P Z *et al.* 1973 *6th Eur. Conf. Contr. Fusion and Plasma Phys.* Moscow **1** 411
Ryutov D D 1988 *Sov. Phys. Usp.* **31** 300
[103] Arzhannikov A V *et al.* 1998 Yokohama **3** 1169, see also:
Arzhannikov A V *et al.* 1990 Washington **2** 547
Burdakov A V *et al.* 1992 Würzburg **2** 659
[104] Mirnov V V and Lichtenberg A J 1996 *Reviews of Plasma Physics* ed. B B Kadomtsev **19** 53
[105] Gross R A 1975 *Nucl. Fusion* **15** 729
Gross R A 1982 *Phys. Fluids* **25** 1042
[106] Oliphant T A 1973 *Nucl. Fusion* **13** 521
[107] Alikhanov S G *et al.* 1980 Brussels **2** 699
Linhart J G 1970 *Nucl. Fusion* **10** 211
Sherwood A R *et al.* 1980 Brussels **2** 691
Ribe F L and Sherwood A R 1981 Fast liner-compression fusion systems *Fusion* **1B** ed. E Teller (New York: Academic) ch 10
[108] Artsimovich L A 1958 Geneva **31** 6
Anderson O A *et al.* 1958 Phys. Rev. **109** 612
Fowler C M, Garn W B and Caird R S 1960 *J. Applied Phys.* **31** 588
Sakharov A D *et al.* 1966 *Sov. Phys. Doklady* **10** 1045
[109] Post R F 1956 *Rev. Mod. Phys.* **28** 338
Rose D J and Clarke M 1961 *Plasmas and Controlled Fusion* (Cambridge: MIT Press) p 20
[110] Miyamoto G, Iwata G, Mori S and Inoue K 1957 *J Phys. Soc. Japan* **12** 438
[111] Barsanti G *et al.* 1958 Geneva **31** 238
[112] Prévot F, Hubert P and Gourdon C 1959 *C. R. Acad. Sci. (Paris)* **249** 997
[113] Maglich B C *et al.* 1971 *Phys. Rev. Lett.* **29** 909
[114] Maglich B C 1987 *Int. Symp. Feasibility of Aneutronic Power* Princeton *(Proceedings in 1988 Nucl. Instr. Meth. in Physics Res.)* **A271** 13
[115] Miyamoto G *et al.* 1993 *7th Int. Conf. Emerging Nuclear Systems* Makahari, Japan p 31
[116] Dawson J M, Furth H P and Tenney F H 1971 *Phys. Rev. Lett.* **26** 1156
[117] See proceedings [114]

Chapter 4

[1] Meyer F and Schmidt H U 1958 *Z. Naturf.* **13a** 1005
Lotz W, Remy E and Wolf G H 1964 *Nucl. Fusion* **4** 335
[2] Weizner H 1971 *Phys. Fluids* **14** 658, Madison **3** 223
Freidberg J P 1971 Madison **3** 215
Grad H 1971 Madison **3** 229
Berge G and Freidberg J P 1974 Tokyo **3** 3
Hernegger F *et al.* 1976 Berchtesgaden **2** 183
Barnes D C *et al.* 1976 Berchtesgaden **2** 203
[3] Osovetz S M 1960 *Sov. Phys. JETP* **12** 221
[4] Burnett S C *et al.* 1971 Madison **3** 201
[5] Bromberg J L 1982 *Fusion; Science, Politics, and the Invention of a New Energy Source* (Cambridge, MA: MIT Press) p 222
[6] Bartsch R R *et al.* 1976 Berchtesgaden **2** 193
[7] Braun W *et al.* 1974 Tokyo **3** 25
[8] Harding G N and Roberts V 1961 Salzburg **3** 883
Jones B B and Wilson R 1961 Salzburg **3** 889
Rusbridge M G, Lees D J and Saunders P A H 1961 Salzburg **3** 895
Burton W M *et al.* 1961 Salzburg **3** 903
Afrosimov V V *et al.* 1961 Salzburg **3** 921 (Engl. transl. **1** 490)
Afrosimov V V *et al.* 1960 *Zh. Techn. Fyz.* **30** 1381
[9] Butt E P *et al.* 1965 Culham **2** 751
[10] Rosenbluth M N 1958 Geneva **31** 85
Suydam B R 1958 Geneva **31** 157
Conner J P *et al.* 1958 Geneva **32** 297
Tuck J 1959 *4th Int. Conf. Ionization Phenomena in Gases* Uppsala (Amsterdam: North-Holland) **2** 920
Colgate S A and Wright R E 1958 Geneva **32** 14
Butt E P *et al.* 1958 Geneva **32** 42
Kerst D W 1958 Geneva **32** 106
Ware A A 1961 Salzburg **3** 869
Allen N L and Liley B S 1959 *4th Int. Conf. Ionization Phenomena in Gases* Uppsala (Amsterdam: North-Holland) **2** 937
Lees D J and Rusbridge M G 1959 *4th Int. Conf. Ionization Phenomena in Gases* Uppsala (Amsterdam: North-Holland) **2** 954
Burtsev V A *et al.* 1961 Salzburg **3** 1230 (Engl. transl. **2** 773)
Babichev A P *et al.* 1961 Salzburg **2** 635 (Engl. transl. **1** 238)
Kadomtsev B B 1961 Salzburg **3** 969 (Engl. transl. **1** 438)
Woltjer L 1958 *Astrophys. J.* **128** 384
Ohkawa T *et al.* 1963 *Phys. Fluids* **6** 846
Laing E W 1958 *AERE Harwell Report* T/M 161
[11] Robinson D C and King R E 1968 Novosibirsk **1** 263
[12] Bodin H A B *et al.* 1971 Madison **1** 225
Baker D A *et al.* 1971 Madison **1** 203
Buffa A *et al.* 1974 Tokyo **3** 431
Ogawa K *et al.* 1974 Tokyo **3** 445
Shimada T *et al.* 1976 Berchtesgaden **1** 463

Hirano K *et al.* 1972 *Proc. 2nd Topical Conf. on Pulsed High-Beta-Plasmas* Garching **1** 127

[13] Taylor J B 1974 *Phys. Rev. Lett.* **33** 1139

[14] Moffat H K 1978 *Magnetic Field Generation in Electrically Conducting Fluids* (Cambridge: University Press)

Berger M A 1999 *26th EPS Conf. Contr. Fusion Plasma Phys. Maastricht*, 1999 *Plasma Phys. Contr. Fusion* **41** B167

Ortolani S 1987 *Varenna Course on Physics of Mirrors, Reversed Field Pinches and Compact Tori* **1** 107

[15] For the history of RFP research, see:

Bodin H A B and Newton A A 1980 *Nucl. Fusion* **20** 1255

Baker D A and Quinn W E 1981 The reversed-field pinch *Fusion* **1A** ed. E Teller (New York: Academic) ch 7

[16] Bodin H A B, James T E and Newton A A 1969 *Proc. Culham Conf. on Nuclear Fusion Reactors*, BNES (1970) **255**, 282

Hancox R *et al.* 1977 *Proc. Madison Meeting on Fusion Reactor Design* (Vienna: IAEA 1978) 319

Hagenson R R L and Krakowski R A 1978 *Santa Fe Meeting on Technology of Controlled Nuclear Fusion*, ANS **1**, 90

[17] A more complete listing may be found in:

Bodin H A B 1990 *Nuclear Fusion* **30** 1725

[18] Baker D A *et al.* 1982 Baltimore **1** 587

[19] The September 1985 anniversary issue of the IAEA journal *Nuclear Fusion* contains brief overviews of the work in these institutes and Bodin's (1990) review paper in the same journal extensively covers the ensemble of RFP research during the 1980s.

[20] Caramara E J and Baker D A 1984 *Nucl. Fusion* **24** 423

[21] Bevir M K and Gray J 1981 *Los Alamos Workshop on RFP Theory*, LA-8944-C 176

[22] Schoenberg K F, Moses R W and Hagenson R L 1984 *Phys. Fluids* **27** 1671

[23] Jacobson A R and Moses R W 1984 *Phys. Rev.* **A29** 3335

[24] Ohkawa T *et al.* 1980 *Nucl. Fusion* **20** 1464

[25] Ortolani S 1985 *Nucl. Fusion* **25** 1291

[26] Connor J W and Taylor J B 1984 *Phys. Fluids* **27** 2676

[27] Bodin H A B 1987 *Varenna Course on Physics of Mirrors, Reversed Field Pinches and Compact Tori* **1** 1 and 307

Alper B *et al.* 1986 Kyoto **2** 399

Alper B, Martini S and Ortolani S, 1986 *Nucl. Fusion* **26** 1256

Sprott J C *et al.* 1990 Washington **2** 519

[28] Antoni V *et al.* 1993 *20th EPS Conf. Contr. Fusion and Plasma Physics* Lisbon **2** 695

[29] Antoni V, Martini S and Ortolani S 1986 *13th EPS Conf. Contr. Fusion and Plasma Heating* Schliersee **1** 385

An Z *et al.* 1984 London **2** 231

[30] Yagi Y *et al.* 1999 *Plasma Phys. Control. Fusion* **41** 255

[31] Rostagni G 1995 *Fusion Eng. Design* **25** 301

Bartiromo R 1999 *26th EPS Conf. Contr. Fusion Plasma Phys. Maastricht*, 1999 *Plasma Phys. Contr. Fusion* **41** B143

For recent work on current profile and plasma rotation control, see also:

Bartiromo R *et al.* 1998 Yokohama **1** 367

Hirano Y *et al.* 1998 Yokohama **1** 375

Fiksel G *et al.* 1998 Yokohama **1** 383
Forest C B *et al.* 2000 Sorrento EX1/1
Martin P *et al.* 2000 Sorrento EX3/5
Yagi Y *et al.* 2000 Sorrento EX4/6

[32] Bobeldijk C *et al.* 1968 Novosibirsk **1** 287
Colgate S A 1958 Geneva **32** 123
Colgate S A *et al.* 1958 Geneva **32** 140

[33] Kolb A C *et al.* 1968 Novosibirsk **2** 567
Zwicker H, Wilhelm R and Krause H 1971 Madison **1** 251
Wilhelm R and Zwicker H 1971 Madison **1** 259
Becker G *et al.* 1974 Tokyo **3** 47
Dippel K H *et al.* 1974 Tokyo **3** 57
Yokoyama J *et al.* 1974 Tokyo **3** 171
Ogawa K *et al.* 1974 Tokyo **3** 445
Hirano K *et al.* 1974 Tokyo **3** 463
Hofmann F, Bighel L and Peiry J M 1975 *7th European Conf. Contr. Fusion and Plasma Physics* Lausanne **2** 137
Aydemir A *et al.* 1978 Innsbruck **2** 105

[34] Becker G *et al.* 1974 Tokyo **3** 47

[35] Goedbloed J P and Sakanaka P H 1974 *Phys. Fluids* **17** 908
d'Ippolito D A *et al.* 1978 *Phys. Fluids* **21** 1600
Galvão R M O *et al.* 1980 Brussels **2** 325
Galvão R M O *et al.* 1982 Baltimore **3** 3

[36] Marshall T C and Lui H C 1976 *Phys. Fluids* **19** 1417

[37] Clarke J F and Sigmar D J 1977 *Phys. Rev. Lett.* **38** 70
Hogan J T 1979 *Nucl. Fusion* **19** 753

[38] Oomens A A M *et al.* 1987 *14th European Conf. Contr. Fusion and Plasma Phys.* Madrid **2** 458

[39] Kiyama H *et al.* 1988 Nice **1** 437

[40] Chu C K *et al.* 1985 *Nucl. Fusion* **25** 1109
Grimm R, Greene J and Johnson J L 1978 *Methods Comput. Phys.* **16** 253
Ling K M and Jardin S C 1985 *J. Comput. Phys.* **58** 300
Sankar M K V *et al.* 1993 *J. Fusion Energy* **12** 303
Ivers T H *et al.* 1994 Seville **2** 287

[41] Eskov A G *et al.* 1978 Innsbruck **2** 187
Bussac M N *et al.* 1978 Innsbruck **3** 249

[42] See e.g. Alfvén H 1958 Geneva **31** 3

[43] Shafranov V D 1957 *Sov. Phys. JETP* **6** 545

[44] Schmidt G 1966 *Physics of High Temperature Plasma* (New York: Academic) see also:
Witalis E A 1990 *J. Meteorology* **15** 121

[45] Andrews M L *et al.* 1971 Madison **1** 169
Phelps D A *et al.* 1974 *Phys. Fluids* **17** 2226

[46] Chen J *et al.* 1980 Brussels **1** 501

[47] Eskov A G *et al.* 1973 *6th Eur. Conf. Contr. Fusion and Plasma Physics* Moscow **1** 595, 599
Eskov A G *et al.* 1974 Tokyo **3** 71 (Engl. transl. 155)

[48] FRC research has been reviewed by:
Tuszewski M 1988 *Nucl. Fusion* **28** 2033

Wright B L 1990 *Nucl. Fusion* **30** 1739; see also the January 1986 issue of the journal *Fusion Technology*, which is devoted to compact torus experiments and reactor studies

[49] Ohi *et al.* 1983 *Phys. Rev. Lett.* **51** 1042
Okada S *et al.* 1994 Seville **2** 441

[50] Okada S *et al.* 1996 Montreal **2** 229

[51] Rej D J *et al.* 1990 Washington **2** 647
Tuszewski M 1988 *Nucl. Fusion* **28** 2033

[52] Hoffman A L and Slough J T 1993 *Nucl. Fusion* **33** 27

[53] Slough J T *et al.* 1992 Würzburg **2** 627
Hoffman and Slough *l.c.*

[54] Rostocker N, Binderbauer M W and Monckhorst H J 1997 *Science* **278** 1419

[55] Bussac M N *et al.* 1978 Innsbruck **3** 249

[56] Rosenbluth M N and Bussac M N 1979 *Nucl. Fusion* **19** 489

[57] Rosenbluth M N and Bussac M N 1979 *l.c*

[58] Yamada M *et al.* in *Proc. US–Japan Symp. on Compact Tori*, Princeton 1979
Grossmann W *et al.* 1980 Brussels **1** 455

[59] Mohri A *et al.* 1978 Innsbruck **3** 311
Narihara K *et al.* 1986 Kyoto **2** 687

[60] FRC and spheromak studies were reviewed by:
Wright B L 1990 *Nucl. Fusion* **30** 1739
and spheromak studies by:
Jarboe T R 1994 *Plasma Phys. Contr. Fusion* **36** 945
Spheromaks have been re-evaluated by:
Fowler T K 1996 *Fusion Techn.* **29** 206
Hooper E B *et al.* 1996 *Fusion Techn.* **29** 191

[61] Baker D A *et al.* 1984 London **2** 439
Bevir M K and Gray J W 1980 *Proc. Reversed Pinch Theory Workshop, LANL* LA-8944-C p 176

[62] Barnes C W *et al.* 1986 *Phys. Fluids* **29** 3415
Much of the work on this subject has been reported in a series of US–Japan Workshops, such as the 11th US–Japan Workshop CT LANL LA-11808-C

[63] Jarboe T R 1989 *Fusion Technol.* **15** 7
Taylor J B and Turner M F 1989 *Nucl. Fusion* **29** 219

[64] Hartman C W and Hammer J H 1982 *Phys. Rev. Lett.* **48** 929
Harned D S *et al.* 1982 Baltimore **2** 293
Brown M R and Bellan P M 1992 *Nucl. Fusion* **32** 1125

[65] Uyama T *et al.* 1990 Washington **2** 621

[66] Ono Y, Yumoyo A and Katsurai M 1987 *IEEE Transactions Plasma Sci.* **15** 418
Ono Y *et al.* 1992 Würzburg **2** 619
Inomoto M *et al.* 1998 Yokohama **3** 927

[67] Kadomtsev B B and Braginsky S I 1958 Geneva **31** 233
Berkowitz J *et al.* 1958 Geneva **31** 171

[68] Allen T K, McWhirter R W P and Spalding I J 1961 Salzburg **1** 67

[69] Kadomtsev B B and Braginsky S I 1958 Geneva **32** 233
Tuck J L 1960 *Nature* **187** 863

[70] Ohkawa T and Kerst D W 1961 *Phys. Rev. Lett.* **7** 41

[71] Kadomtsev B B 1957 in 1958 Leontovich **4** 17

Rosenbluth M N and Longmire C 1957 *Ann. Phys. (NY)* **1** 120
Bernstein I B, Frieman E A, Kruskal M D and Kulsrud R M 1958 *Proc. R. Soc.* **A244** 17
Johnson J L, Oberman C R, Kulsrud R M and Frieman E A 1958 Geneva **31** 198
Furth H P and Rosenbluth M N 1964 *Phys. Fluids* **7** 764
Furth H P, Killeen J, Rosenbluth M N and Coppi B 1965 Culham **1** 103
The literature on this subject has been reviewed by:
Jukes J D 1967 *Rep. Prog. Phy.* **30** 333
Furth H P 1968 *Advances in Plasma Physics* ed. Simon A and Thompson W B (New York: Wiley) **1** 67
Bodin H A B and Keen B E 1977 *Rep. Prog. Phys.* **40** 1415
[72] Eckhartt D, von Gierke G and Grieger G 1965 Culham **2** 719
[73] Ohkawa T, Yoshikawa M and Schupp A A 1968 Novosibirsk **1** 329
Forsen H *et al.* 1968 Novosibirsk **1** 313
[74] Roberts M *et al.* 1968 Novosibirsk **1** 351
[75] Ohkawa T 1970 *Nucl. Fusion* **10** 185
Ohkawa T *et al.* 1971 Madison **1** 15, 1972 *Phys. Rev. Lett.* **28** 1107
[76] Meade D M 1965 *Phys. Rev. Lett.* **17** 677
[77] Tamano T *et al.* 1974 Tokyo **2** 97
[78] Freeman R *et al.* 1971 Madison **1** 27
[79] Ando K *et al.* 1974 Tokyo **2** 103
Drake J R *et al.* 1976 Berchtesgaden **2** 333
[80] Ainsworth N R *et al.* 1975 *7th European Conf. Contr. Fusion Plasma Physics* Lausanne **2** 127
Alcock M W *et al.* 1976 Berchtesgaden **2** 305
[81] Colgate S A and Furth H P 1960 *Phys. Fluids* **3** 982
Anderson O A *et al.* 1958 *Phys. Fluids* **1** 489
Birdsall D H, Colgate S A and Furth H P 1959 *4th Int. Conf. Ionization Phenomena in Gases* Uppsala (Amsterdam: North-Holland) **2** 888, 892
Birdsall D H *et al.* 1961 Salzburg **3** 955
[82] Lehnert B 1959 *J. Nucl. Energy* C **1** 40
Sakharov A D 1951 in Leontovich **1** 21
Tamm I E 1951 in 1958 Leontovich **1** 35
[83] Ware A A 1964 *Phys. Fluids* **7** 2006
Yoshikawa S 1966 *Phys. Fluids* **9** 2295
Furth H P *et al.* 1965 Culham **1** 103
[84] Anderson O A *et al.* 1968 Novosibirsk **1** 443
[85] Freeman R *et al.* 1971 Madison **1** 27
Anderson O A *et al.* 1971 Madison **1** 103
Rivière A C and Sweetman D R 1972 *5th Euro. Conf. Contr. Fusion Plasma Phys.* Grenoble 95
Alcock M W *et al.* 1974 Tokyo **2** 77
[86] Sinnis J *et al.* 1972 *Phys. Rev. Lett.* **29** 1214
Ejima S, Okabayashi M and Schmidt J 1974 *Phys. Rev. Lett.* **32** 872
[87] For reviews of experiments on plasma confinement in internal-ring devices, we refer to:
Yoshikawa S 1973 *Nucl. Fusion* **13** 433
Bodin H A B and Keen B E 1977 *Rep. Prog. Phys.* **40** 95

[88] Futch A H *et al.* 1965 Culham **2** 3
Perkins W A and Barr W L 1965 Culham **2** 115
[89] Hasegawa A, Chen L and Mauel M, 1990 *Nucl. Fusion* **30** 2405
Hasegawa A *et al.* 1992 *Fusion Technol.* **22** 27
Kesner J *et al.* 1998 Yokohama **3** 1165
[90] Kadomtsev B B 1956 in 1958 Leontovich **3** 340
[91] Dandl R A *et al.* 1971 Madison **2** 607
Ard W B *et al.* 1971 Madison **2** 619
Dandl R A *et al.* 1968 Novosibirsk **2** 435
Berry L A *et al.* 1984 London **2** 54
Hiroe S *et al.* 1988 *Nucl. Fusion* **28** 2249
[92] Fujiwara M *et al.* 1980 Brussels **1** 845
Ikegami H *et al.* 1986 Kyoto **2** 489
[93] Alfvén H and Smårs E A 1960 *Nature* **188** 801
Smårs E A 1965 *Arkiv Fysik* **29** 97
[94] Braams C M 1966 *Phys. Rev. Lett.* **17** 470
[95] Wienecke R 1963 *Z. Naturf.* **18a** 1151
K Berghahn H H *et al.* 1968 Novosibirsk **2** 113
[96] Alikhanov S G, Konkashbaev I K and Chebotaev P Z 1970 *Nucl. Fusion* **10** 13
[97] Verboom G K and Rem J 1973 *Nucl. Fusion* **13** 69
[98] Lehnert B 1968 *Nucl. Fusion* **8** 173
Lehnert B 1973 *Nucl. Fusion* **19** 958
[99] Bergström J, Holmberg S and Lehnert B 1965 Culham **1** 341
Lehnert B 1971 *Nucl. Fusion* **11** 485
[100] Schüller FC *et al.* 1974 Tokyo **2** 725
Brandt B and Braams C M 1974 Culham 385
[101] Engelmann F *et al.* 1976 *Proc. Int. Symp. Plasma–Wall Interaction* Jülich (Pergamon for CEC 1977) 627
Volkov T F, Igitkhanov Y L and Tokar M Z 1976 Berchtesgaden **3** 359

Chapter 5

[1] Kadomtsev B B and Nedospasov A V 1960 *J. Nucl. Energy* C **1** 230
Hoh F C and Lehner B 1961 *Phys. Rev. Lett.* **7** 75
[2] Grieger G, Renner H and Wobig H 1985 *Nucl. Fusion* **25** 1231
[3] Eckhartt D, von Gierke G and Grieger G 1965 Culham **2** 719
Akulina D K *et al.* 1965 Culham **2** 733
[4] Bishop A S and Hinnov E 1965 Culham **2** 673
Stodiek W, Grove D J and Kessler J O 1965 Culham **2** 687
[5] Lees D J 1985 *Nucl. Fusion* **25** 1259
[6] Gibson A *et al.* 1968 Novosibirsk **1** 465
[7] Gourdon C *et al.* 1968 Novosibirsk **1** 847
[8] Uo K *et al.* 1962 *Phys. Fluids* **5** 1293
[9] Bezbatchenko A L *et al.* 1956–1957 in 1958 Leontovich **4** 135
[10] Strelkov V S 1985 *Nucl. Fusion* **25** 1189
[11] Dolgov-Saveliev G G *et al.* 1958 Geneva **32** 82
[12] Dolgov-Saveliev G G *et al.* 1959 Uppsala **2** 947

[13] Early tokamaks in Moscow are described in:
Bezbatchenko A L *et al.* 1956–1957 *l.c.*
Artsimovich L A *et al.* 1965 Culham **2** 595 (Engl. transl. 417)
Ivanov D P and Parfenov V S 1965 Culham **2** 659 (Engl. transl. 484)
Dolgov-Saveliev G G *et al.* 1958 Geneva **32** 82
Artsimovich L A *et al.* 1968 Novosibirsk **1** 157 (Engl. transl. **1** 17)
Artsimovich L A 1961 *Controlled Thermonuclear Reactions* (Engl. transl. Edinburgh: Oliver & Boyd, 1964)
Gorbunov E P *et al.* 1961 Salzburg **3** 941 (Engl. transl. **1** 412)
Strelkov V S 1985 *Nucl. Fusion* **25** 1189 and private communication
Gorbunov E P and Rasumova K A 1963 *At. Energ.* **15** 363
Gorbunov E P 1965 Culham **2** 629 (Engl. transl. 440)
Mukhovatov V S 1965 Culham **2** 577 (Engl. transl. 399)
Artsimovich L A 1971 Madison **1** 443 (Engl. transl. 41)

[14] Mukhovatov *l.c.*; further diagnostic work in Gorbunov 1965 *l.c.* and in:
Vinogradova N D and Razumova K A 1965 Culham **2** 617 (Engl. transl. 440)
Gorelik L L, Razumova K A and Sinitsyn V V 1965 Culham **2** 647 (Engl. transl. 471)
Ivanov D P and Parfenov D S 1965 Culham **2** 659

[15] Shafranov V D 1962 *At. Energ.* **13** 521

[16] Artsimovich L A, Mirnov S V and Strelkov V S 1964 *At. Energ.* **17** 170

[17] Goldberg R and Liley B S 1986 *Vacuum* **18** 631
Liley B S, Vance C F and Smith J M 1968 *Proc. Inst. Radio Electron. Eng.* **29** 221

[18] Artsimovich L A *et al.* 1968 Novosibirsk **1** 157 (Engl. transl. 17)

[19] Kadomtsev B B and Pogutse O P 1967 *Sov. Phys. JETP* **24** 1172
Dreicer H 1958 Geneva **31** 57

[20] Artsimovich L A 1961 Salzburg **1** 9 (Engl. transl. **1** 15)

[21] Afrosimov V V and Petrov M P 1967 *Zh. Tekh. Fiz.* **37** 1995 (Engl. trans. *Sov. Phys.–Tech. Phys.* **12** 1467); see also:
Artsimovich L A 1972 *Nucl. Fusion* **12** 215

[22] Furth H P 1981 The tokamak *Fusion* **1A** ed. E Teller (New York: Academic) ch 3, pp 146–147

[23] Peacock N J *et al.* 1969 *Nature* **224** 448

[24] Stott P E 1992 *Rep. Prog. Phys.* **55** 1715

[25] Artsimovich L A *et al.* 1969 *Sov. Phys.–JETP* **10** 130

[26] Artsimovich L A *et al.* 1971 Madison **1** 443 (Engl. transl. 41)

[27] Young K M 1974 *Plasma Phys.* **16** 119

[28] Miyamoto K 1978 *Nucl. Fusion* **18** no. 2

[29] Uo K *et al.* 1981 *Proc. 10th European Conf. Cont. Fusion and Plasma Physics* Moscow, Europhysics Conference Abstracts vol 5G, part 1, p 357
Uo K *et al.* 1982 Baltimore **2** 209

[30] Fujita J and Matsuoka K *Nucl. Fusion* **25** 1253

[31] Uo K 1985 *Nucl. Fusion* **25** 1243

[32] Seki H *et al.* 1988 Nice **2** 645

[33] Grieger G *et al.* 1992 *Phys. Fluids* **B4** 2081
Nührenberg J *et al.* 1995 *Trans. Fusion Technology* **27** 71
Wagner F 1998 *Trans. Fusion Technology* **33** 67
Hartmann D A 2000 *Trans. Fusion Technology* **37** 71

[34] Palumbo D 1968 *Nuovo Cimento* **B53** 507

[35] Stroth U *et al.* 1996 *Nucl. Fusion* **36** 1063
[36] Fujiwara M, 1999 *Plasma Phys. Contr. Fusion* **41** B157
[37] Carreras B A *et al.* 1988 *Nucl. Fusion* **28** 1613

Chapter 6

[1] Bromberg J L 1982 *Fusion; Science, Politics, and the Invention of a New Energy Source* (Cambridge, MA: MIT Press) p 28
[2] Heppenheimer T A 1984 *The Man Made Sun: the Quest for Fusion Power* (Boston: Little, Brown)
[3] Ohkawa T and Voorhies H G 1969 *Phys. Rev. Lett.* **22** 1275
[4] Ohkawa T *et al.* 1974 Tokyo **1** 281
[5] Based on data in Wesson J 1997 *Tokamaks* (Oxford: Science Publications)
[6] Parker R R *et al.* 1985 *Nucl. Fusion* **25** 1127
[7] Ellis R A 1985 *Nucl. Fusion* **25** 1145
[8] Hosea J *et al.* 1985 *Nucl. Fusion* **25** 1155
[9] Bol K *et al.* 1985 *Nucl. Fusion* **25** 1149
[10] ORMAK-ISX Group 1985 *Nucl. Fusion* **25** 1137
[11] Maisel M W 1985 *Nucl. Fusion* **25** 1113
[12] TFR Group 1985 *Nucl. Fusion* **25** 1025
[13] Briffod G *et al.* 1985 *Nucl. Fusion* **25** 1033
[14] Pulsator Team 1985 *Nucl. Fusion* **25** 1059
[15] Alladio F *et al.* 1985 *Nucl. Fusion* **25** 1069
[16] Robinson D C 1985 *Nucl. Fusion* **25** 1101
[17] Paul J W M 1985 *Nucl. Fusion* **25** 1097
[18] Tanaka M 1985 *Nucl. Fusion* **25** 1073
[19] Strelkov V S 1985 *Nucl. Fusion* **25** 1189 and private communication
[20] Golant V E 1985 *Nucl. Fusion* **25** 1183 and private communication
[21] See 1982, 1986 *World Survey of Major Activities in Controlled Fusion Research* (Vienna: IAEA)
[22] Coppi B, in 1976 *Fusion Reactors for Break-Even*, Erice–Trapani (published by Pergamon for CEC) p 303
[23] Berry L A 1974 Tokyo **1** 113
[24] Cordey J G *et al.* 1974 *Nucl. Fusion* **14** 441
Barnett C F *et al.* 1973 *Nucl. Fusion* **13** 441
Bol K *et al.* 1973 *Phys. Rev. Lett.* **32** 661
[25] Equipe TFR 1976 Berchtesgaden **1** 69
[26] Vlasenkov V S *et al.* 1976 Berchtesgaden **1** 85
[27] Eubank H *et al.* 1978 Innsbruck **1** 167
[28] Gondhalekar A *et al.* 1978 Innsbruck **1** 199
[29] Mirnov S V and Semenov I B 1971 *Sov. Phys. JETP* **33** 1143
Mirnov S V and Semenov I B 1971 Madison **2** 401
[30] Hosea J C, Bobeldijk C and Grove D J 1971 Madison **2** 425
Bol K *et al.* 1978 Innsbruck **1** 11
Callen J D *et al.* 1978 Innsbruck **1** 415
Furth H P 1981 The tokamak *Fusion* **1A** ed. E Teller (New York: Academic) ch 3, section V A

[31] Wesson J A *et al.* 1989 *Nucl. Fusion* **29** 641
Schüller F C 1995 *Plasma Phys. Contr. Fusion* **37** A135
[32] Murakami M, Callen J D and Berry L A 1976 *Nucl. Fusion* **16** 347
[33] Paul J W M *et al.* 1977 *Proc. 8th European Conference on Controlled Fusion and Plasma Physics,* Prague vol II p 49
[34] von Goeler S 1974 *Phys. Rev. Lett.* **33** 1201
[35] Kadomtsev B B 1975 *Fiz. Plazmy* **1** 710, 1976 Berchtesgaden **1** 555
[36] Reviewed by Jahns *et al.* 1978 *Nucl. Fusion* **18** 609 and more recently by Wesson J D 1997 *Tokamaks* (Oxford: Science Publications)
[37] Equipe TFR 1980 Brussels **2** 547
Dubois M, Marty D and Pochelon A 1980 *Nucl. Fusion* **20** 1355
[38] Dubois M and Samain A 1980 *Nucl. Fusion* **20** 1355
[39] Campbell D J *et al.* 1986 Kyoto **1** 433
[40] Soltwisch H *et al.* 1986 Kyoto **1** 263
[41] Equipe TFR 1974 Tokyo **1** 127
Fujsawa N *et al.* 1974 Tokyo **1** 3
Berry L A *et al.* 1974 Tokyo **1** 101
Colchin R J *et al.* 1976 *J. Nucl. Mater.* **63** 74
[42] See the proceedings of the series of *International Conferences on Plasma Surface Interactions* published in *J. Nucl. Mater.* **63** (1976), **76 + 77** (1978), **93 + 94** (1980), **111 + 112** (1982), **128 + 129** (1984), **145–147** (1987), **162–164** (1989), **176 + 177** (1990), **196–198** (1992), **220–222** (1995), **241–243** (1997), **266–269** (1999)
[43] Meade D M 1974 *Nucl. Fusion* **14** 289
[44] Taylor R J 1978 *J. Nucl. Mater.* **76 + 77** 41
[45] Stott P E, Daughney C C and Ellis R A 1975 *Nucl. Fusion* **15** 431
[46] Winter J 1990 *J. Nucl. Mater.* **176 + 177** 14
Fielding S J *et al.* 1977 *Nucl. Fusion* **17** 1382
Wootton A J *et al.* 1982 *J. Nucl. Mater.* **111 + 112** 479
Suzuki N *et al.* 1980 *J. Nucl. Mater.* **93 + 94** 282
Bol K *et al.* 1978 Innsbruck **1** 11
Dylla H F *et al.* 1987 *J. Nucl. Mater.* **145–147** 48
[47] McCracken G M and Stott P E 1979 *Nucl. Fusion* **19** 889
[48] Milora S L *et al.* 1995 *Nucl. Fusion* **35** 657
[49] Milora S L 1989 *Vac. Sci. Technol.* A7 925
[50] Lang P *et al.* 1997 *Phys. Rev. Lett.* **79** 1487
[51] Spitzer L 1951 *US Atomic Energy Commission Report* NYO-993
[52] Pitcher C S and Stangeby P C 1997 *Plasma Phys. Control. Fusion* **39** 779
[53] The bundle divertor concept is described by
Stott P E, Gibson A and Wilson C 1977 *Nucl. Fusion* **17** 481
The ergodic divertor is reviewed by
Ghendrih Ph, Grosman A and Capes H 1996 *Plasma Phys. Control. Fusion* **38** 1653
[54] Galeev A A and Sagdeev R Z 1968 *Sov. Phys. JETP* **26** 233
[55] Heuristic derivations of these effects are given by:
Kadomtsev B B 1992 *Tokamak Plasma, a Complex Physical System* (Bristol: IOP Publishing)
Rosenbluth M N and Rutherford P H 1981 Tokamak plasma stability *Fusion* **1A** ed. E Teller (New York: Academic) ch 2, section VIII A, p 104
[56] Ware A A 1970 *Phys. Rev. Lett.* **25** 15

Galeev A A and Sagdeev R Z 1971 *JETP Lett.* **13** 113
Kadomtsev B B and Shafranov V D 1971 Madison **2** 479
[57] Bickerton R J, Conner J W and Taylor J B 1971 *Nature* **229** 110
Galeev A A and Sagdeev R Z 1971 Madison **1** 481 (Engl. transl. 45)
[58] Hinton F L and Hazeltine R D 1976 *Rev. Mod. Phys.* **48** 239
[59] Artsimovich L A 1971 *Sov. Phys. JETP Lett.* **13** 101
Yoshikawa S 1970 *Phys. Rev. Lett.* **25** 353
Yoshikawa S 1970 *Phys. Fluids* **13** 2300
Yoshikawa S and Christofilos N C 1971 Madison **2** 357
[60] Jassby D L, Cohn D R and Parker R R 1976 *Nucl. Fusion* **16** 1045
[61] Hugill J 1983 *Nucl. Fusion* **23** 331
[62] Furth H P 1981 The tokamak *Fusion* **1A** ed. E Teller (New York: Academic) ch 3, section VI E

Chapter 7

[1] Based on data in Wesson J D 1997 *Tokamaks* (Oxford: Science Publications)
[2] Itoh S *et al.* 1986 Kyoto **3** 321
[3] Tanaka S *et al.* 1986 Kyoto **1** 553
[4] Keilhacker M *et al.* 1985 *Nucl. Fusion* **25** 1045
[5] TCA Team 1985 *Nucl. Fusion* **25** 1041 and Lister J, private communication
[6] Callen J D *et al.* 1982 Baltimore **1** 297
[7] Hawryluk R J *et al.* 1986 Kyoto **1** 51
Navratil G A 1990 Washington **1** 209
JT-60 Team, 1990 Washington **1** 53
[8] Ohkawa T 1970 *Nucl. Fusion* **10** 185
Start D F H *et al.* 1978 *Phys. Rev. Lett.* **40** 1497
Axon K B *et al.* 1980 Brussels **1** 413
Start D F H, Cordey J G and Jones E M 1980 *Plasma Phys.* **22** 303
[9] Thonemann P C, Cowhig W T and Davenport P A 1952 *Nature* **169** 34
Blevin H A and Thonemann P C 1961 Salzburg **1** 55
[10] Wort D J H 1970 *Plasma Phys.* **13** 258
[11] Yoshikawa S and Yamato H 1966 *Phys. Fluids* **9** 1814
[12] Ohkawa T 1970 *Nucl. Fusion* **10** 185
[13] Fisch N J 1978 *Phys. Rev. Lett.* **41** 873 (see also **42** 410)
Karney C F F and Fisch N J 1979 *Phys. Fluids* **22** 1817
[14] Fisch N J and Boozer A H 1980 *Phys. Rev. Lett.* **45** 720
[15] Yamamoto T *et al.* 1980 *Phys. Rev. Lett.* **45** 716
Maekawa T *et al.* 1981 *Phys. Lett.* **85A** 339
[16] Hooke W *et al.* 1990 Baltimore **1** 239
[17] JT-60 team 1986 Kyoto **1** 11
[18] Ikeda Y *et al.* 1994 Seville **1** 415
[19] Itoh S *et al.* 1990 Washington **1** 733, 1998 Yokohama **1** 99
[20] ITER 1989 *Concept Definition* Vol 1 (Vienna: IAEA)
[21] Eubank H *et al.* 1978 Innsbruck **1** 167
[22] TFR Group 1977 EUR-CEA-875
[23] Vorobyev A V *et al.* 1978 *Fyz. Plazmy* **4** 982

[24] Behrisch R *et al.* 1984 *J. Nucl. Mater.* **128 + 129** 470
[25] Behrisch R *et al.* 1987 *J. Nucl. Mater.* **145–147** 731
[26] Chicherov V M 1987 *J. Nucl. Mater.* **145–147** 117
[27] TFR Group 1987 *J. Nucl. Mater.* **145–147** 681
[28] Dobrokhotov E I *et al.* 1982 Baltimore **3** 229
[29] Rebut P H *et al.* 1989 *J. Nucl. Mater.* **162–164** 172
[30] Sengoku S 1980 *J. Nucl. Mater.* **93/94** 178
[31] Winter J 1996 *Plasma Phys. Contr. Fusion* **38** 1503
[32] Furth H P 1981 The tokamak *Fusion* **1A**, ed. E Teller (New York: Academic) ch 3
[33] Raeder J *et al.* 1981 *Kontrollierte Kernfusion* (Stuttgart: Teubner)
[34] Bortnikov A V *et al.* 1974 Tokyo **1** 147 (Engl. transl. 13)
[35] Maeda H *et al.* 1978 Innsbruck **1** 377
Vlasenkov V S *et al.* 1978 Innsbruck **1** 211
[36] Nagami M, Overskei D *et al.* 1982 Baltimore **1** 27
[37] Troyon F *et al.* 1983 *11th EPS Plasma Physics Division Conf.* Aachen, 1984 *Plasma Phys. Contr. Fusion* **26** no 1A 209
[38] Lazarus E A, Lister J B and Neilson G H 1990 *Nucl. Fusion* **30** 111
[39] Swain D W *et al.* 1981 *Nucl. Fusion* **21** 1409
[40] Goldston R J 1983 *11th EPS Plasma Div. Conf.* Aachen, 1984 *Plasma Phys. Contr. Fusion* **26** no 1A p 87
[41] Yushmanov P N *et al.* 1990 *Nucl. Fusion* **30** 1999
[42] Hugill J and Sheffield J 1978 *Nucl. Fusion* **18** 15
[43] Wagner F *et al.* 1982 Baltimore **1** 43
[44] Odajima K *et al.* 1986 Kyoto **1** 151
[45] Itoh S-I, Itoh K and Fukuyama A 1995 *J. Nucl. Mater.* **220–222** 117
[46] Connor J W and Wilson H R 2000 *Plasma Phys. Contr. Fusion* **42** R1
Hugill J 2000 *Plasma Phys. Contr. Fusion* **42** R75
[47] Greenwald M *et al.* 1984 London **1** 45
[48] Coppi B and Sharky N 1981 *Nucl. Fusion* **21** 1363
[49] Callen J D *et al.* 1987 *Nucl. Fusion* **27** 1857
[50] Luce T C, Petty C C and de Haas J C M 1992 *Phys. Rev. Lett.* **68** 52
[51] Coppi B 1980 *Comm. Plasma Phys. Contr. Fusion* **5** 261
[52] Rebut P H *et al.* 1986 Kyoto **2** 187
Rebut P H, Lallia P P and Watkins M L 1988 Nice **2** 191
[53] Furth H P 1986 *Plasma Phys. Contr. Fusion* **28** 1305
[54] Janes G S 1959 *Plasma Acceleration, 4th Lockheed Symp. on Magnetohydrodynamics* (Stanford: University Press)
Beiser A and Raab B 1961 *Phys. Fluids* **4** 177
Braams C M 1961 *High Magnetic Fields* ed. H Kolm *et al.* p 109 (New York: Wiley)
[55] Kadomtsev, like Beiser and Raab, emphasized the dimensionless quantities that characterize the plasma in its magnetic confinement, while Connor and Taylor, like Janes, focused on the structure of the equations:
Kadomtsev B B 1975 *Sov. J. Plasma Phys.* **1** 295
Connor J W and Taylor J B 1977 *Nucl. Fusion* **17** 1047
Connor J W 1988 *Plasma Phys. Contr. Fusion* **30** 619
Kadomtsev B B 1992 *Tokamak Plasma, a Complex Physical System* (Bristol: IOP Publishing)
[56] Wagner F and Stroth U 1993 *Plasma Phys. Control. Fusion* **35** 1321

[57] Düchs D F, Post D E and Rutherford P H 1977 *Nucl. Fusion* **17** 565
[58] Hawryluk RJ 1980 *Physics of Plasmas Close to Thermonuclear Conditions* ed. B Coppi *et al.* (London: Pergamon) **1** 19

Chapter 8

[1] Bromberg J L 1982 *Fusion; Science, Politics, and the Invention of a New Energy Source* (Cambridge, MA: MIT Press) p 28
[2] *The JET Project—Design Proposal* 1975 EUR-JET-R5 (Brussels: CEC)
[3] Grove D *et al.* 1976 Berchesgaden **1** 21
[4] Willson D 1981 *A European Experiment* (Bristol: Hilger)
Shaw E N 1990 *Europe's Experiment in Fusion* (Amsterdam: North-Holland)
[5] *TFTR Conceptual Design* 1976 Princeton Plasma Physics Laboratory Report PPPL-1275, PH-R-001
[6] Dawson J M, Furth H P and Tenney F H 1971 *Phys. Rev. Lett.* **26** 1156
[7] Furth H P and Jassby D L 1974 *Phys. Rev. Lett.* **32** 1167
Berk H L *et al.* 1974 Tokyo **3** 569
[8] Dylla H F *et al.* 1989 *J. Nucl. Mater.* **162–164** 128
[9] Strachan J D *et al.* 1987 *Phys. Rev. Lett.* **58** 1004
Goldston R J *et al.* 1986 Kyoto **1** 75
[10] JT-60 Team 1990 Washington **1** 53
[11] Tanga A *et al.* 1986 Kyoto **1** 65
[12] Rebut P H *et al.* 1986 Kyoto **1** 31
[13] Meade D M *et al* 1990 Washington **1** 9
[14] JET Team 1990 Washington **1** 27
[15] JT-60 Team 1994 Seville **1** 31
[16] Hawryluk R J 1998 *Rev. Mod. Phys.* **70** 537
Strachan J D *et al.* 1997 *Plasma Phys. Control. Fusion* **39** B103
[17] Schneider R *et al.* 1998 Yokohama **4** 1525
ITER Team 1999 *Nucl. Fusion* **39** 2391
[18] Lipshultz B *et al.* 1995 *J. Nucl. Mater.* **220–222** 50
Petrie T W *et al.* 1997 *Nucl. Fusion* **37** 321
Asakura N *et al.* 1997 *J. Nucl. Mater.* **241–243** 559
Loarte A *et al.* 1998 *Nucl. Fusion* **38** 331
[19] Rebut P H and Lallia P P 1988 *15th Symp. on Fusion Technology* Utrecht (Amsterdam: North-Holland 1989) Vol 1 p 1
[20] Litaudon X 1998 *Plasma Phys. Control. Fusion* **40** A251
[21] Peng Y-K and Strickler D J 1986 *Nucl. Fusion* **26** 769
[22] Robinson D C 1999 *Plasma Phys. Control. Fusion* **41** A143
[23] Sykes A *et al.* 1992 *Nucl. Fusion* **32** 694
[24] Sykes A 1999 *Plasma Phys. Control. Fusion* **39** B247
[25] Sykes A *et al.* 2000 Sorrento IAEA-CN-77/OV4/1
[26] Coppi B 1976 in *Tokamak Reactors for Breakeven* Erice-Trapani (published by Pergamon for CEC) p 303
[27] Coppi B and Taroni A in 1979 *Physics of Plasmas Close to Thermonuclear Conditions* Varenna (published by Pergamon for CEC) vol 2 p 687 (see p 832 for a summary of the objectives of the Ignitor experiment)

[28] Meade D M *et al.* 2000 Sorrento IAEA-CN-77/FTP2/16
[29] Colombo U *et al.* 1990 *Fusion Programme Evaluation* (Brussels: CEC)
[30] Coppi B *et al.* 1996 Montreal **3** 579
[31] Azizov E A *et al.* 1980 Brussels **2** 599
Azizov E A, Kovan I A and Mirnov S V 1992 Würzburg **1** 153

Chapter 9

[1] Thomson G P and Blackman M 1946 British Patent No 817 681. The patent was reproduced in *Plasma Phys. Control. Fusion* **38** (1996) 643 to commemorate the 50th aniversary of its filing. See also 1952/1953 British Patent No 822 462
[2] Tamm I E 1951 in 1958 Leontovich **1** 1
Sakharov A D 1951 in 1958 Leontovich **1** 21
Tamm I E 1951 in 1958 Leontovich **1** 35
[3] Bishop A S 1958 *Project Sherwood* (Reading, MA: Addison-Wesley)
Bromberg J L 1982 *Fusion: Science, Politics, and the Invention of a New Energy Source* (Cambridge, MA: MIT Press)
[4] Post R F 1960 *Int. Summer Course in Plasma Physics, Risö Laboratory Report* 18 (Roskilde: Danish AEC) p 367
Post R F 1961 Salzburg **1** 99
[5] Mills R G 1967 *Nucl. Fusion* **7** 223; 1969 PPPL Report Matt-658
[6] Ribe F L, Oliphant T A and Quinn W E 1965 *Los Alamos Report* LA-3294-MS
[7] Carruthers R, Davenport P A and Mitchell T A D 1967 *UKAEA Report* CLM-R 85
[8] Rose D J 1968 *Oak Ridge Report* ORNL-TM-2204
[9] Post R F 1969 Culham 88
[10] Bell G I, Borkenhagen W H and Ribe F L 1969 Culham 242
Bodin H A B, James T E and Newton A A 1969 Culham 255
[11] Golovin I N, Dnestrovsky Yu N and Kostomarov D P 1969 Culham 194
[12] Kadomtsev B B 1975 *Sov. J. Plasma Phys.* **1** 295
[13] Bishop A S 1969 Culham 571
[14] Fraas A P 1969 Culham 1
Johnson E F 1969 Culham 441
[15] Morley F and Kennedy J W 1969 Culham 54
[16] Tuck J L 1969 Culham 31
[17] Eastllund and Gough 1969 Culham 410
[18] Lidsky L M 1969 Culham 41; hybrid reactors had been considered already in 1954–1955, see:
Moir R W 1981 The fusion–fission fuel factory *Fusion* **1B** ed. E Teller (New York: Academic) ch 15, who mentions a 1954 study by D H Imhoff *et al.* and:
Hancox 1980 *Nucl. Fusion.* **20** 1065, who mentions a Harwell proposal from 1955
[19] Joint Committee on Atomic Energy 1971 *Controlled Thermonuclear Research* (Washington: US Government Printing Office 1972)
[20] Bromberg J L 1982 *Fusion: Science, Politics, and the Invention of a New Energy Source* (Cambridge, MA: MIT Press)
[21] Fraas A P 1973 *Oak Ridge Report* ORNL-TM-3096
[22] Mills R G (ed.) 1974 *Princeton Plasma Phys. Lab. Report* MATT-1050
[23] Werner R W *et al.* 1974 Culham 171

For later tandem mirror reactor studies, we refer to:
Logan B G 1983 *Varenna Course on Mirror-Based and Field-Reversed Approaches to Magnetic Fusion* **1** 347. TASKA-Team *ibid.* 365
[24] Conn R W 1981 Magnetic fusion reactors *Fusion* **1B** ed. E Teller (New York: Academic) ch 14 on p 195 refers to 1974 Los Alamos Report LA-5336
[25] Badger B *et al.* 1974 *University of Wisconsin Report* UWFDM-68
[26] Sako K *et al.* 1974 Culham 27
[27] Hatori T *et al.* 1974 Culham 189
[28] Cordey J C and Watson C J H 1974 Culham 199
[29] Bertolini E *et al.* 1974 Culham 79
Bertolini E *et al.* 1974 Tokyo **3** 485
[30] Joint Committee on Atomic Energy 1971 *Controlled Thermonuclear Research* (Washington: US Government Printing Office 1972) pp 195, 226
[31] Häfele W and Starr C 1974 *J. Brit. Nucl. En. Soc.* **13** 131 see also the ensuing discussion:
von Hippel F 1975 *ibid* **14** 119, Holdren J P 1975 *ibid* **14** 120, Häfele W and Starr C 1975 *ibid* **14** 265
[32] Häfele W *et al.* 1977 International Institute for Applied Systems Analysis RR-77-8 (Laxenburg: IIASA)
[33] Holdren JP 1978 *Science* **200** 168
Holdren JP 1980 *Env. Cons.* **7** 289
[34] Lovins A 1976 *Foreign Affairs* **55** 65
[35] Metz W D 1976 *Science* **192** 1320, **193** 38, 76 and **193** 307
[36] Conn RW 1981 Magnetic fusion reactors *Fusion* **1B** ed. E Teller (New York: Academic) ch 14
[37] Abdou M A *et al.* 1980 Brussels **2** 119
[38] Sheffield J *et al.* 2000 Joint Institute for Energy and Environment Knoxville USA JIEE 2000-06
[39] Wootton A J *et al.* 1997 *Nucl. Fusion* **37** 927
[40] Stambaugh R D *et al.* 1996 Montreal **3** 395
[41] Estimates for $\sim$10 MA experiments presented in:
Sykes A 1994 *Plasma Phys. Control. Fusion* **36** B 93
Peng M *et al.* 1998 Yokohama **4** 1243
Robinson D *et al.* 1998 Yokohama **4** 1247
would extrapolate to around 200 MW for a 30–40 MA reactor.
[42] The Starlite Study: *Assessment of Options for Tokamak Power Plants*, Final Report 1997, UCSD-ENG-005.
[43] Raeder J *et al.* 1995 *Safety and Environmental Assessment of Fusion Power* EURFUBRU XII-217/95 (Brussels: European Commission)
This work has developed into two ongoing activities under EFDA; see
Borrelli G *et al.* 2001 *Socio-Economic Research on Fusion* EFDA-RE-RE-1
Cook I *et al.* 2001 *Safety and Environmental Impact of Fusion* EFDA-S-RE-1
[44] D'haeseleer W D 1999 *Plasma Phys. Control. Fusion* **41** B25
[45] INTOR 1980, 1982, 1983, 1986, 1988a, 1988b *Workshop Reports* (Vienna: IAEA)
[46] INTOR 1988 *Workshop Report* vol. 2 (Vienna: IAEA)
[47] INTOR 1988b *Workshop Report* (Vienna: IAEA)
The NET design also allowed for a regime with a pulsed plasma current of 15.2 MA. See the following, as well as accompanying papers in the same issue of *Fusion Technology*:

Toschi *et al.* 1988 *Fusion Technol.* **14** 19
Engelmann F *et al.* 1988 *Fusion Technol.* **14** 30
[48] INTOR 1988a *Workshop Report* (Vienna: IAEA)
[49] Dean S O 1988 *J. Fusion Energy* **17** 155
[50] Stacey W M Jr *et al.* 1986 Kyoto **3** 199
[51] Tomabechi K 1990 Washington **3** p 217
[52] ITER 1997 *Final Design Report* (Vienna: IAEA)
[53] ITER 2001 *Final Design Report* (Vienna: IAEA)
[54] ITER Physics Basis Editors 1999 *Nucl. Fusion* **39** 2391
[55] ITER Physics Basis Editors 1999 *Nucl. Fusion* **39** 2137
[56] Panel Report to FESAC 'Review of the ITER Detailed Design Report'
[57] *National Energy Strategy—Powerful Ideas for America, US DOE Report* 1991
[58] Fusion Programme Evaluation 1996 EUR 17521

Chapter 10

[1] Fowler K T 1997 *The Fusion Quest* (Baltimore: Johns Hopkins University Press)
[2] Conclusions of the Fusion Fast Track Experts Meeting held on 27 November 2001 under the chairmanship of Prof. D King, on the initiative of Mr De Donnea, President of the Research Council of the European Union.
[3] Artsimovich wrote in 1972, the year before his death: '*. . . the parameters in our research . . . grow, and this means our gradual advance towards the accomplishment of the task. It is bound to be accomplished when thermonuclear energy becomes absolutely necessary for mankind, because there are no insurmountable obstacles in our path*'. This translation appeared in: Anonymous 1982 *Tokamak: Towards Thermonuclear Energy* (Moscow: Novosti)
For a view from the developing world, we may refer to
Kaw P K 1992 Würtzburg *Artsimovich Memorial Lecture, 'Fusion power: who needs it?'* **1** 3

Symbols

Consistency in the use of symbols is difficult because there are some clashes between usage in open-ended and closed systems and also conventions have changed during the period covered by this book.

For open-ended systems we use a cylindrical coordinate system

(r, θ, z)	radial, azimuthal and axial coordinates
a	diameter of the plasma
L	length of the plasma column

However, in section 4.5 we use L for inductance and R for resistance, together giving the L/R time.

I, I_p, I_z	axial plasma current
I_θ	azimuthal plasma current
B_z	axial magnetic field
B_θ	azimuthal magnetic field
M	mirror ratio $M = B_{max}/B_{min}$
m, z	azimuthal and axial mode numbers

For closed systems we use a toroidal coordinate system

(R, ϕ, z)	cylindrical coordinates of the torus (z is the major axis and $z = 0$ the mid-plane)
(r, θ)	the polar coordinates in the minor (poloidal) cross-section
R_0	major radius of the plasma
a	horizontal minor radius of the plasma
b	vertical minor radius of the plasma
A	aspect ratio $A = R_0/a$ of the plasma
ε	inverse aspect ratio $\varepsilon = r/R$
κ	elongation of the plasma cross-section $\kappa = b/a$
I, I_p	plasma current (in the toroidal direction)
B_t	toroidal magnetic field (B_z and B_ϕ are used in some early papers)
B_p	poloidal magnetic field (B_θ is used in many early papers)
ι	rotational-transform angle

q	safety factor $q(r) = 2\pi/\iota$
q_a	safety factor at the plasma edge $q_a = q(a) \approx 5a^2B_t/R_0I_p$
q_0	safety factor at the plasma centre $q_0 = q(0)$
l, m	periodicity and number of periods of a stellarator helical field
m, n	poloidal and toroidal mode numbers

Plasma quantities

Species are denoted by subscripts e (electron), i (ion), j (= e, i)

m_j	mass of a particle
Z_i	charge/e
M_i	mass number
μ_j	magnetic moment of a particle $\mu = m_j v_\perp^2/2B$
v_j	velocity of a particle
$v_\perp^2$	velocity perpendicular to magnetic field
T_j	plasma temperature
T	temperature when $T_e = T_i$
n_j	particle density
N_e	number of electrons per metre length of a plasma column
λ_d	Debye length $\lambda_d = (\varepsilon_0 kT_e/n_e e^2)^{1/2}$
V_A	Alfvén velocity $V_A = (B^2/\mu_0\rho)^{1/2}$
P	power
τ_E	energy confinement time
τ_p	particle confinement time
H	the enhancement of confinement time compared to L-mode $H = \tau_E/\tau_L$
β	beta ratio of plasma kinetic and magnetic pressures $\beta = 2\mu_0 p/B^2$
β_N	normalized beta defined by $\beta = \beta_N I_N$ where $I_N = I/aB_T$ with I in MA
R	ratio of fusion energy yield to input or loss
Q	ratio of fusion energy output to external energy supplied to the plasma
Q_{DT}	the value of Q in a D–T plasma
ω_p	plasma or Langmuir frequency $\omega_p = (n_e e^2/\varepsilon_0 m_e)^{1/2}$
ω_{Bj}	Larmor or gyro frequency $\omega_{Bj} = eB/m_j$
ρ_j	gyroradius of species j
ρ^*	ρ_i/a dimensionless ion gyroradius
ρ	mass density of the plasma
η	the resistivity of the plasma
Z_{eff}	the effective ion charge
τ_c	the collision time
χ	the heat diffusivity $\chi = K/n_e$
K	the heat conductivity
$D_{\perp B}$	Bohm diffusion coefficient $D_{\perp B} = kT/16eB$

$\langle \sigma v_i \rangle$	fusion reaction cross section averaged over the relative velocity of colliding ions
A_{eq}	equivalent current of a neutral beam flux

General

e	electronic charge
k	Boltzmann's constant
ε_0	permittivity of free space
μ_0	permeability of free space

Glossary

Abbreviations, acronyms and fusion terminology

2XIIB: mirror machine at Livermore, the first to produce high-energy, high-density, stable mirror plasmas.

adiabatic compression: heating a plasma by compression in a magnetic field that rises on a time scale short compared with the energy-loss time, but not so short as to induce shock waves.

AEC: (US) Atomic Energy Commission, responsible for fusion research (from 1974 **ERDA**: Energy Research and Development Agency and from 1977 **DOE**: Department of Energy).

Alcator, Alcator A; **Alcator B**; **Alcator C-mod:** series of tokamaks at MIT with very high toroidal fields and plasma current densities producing enhanced Ohmic heating.

Aldermaston: site of UK nuclear weapons laboratory engaged in fusion research until about 1960 when fusion was transferred to Culham. Also site of research laboratory of the AEI (Associated Electrical Industries).

Al: mirror machine at Livermore.

Alice-Baseball: modified version of Alice with a magnetic well.

Alpha: replica of the ZETA toroidal pinch experiment; built in Leningrad.

Alfvén waves: plasma waves in which the plasma is tied to the magnetic field.

Ambal-M: tandem-mirror experiment at the Budker Institute, Novosibirsk.

ambipolar potential: the electrostatic potential which develops spontaneously to equalize the loss rates of ions and electrons.

ASDEX, ASDEX-U: Axially Symmetric Divertor Experiment (Upgrade)—tokamaks at Garching.

aspect ratio: the ratio (R_0/a) of the major and minor radii in a toroidal system.

Astron: a configuration based on the mirror machine in which a layer of high-energy electrons was intended to modify the basic mirror field sufficiently to reverse the magnetic field on axis.

ATC: Adiabatic Toroidal Compressor—small tokamak at Princeton University, built to study plasma heating by adiabatic compression and neutral beam injection heating.

ATF: Advanced Toroidal Facility—a torsatron at Oak Ridge.
axisymmetric: having rotational symmetry about an axis.
B-3: early stellarator at Princeton.
Baltimore: venue of 9th IAEA Fusion Conference in 1982.
baseball coil: coil combining the basic mirror and Ioffe coils in a mirror machine.
Berchtesgaden: venue of 6th IAEA Fusion Conference in 1976.
Berkeley: site of US National Laboratory (responsible for developing the neutral beam sources used for plasma heating).
beta: the parameter β is the plasma kinetic pressure divided by the pressure of the magnetic field.
beta limit: the value of beta at which an instability starts to develop.
blanket: annular region in a fusion reactor surrounding the plasma in which neutrons would be slowed down to extract their energy as heat and would interact with lithium to generate tritium.
Bohm diffusion: anomalous loss mechanism described by a semi-empirical formula that threatened fusion research during the 1960s.
Bohm scaling: scaling relation for energy confinement that is determined by turbulent fluctuations with scale lengths of the order of the plasma minor radius (see gyro-Bohm scaling).
bootstrap current: a self-generated toroidal electric current maintained by the radial pressure gradient.
boronization: technique to deposit a layer of boron on plasma-facing surfaces in order to reduce impurities and recycling.
BPX: Burning Plasma Experiment—proposal for an ignition experiment at Princeton which failed to gain approval.
break-even: the point ($Q = 1$) when the fusion power produced is equal to the externally applied heating power.
bremsstrahlung: electromagnetic radiation emitted by a plasma due to the acceleration of electrons by collisions with other charged particles.
Brussels: venue of 8th IAEA Fusion Conference in 1980.
bumpy torus: toroidal confinement scheme which may be regarded either as a set of joined mirrors or as a torus with a corrugated field.
burnout: a state when the plasma in a mirror machine becomes sufficiently dense to be opaque to neutral particles from the wall.
Cadarache: Site of French nuclear research establishment near Aix en Provence where the French fusion programme was relocated in 1985.
carbonization: technique to deposit a layer of carbon on plasma-facing surfaces in order to reduce metallic impurities.
CERN: Centre Européene pour la Recherche Nucléaire—European laboratory in Geneva for high energy particle physics.
charge-exchange: a process whereby two particles exchange an electron. It is a loss mechanism when fast plasma ions are neutralized and a plasma forming or heating mechanism when fast atoms are injected into the plasma.

CHS: Compact Helical System—a torsatron at Nagoya, Japan.

CIT: Compact Ignition Tokamak—a proposal for an ignition experiment at Princeton which failed to gain approval.

CLASP: stellarator at Culham which demonstrated single particle confinement.

CLEO: stellarator at Culham. Operated initially without the helical coils as a tokamak making the first demonstration of neutral beam heating.

compact tori: confinement systems including RFC and spheromaks.

Compass-D: small tokamak at Culham.

C-stellarator: largest of a series of stellarators built at Princeton University—later converted into a tokamak (ST).

Culham: venue of 2nd IAEA Fusion Conference in 1965.

Culham Laboratory: UK fusion research laboratory near to Oxford established in 1960 by the integration of groups previously at Harwell and Aldermaston.

cyclotron radiation: radiation emitted by the plasma electrons (or ions) due to their motion in the magnetic field (this was called synchrotron radiation in the early days of fusion research).

DCLC: Drift Cyclotron Loss Cone—instability in mirror machines.

DCX: Direct Current Experiment—early mirror machine at Oak Ridge where the plasma was produced by dissociation of an injected molecular ion beam.

D–D: deuterium–deuterium fusion reaction.

Debye length: plasma parameter characterizing the range over which collective phenomena occur.

DEMO: generic name for a prototype fusion reactor demonstrating power production.

density limit: an empirical limit for the maximum stable density in a tokamak.

disruption: a catastrophic instability in a tokamak resulting from either too high a current or too high a density.

DITE: Divertor Injection Tokamak Experiment at Culham Laboratory, equipped with neutral injection heating and with a bundle divertor.

DIVA: first generation divertor tokamak experiment at JAERI, Japan.

divertor: a modification of the basic toroidal confinement system which deflects field lines at the periphery so that particle and energy fluxes can be controlled.

DOE: US Department of Energy [successor of ERDA (Energy Research and Development Agency), which in turn succeeded the AEC (Atomic Energy Commission)].

Doublet I, II and **III**: tokamaks at General Atomics laboratory in San Diego, USA, with elongated and strongly indented plasma cross-section; Doublet III was rebuilt as the divertor tokamak **D III-D**.

D–T: deuterium–tritium fusion reaction.

dynamic stabilization: a method that relies on modulating the system at a faster rate than the instability can grow.

ECE: Electron Cyclotron Emission—radiation at the electron cyclotron frequency and its multiples (harmonics); used to measure the electron temperature.

ECRH: Electron Cyclotron Resonance Heating—a heating method using microwave radiation tuned to the resonance gyro-frequency of the plasma electrons in the confining magnetic field.

ELM: Edge Localized Mode—a relaxation instability of the steep edge density gradient in the H-mode.

ELMO: bumpy torus experiment at Oak Ridge.

ELMy H-mode: H-mode in which ELMs have been induced to control the density.

elongation: (b/a) the ratio of the vertical and horizontal minor radii in a torus with non-circular poloidal cross-section.

EPS: European Physical Society; organizes an annual European conference on nuclear fusion and plasma physics.

ERDA: (US) Energy Research and Development Agency, responsible for fusion research from 1974 to 1977 when it was succeeded by **DOE**.

Eta-Beta-I, Eta-Beta-II: reversed field pinch experiments at Padua, Italy.

Etude: early stellarator at Princeton.

EURATOM: one of the three European Communities that merged into the European Union, seated in Brussels; responsible for coordinating fusion research in Europe.

EXTRAP: series of experiments in the Royal Institute of Technology, Stockholm, Sweden.

feedback stabilization: method to detect an instability at low amplitude and to apply a corrective measure.

FIRE: Fusion Ignition Research Experiment; US proposal for a 'next-step' experiment.

flute mode: pressure-driven Rayleigh–Taylor instability where the plasma tries to change place with the magnetic field (also called an **interchange mode**) developing distortions like a fluted Greek column.

FOM: research organization in the Netherlands; responsible for the Institute for Plasma Physics at Nieuwegein.

Fontenay-aux-Roses: site of French nuclear research institute near Paris (fusion research moved from FaR to Cadarache).

Franck–Condon neutrals: hydrogenic atoms emitted with some 2 eV kinetic energy when excited molecular ions disintegrate into atomic ions and atoms.

Frascati: site of nuclear research laboratory near Rome.

FRC: Field Reversed Configuration—class of magnetic confinement schemes where the field of an open-ended system is reversed to form closed magnetic field lines.

FT, FT-U: Frascati Tokamak (Upgrade)—high field tokamaks at Frascati.

GA: General Atomics research laboratory at San Diego, California.

Gamma-6: tandem-mirror experiment in Tsukuba, Japan.

Garching: site of the Max-Planck Institut für Plasmaphysik near Munich, Germany.

gas blanket: confinement scheme where the plasma is surrounded by a layer of neutral gas.

Geneva: venue of first (1955) and second (1958) UN Conference on the Peaceful Uses of Atomic Energy. The second provided the public release of the hitherto secret work on fusion in the Soviet Union, the USA and Europe.

gettering: technique to control impurities and recycling by depositing a thin active layer (e.g. titanium or beryllium) on to plasma-facing surfaces.

Goldston scaling: empirical scaling for the tokamak energy confinement time.

Greenwald limit: empirical limit for the maximum stable density in a tokamak.

Grenoble: site of French research laboratory to which the fusion programme from Saclay moved before being finally relocated to Cadarache.

gyro-Bohm scaling: scaling relation for energy confinement that is determined by turbulent fluctuations with scale lengths of the order of the ion gyroradius (see Bohm scaling).

gyrofrequency: the frequency with which the ions or electrons rotate in the magnetic field.

gyrotron: a high-power microwave source. Requires a frequency in the range 60–180 GHz for ECRH in fusion experiments.

H-1: heliac experiment at Canberra, Australia.

Harwell: site of UK nuclear research laboratory engaged in fusion research until 1960 when fusion was transferred to Culham (ZETA remained working at Harwell until 1969).

HBT, HBT(EP): High Beta Tokamaks at Columbia University, New York, USA.

HBTX: High Beta Toroidal Experiment—first generation reversed field pinch at Culham.

helical winding: the twisted coils that generate the rotational transform in a stellarator.

heliac: a variant of the stellarator family with a twisted magnetic axis.

Heliotron: a series of toroidal experiments at Kyoto which evolved towards a torsatron-like configuration in **Heliotron E**.

high-beta stellarator: a toroidal version of the theta pinch with helical symmetry.

high-*Z*: materials and impurities like tungsten with large atomic number.

HL-2A: tokamak at the South Western Institute of Physics, China (reconstructed Asdex).

H-mode: a regime of improved confinement characterized by steep density and temperature gradients at the plasma edge.

Homopolar: early mirror machine with rotating plasma.

IAEA: International Atomic Energy Agency—a United Nations organization based in Vienna.

ICF: Inertial Confinement Fusion where the fusion fuel is heated and compressed by laser or particle beams.

ICRH: Ion Cyclotron Resonance Heating—a heating method using radio frequencies tuned to the gyro-resonance of a plasma ion species in the confining magnetic field.

ICSE: Intermediate-Current Stability Experiment; a large successor to ZETA proposed around 1960 but never built.

IEA: International Energy Agency based in Paris.

ignition: the point where a fusion plasma becomes energetically self-sustaining through internal heating by reaction products.

Ignitor: Ignited Torus, concept for a compact ignition experiment based on high toroidal magnetic fields.

Innsbruck: venue of 7th IAEA Fusion Conference in 1978.

interferometry: a diagnostic method which determines the electron density by measuring the plasma refractive index with beams of microwaves or infrared radiation.

INTOR: a conceptual design study of an international tokamak fusion reactor.

Ioffe Institute: research laboratory in St Petersburg, Russia.

Ioffe winding: a set of multipolar conductors added to the basic mirror machine to provide a magnetic well.

ISX-B: Impurity Studies Experiment—tokamak at Oak Ridge National Laboratory.

ITER: International Thermonuclear Experimental Reactor—international collaboration started in 1988 towards a prototype fusion reactor based on the tokamak.

ITER-CDA: Conceptual Design Activity.

ITER-EDA: Engineering Design Activity.

Ixion: early mirror machine with rotating plasma.

JAERI: Japan Atomic Energy Research Institute, whose principal fusion research facility (JT-60U) is located at Naka. Residing under the Ministry of International Trade and Industry (see NIFS).

JET: Joint European Torus—world's largest tokamak, built at Culham, UK, as a European collaboration.

JFT-2: Jaeri Fusion Torus—tokamak experiment at JAERI, later rebuilt with elongated cross-section and open poloidal divertor as **JFT-2M**.

JT-60: Japan Tokamak—largest Japanese tokamak, later rebuilt as **JT-60U**.

keV: kilo-electron volt—used as a measure of plasma temperature (1 keV corresponds to 11.8 million degrees Celsius).

Kharkov: the Institute of Plasma Physics in Kharkov, Ukraine.

Kruskal–Shafranov limit: upper limit for the current in a tokamak or a stabilized toroidal pinch. The safety factor must satisfy $q > 1$ for MHD stability.

Kurchatov Institute: the Institute of Atomic Energy in Moscow, Russia's leading centre for fusion research, named after Academician Igor Kurchatov.

Kyoto: venue of 11th IAEA Fusion Conference in 1986; also site of the fusion laboratory of the University of Kyoto.

***L/R* time:** characteristic time of the exponential decay of an electric current in a closed loop with inductance *L* and resistance *R*.

L-1, L-2: stellarator experiments at Lebedev Institute, Moscow.

Lausanne: site of Swiss National Centre for Research in Plasma Physics.

Lawson criterion: the condition for a fusion reactor to become energetically self-sustaining. This is usually quoted as the condition where the fusion fuel ignites due to the self-heating of the alpha particles.

Lebedev Institute: an institute in Moscow of the Academy of Sciences; after 1985 the fusion research became part of the newly formed General Physics Institute on the same site.

Levitated Toroidal Octupole: a toroidal experiment at the University of Wisconsin in Madison with four internal current-carrying rings.

Levitron: a toroidal configuration where the poloidal field is generated by a current in a levitated (normally superconducting) internal ring conductor.

LHCD: Lower Hybrid Current Drive—a technique to drive a non-inductive current in a toroidal plasma using microwave radiation at the lower hybrid resonance frequency.

LHD: Large Helical Device—large torsatron-type experiment at NIFS, Toki, Japan.

limiter: a device to restrict the plasma aperture and to concentrate the particle and heat fluxes on to a well defined material surface. Originally of refractory metal (tungsten or molybdenum), later usually of carbon.

linear pinch: also known as Z-pinch—open-ended system where the plasma current flows axially and generates an azimuthal magnetic field which constricts ('pinches') the plasma.

Livermore: National Laboratory—US nuclear weapons laboratory, which was the USA's leading centre for mirror machines, now mainly devoted to inertial confinement fusion.

LMFBR: liquid metal fast breeder (fission) reactor.

L-mode: denotes the normal (degraded) confinement in tokamaks.

London: venue of 10th IAEA Fusion Conference in 1984.

low-*Z*: materials and impurities like carbon and oxygen with relatively small atomic number.

Los Alamos: National Laboratory—US nuclear weapons laboratory also engaged in fusion research.

loss-cone: the region in velocity space where a charged particle will escape the magnetic confinement in a mirror machine.

LT-3: toroidal pinch experiment at Australian National University, Canberra, which also operated as a tokamak; replaced later by **LT-4**.

***l*-windings:** the external helical coils on a stellarator that produce the poloidal magnetic field.

Madison: venue of 4th IAEA Fusion Conference in 1971; also the location of a fusion laboratory at the University of Wisconsin.

magnetic axis: the field line into which the inner surface of a set of nested toroidal magnetic surfaces degenerates.

magnetic surfaces: toroidal surfaces formed by magnetic field lines (see box 2.2).

magnetic well: property of the magnetic field strength to increase in all directions away from the plasma (see minimum-*B*).

MAGPIE: a fibre pinch experiment at Imperial College in London.

MARFE: a poloidally localized radiation instability.

MAST: Mega-Amp Spherical Tokamak; an experiment at Culham.

MFTF-B: Mirror Fusion Test Facility—the largest tandem mirror device built, but never operated, at Livermore.

MHD: magnetohydrodynamics—the set of equations that describe a conducting fluid or plasma in a magnetic field.

Microtor and Macrotor: small tokamaks at MIT.

minimum-*B*: a property of a confinement system where the field forms a magnetic well.

mirror machine: an open-ended system where the plasma is confined in a solenoidal field which increases in strength at the ends—the 'magnetic mirrors'.

MIT: Massachusetts Institute of Technology.

modular stellarator: a version of the stellarator where the toroidal and poloidal magnetic fields are both produced by the same set of non-interlinking coils.

Montreal: venue of 16th IAEA Fusion Conference in 1996.

MST: Madison Symmetric Torus—reversed field pinch experiment at University of Wisconsin, Madison.

NBI: Neutral Beam Injection for heating and refuelling.

neoclassical theory: theory describing particle and energy transport, taking toroidal effects into account.

NET: Next European Torus—a study for a post-JET European experiment which became integrated with the European contribution to ITER.

Nice: venue of 12th IAEA Fusion Conference in 1988.

Nieuwegein: site of FOM Institute for Plasma Physics in the Netherlands.

NIF: National Ignition Facility, laser-driven ICF experiment at Livermore.

NIFS: National Institute for Fusion Science, Japanese centre for fusion research under the Ministry of Science and Education, sited near Nagoya. Its principal research facility is the LHD stellarator (see JAERI).

Novosibirsk: venue of 3rd IAEA Fusion Conference in 1968; also the location of a major fusion laboratory (the Budker Institute).

NRL: Naval Research Laboratory, Washington, DC.

Oak Ridge: US National Laboratory at Oak Ridge, Tennessee.

octopole: toroidal device with four internal current-carrying rings which may be mechanically supported or levitated.

OGRA: *odin gram neitronov v sutki* ('one gram of neutrons per day')—an early mirror machine at the Kurchatov Institute where the plasma was produced by dissociation of an injected molecular ion beam.

Ohmic heating: the process which heats a plasma by an electric current (known as Joule heating in some other branches of physics).

OHTE: reversed-field pinch at GA laboratories in San Diego, USA.

ORMAK: Oak Ridge Tokamak.

PBX: Princeton Beta Experiment, a modified version of PDX.

PDX: Princeton Divertor Experiment.

pellet refuelling: method to refuel a fusion reactor by injecting small pieces of solid hydrogen accelerated to velocities in the range 1 to 10 km s^{-1}.

Perhapsatron: series of early toroidal pinch experiments at Los Alamos.

Petula: small tokamak at Grenoble devoted to studies of radio frequency heating. Later transferred as RTP to Nieuwegein, the Netherlands.

Phaedrus-B: tandem mirror experiment at the University of Wisconsin in Madison.

pinch effect: the compression of a plasma due to the inward radial force of the azimuthal magnetic field associated with a longitudinal current.

plasma focus: intense plasma source developed from the z-pinch.

plasma frequency: the natural oscillation frequency of a plasma electron.

PLT: Princeton Large Torus—the first tokamak with a current larger than 1 MA.

poloidal: when a straight solenoid is bent into a torus, the azimuthal coordinate θ becomes the poloidal coordinate. Thus in a toroidal pinch or tokamak, the poloidal magnetic field is the self-field of the toroidal current.

Princeton: the Plasma Physics Laboratory (**PPPL**) at Princeton University is the leading fusion laboratory in the USA.

Proto-Cleo: small stellarator experiment at Culham—later rebuilt at University of Wisconsin at Madison.

Pulsator: early small tokamak at Garching.

quadrupole: toroidal device with two internal current-carrying rings, which may be mechanically supported or levitated.

***q*-value:** the magnitude of the parameter q known as the safety factor.

***Q*-value:** the ratio of fusion power to heating power; $Q = 1$ is referred to as break-even.

recycling: loss of particles from the plasma to the wall and their subsequent return to the plasma.

REPUTE: series of reversed field pinch experiments at the University of Tokyo.

RF heating: Radio Frequency heating—methods of heating the plasma by means of electromagnetic waves at various plasma resonance frequencies.

RFP: Reversed-Field Pinch—version of the toroidal pinch where the stabilizing toroidal magnetic field reverses on the outside of the torus.

RFX: Reversed Field Experiment—a high-current RFP in Padua.

rotational transform: ι is the poloidal angle through which a field line in a toroidal confinement system rotates in the course of one complete transit in the toroidal direction (see section 2.4).

RTP: Rijnhuizen Tokamak Project at Nieuwegein (reconstructed Petula).

runaways: electrons that have been accelerated to such a high velocity that friction with the ions is too small to balance their continued acceleration to MeV energies.

safety-factor: a measure of the helical twist of magnetic field lines in a torus denoted by the parameter $q = 2\pi/\iota$ (see rotational transform).

Saclay: French nuclear research centre near Paris; fusion research later transferred to Grenoble and then to Cadarache.

Salzburg: venue of 1st IAEA Fusion Conference in 1961.

sawteeth: periodic relaxation oscillations of the central temperature.

scaling: an empirical relation between a quantity such as the energy confinement time and various plasma parameters.

SCEPTRE: Stabilized Controlled (E) Pinch Thermonuclear Reaction Experiment, a series of toroidal pinches at Associated Electrical Industries, Aldermaston, UK.

scrape-off layer: the region at the plasma edge where particles and energy flow along field lines to the limiter or divertor.

screw pinch: toroidal configuration where the poloidal and toroidal fields are rapidly pulsed in phase.

Scylla: series of linear theta pinches at Los Alamos.

Scyllac: toroidal versions of Scylla.

separatrix: limiting magnetic surface separating closed and open surfaces.

Seville: venue of 15th IAEA Fusion Conference in 1994.

shock-heat: method to heat a plasma by dissipation of a magneto-hydrodynamic (or hydromagnetic) shock wave, driven by a rapidly rising magnetic field.

Sorrento: venue of 18th IAEA Fusion Conference in 2000.

spherator: a series of toroidal devices at Princeton (**Spherator**, **SP-3**, **LSP** and **FM-1**) where the plasma was detached from a single internal ring conductor (mechanically supported or levitated).

spherical tokamak: a toroidal configuration with an extremely small aspect ratio where the toroidal field coils of a conventional tokamak are replaced by a straight rod.

spheromak: a compact system with a trapped toroidal field and purely external poloidal field.

SPICA: screw pinch experiment at Nieuwegein.

ST: Symmetric Torus—first tokamak in the USA—converted from the C-stellarator.

Starfire: US fusion reactor design study based on the tokamak.

START: Spherical Tight Aspect Ratio Tokamak—spherical tokamak at Culham.

stellarator: a generic class of toroidal confinement systems where rotational transform of the basic toroidal magnetic field is produced by twisting the magnetic axis or by adding an external helical field. Proposed in 1951 by Lyman Spitzer at Princeton University. The class includes the early figure-eight and helical-winding ('classical') stellarators at Princeton as well as torsatrons, heliacs and modular stellarators.

STP: series of reversed field pinch experiments at Nagoya, Japan.

superconducting: property of a conductor to lose its resistivity at low temperature; refers to a machine with magnetic field coils cooled by liquid helium to become superconducting.

supershots: a regime of enhanced energy confinement developed in TFTR by powerful neutral beam injection heating combined with reduced particle recycling.

SWIP: South Western Institute of Physics, research institute at Sichuan, China.

T-3: tokamak at the Kurchatov Institute in Moscow whose results prompted a world-wide emphasis on tokamaks; **T-4** to **T-15** are later experiments.

tandem mirror: modified mirror machine with end cells to reduce particle losses.

TCA: Tokamak Chauffage Alfvén—experiment at Lausanne, Switzerland.

TCV: Tokamak à Configuration Variable—experiment at Lausanne, Switzerland.

TEXT: Texas Tokamak—a small experiment at University of Texas, Austin mainly devoted to detailed physics studies.

TEXTOR: tokamak at the Jülich fusion laboratory, Germany.

TFR: Tokamak à Fontenay-aux-Roses, France.

TFTR: Tokamak Fusion Test Reactor—largest US tokamak experiment, located at Princeton University.

theta pinch: open-ended confinement system where a rapidly pulsed axial magnetic field generates a plasma current in the azimuthal (i.e. theta) direction.

Thomson scattering: method to measure the electron temperature and density, based on scattering laser light from the electrons; **collective Thomson scattering** measures the ion properties by scattering off the associated electron cloud.

TMX and **TMX-U:** Tandem Mirror Experiment (and Upgrade) at Livermore.

tokamak: *toroidalnaya kamera magnitnaya katushka*—a toroidal confinement system with strong stabilizing toroidal magnetic field and a poloidal field generated by a toroidal current in the plasma. Proposed in 1951 by Andrei Sakharov and Igor Tamm in the Soviet Union and now the front runner in magnetic confinement systems.

Tokyo: venue of 5th IAEA Fusion Conference in 1974. Site of Tokyo Institute of Technology and University of Tokyo.

Tore-Supra: superconducting tokamak at the French fusion laboratory in Cadarache.

toroidal pinch: the first (studied since about 1947) toroidal confinement scheme where a plasma current in the toroidal direction both heats the plasma and provides the poloidal ('pinch') magnetic field.

torsatron: a variant of the stellarator where a single set of helical coils with unidirectional currents supplies both the poloidal and toroidal magnetic fields.

TPE: Toroidal Pinch Experiment—series of reversed-field pinch experiments at Tsukuba, Japan.

TPX: Tokamak Physics Experiment—proposal for an experiment at Princeton to study α-particle heating which failed to gain approval.

TRANSP: a set of computational codes used to analyse and extrapolate data from tokamak experiments.

TRIAM-1M: small superconducting tokamak at Kyushu University which has been kept in a steady state for several hours with lower hybrid current drive.

triangularity: parameter characterizing the 'D-shape' in a tokamak cross-section.

Troitsk: site on the outskirts of Moscow of what was originally a branch of the Kurchatov Institute, subsequently an independent institution named **TRINITI**.

Troyon limit: empirical limit on β (the ratio of plasma kinetic pressure and magnetic field pressure) for stable operation.

TSP: Tokamak Silnoye Pole ('with a strong field')—a tokamak at the Troitsk laboratory near Moscow with high magnetic fields and compression heating intended to operate with tritium and reach ignition.

TUMAN: a series of tokamaks at Ioffe Institute, St Petersburg.

UKAEA: United Kingdom Atomic Energy Authority.

Uragan: series of stellarator and torsatron experiments at Kharkov.

UWMAK-I: a fusion reactor design study at University of Wisconsin.

Washington: venue of 13th IAEA Fusion Conference in 1990.

Ware pinch: an inward plasma flow resulting from neoclassical effects of the toroidal electric field.

Wendelstein: series of stellarators **(W IA, W IB, W II, W VII-A, W VII-AS)** at Garching; **W VII-X** is a new machine under construction at Greifswald.

Würzburg: venue of 14th IAEA Fusion Conference in 1992.

yin–yang coils: coils producing a minimum-B mirror field.

Yokohama: venue of 17th IAEA Fusion Conference in 1998.

ZETA: Zero Energy Thermonuclear Assembly—a large toroidal pinch experiment at Harwell first operated in 1957 and for many years the largest fusion experiment in the world.

Z-pinch: device where the plasma is confined by the azimuthal magnetic field generated by an axial current flowing in the plasma. (In a straight cylindrical pinch, z is the axial and θ is the azimuthal coordinate; in a toroidal pinch the current is toroidal and the field poloidal.)

ZT-1: first generation reversed field pinch at Los Alamos.
ZT-40: reversed field pinch at Los Alamos.
ZTH: high-current RFP experiment in Los Alamos, whose construction was not completed.

Index